Mapping Forest Landscape Patterns

Tarmo K. Remmel • Ajith H. Perera
Editors

Mapping Forest Landscape Patterns

Springer

Editors
Tarmo K. Remmel
Department of Geography
York University
Toronto, ON, Canada

Ajith H. Perera
Ontario Forest Research Institute
Ministry of Natural Resources and Forestry
Sault Ste. Marie, ON, Canada

ISBN 978-1-4939-7329-3 ISBN 978-1-4939-7331-6 (eBook)
DOI 10.1007/978-1-4939-7331-6

Library of Congress Control Number: 2017951195

Printed on acid-free paper

This Springer imprint is published by Springer Nature
The registered company is Springer Science+Business Media LLC
The registered company address is: 233 Spring Street, New York, NY 10013, U.S.A.

Preface

Forest landscape maps provide the foundational information for managing, conserving, and utilizing forests—one of the most important global resources. Use of the spatially and temporally explicit knowledge generated by these maps, once limited to professionals engaged in research, management, and education, is now possible for nonprofessionals, including the general public. Recent interest in global issues such as conserving biodiversity, the impacts of anthropogenic activities, and the effects of climate change have further increased the demand for mapped information about forest landscapes. This demand is being met by a rapidly expanding capability to supply that information readily and inexpensively. Given this positive and dynamic scenario, and the growing enthusiasm of academics and students, it is perhaps time to pause and ponder the status quo for using mapped information. Thus, in this book, we examine the mapping of forest landscapes with the goals of providing a broad overview of the current status of mapping and highlighting some advances in the techniques available for obtaining and visualizing this information. It is not intended to be an exhaustive review of the state of our current knowledge of forest landscape mapping, but rather a recap of some key aspects related to the utility and use of these maps.

The book is composed of eight chapters. We begin with a broad introduction to mapping forest landscapes, which will serve as a primer for readers who are either not well versed in this topic or who need a refresher course in the basics. Chapter 1 reminds the reader that maps are abstractions of reality, and that the degree of the abstraction is influenced by the map's scale and the developers' decisions regarding the geometric projection and representation; forest landscapes represent a high level of ecological organization that includes many hierarchically structured elements at multiple scales, and the elements we map are often fuzzy and heterogeneous with respect to their attributes and geometry. Data for maps are gathered from numerous sources at multiple scales; the collection can span many sources, so assessments of the inherent errors and of map accuracy are essential for defining the validity and credibility of maps.

Chapter 2 describes the concept of mapping forest landscapes as fuzzy elements, since both classification themes and the boundaries of mapped entities are often unknown or imprecise. This chapter describes the approaches that can be used to obtain probabilistic and membership values for mapped ecosystems, how the membership functions that provide these values can take many forms that depend on the spatial resolution, and the consequences for the uncertainty of mapped themes and spatial details.

Chapter 3 addresses the specific case of mapping wildfires in forest landscapes as discrete but complex objects. It describes how wildfire footprints, despite commonly being mapped as simple polygons with a uniform interior and a definitive boundary, are highly complex with respect to their internal composition and spatial attributes. This complexity results from the heterogeneous and stochastic nature of the processes that underlie fire behavior. Accurate portrayal of these processes is scale-dependent.

Chapter 4 describes the utility of three-dimensional mapping of forest landscapes using airborne light detection and ranging (LiDAR). This chapter compares the uses of discrete and full-waveform LiDAR data collection and discusses the types of summary statistics about forests that each can provide, including various characterizations of tree crowns and the development of canopy and terrain models. Having access to 3D forest landscape data has implications for mapping biomass and stem density that go beyond traditional two-dimensional stand-level genus or species classifications.

Chapter 5 explores the mapping of outputs from spatial models to understand the interactions among processes from the resulting spatial patterns. Such mapping often extends into the construction of mathematical surfaces, for which the mapped quantities represent the outputs of a model (e.g., CART, Random Forests) and are themselves abstractions of a landscape. In such cases, the spatial patterns can be used as a surrogate for studying complex ecological processes, and ensemble methods can be used to integrate data within analyses of growing archives of spatial data.

Chapter 6 continues the discussion of landscape abstraction through the use of landscape metrics in two, three, and four dimensions. Although numerous metrics exist, most capture some aspect of the landscape's composition or configuration by summarizing the size, density, shape, core, edge, or connectivity characteristics of specific land cover types. Transition zones (e.g., ecotones, edges) can be visualized as concentric bands, which are conceptually simpler than fuzzy membership functions in terms of how they describe changes with respect to the distance across an interface, since fuzzy membership functions can take on highly complex forms.

Chapter 7 explores the automation of data processing to standardize the production of forest maps and to ensure both consistency and more rapid map development. Ensuring the effectiveness of such workflows will require consistent terminology in the context of an increasingly automated environment. In this chapter, we reiterate the distinction between land *use* and land *cover*, particularly since the important information obtained from land use classification remains difficult to extract from remote-sensing imagery. The chapter concludes with a discussion of tools that can integrate this imagery with LiDAR data in a logistic regression model to allow interpretation of fractional tree cover, which connects neatly with the concept and fuzziness or mixed pixels that contain various combinations of land cover types.

The book concludes with a brief synopsis of the need to improve the application of forest landscape maps, particularly in terms of the efficiency and effectiveness of using the mapped information. In this epilogue, we remind readers that all applications of maps are scale-related; the best information is not necessarily the most detailed or presented at the highest resolution, but rather uses the optimal (most appropriate) scale for a specific application. We also reiterate that maps are abstractions and simplifications of complex natural systems, and that these abstractions depend on many assumptions and are undermined by the many sources of error that are associated with any map.

Our target audience for this book is readers who are involved with generating and using forest landscape maps. We anticipate this volume to benefit readers from the communities of developers and users of geospatial data about forest landscapes. Throughout the book, we emphasize how creators and users of maps must actively and continuously communicate their needs and interact to achieve those needs, with the ultimate goal being to improve the efficiency and effectiveness of creating and using forest landscape maps.

On behalf of all authors who contributed to this book, we thank the colleagues who critically reviewed drafts of the chapter manuscripts and suggested improvements: Raivo Aunap, Den Boychuk, Nicholas Coops, Christoph Fischer, Curtis Gautschi, Wendy Goetz, Emilie Henderson, John Lindsay, Jed Long, Kevin McGarigal, Don McKenzie, Scott Mitchell, Achilleas Psomas, Marc Simard, Christoph Straub, and Hannah Wilson. Finally, we gratefully acknowledge Geoff Hart for improving the clarity of the messages in several chapters, Marc Ouellette for refining many images and illustrations, and Janet Slobodien for guiding the publication process.

Tarmo K. Remmel
Toronto, ON, Canada
Ajith H. Perera
Sault Ste. Marie, ON, Canada

Contents

Contributors

Ruedi Boesch Swiss Federal Institute for Forest, Snow and Landscape Research, Birmensdorf, Switzerland

Mathieu L. Bourbonnais Spatial Pattern Analysis and Research (SPAR) Laboratory, Department of Geography, University of Victoria, Victoria, BC, Canada

Christian Ginzler Swiss Federal Institute for Forest, Snow and Landscape Research, Birmensdorf, Switzerland

Connie Ko Department of Geography, York University, Toronto, ON, Canada

Trisalyn A. Nelson School of Geographical Sciences and Urban Planning, Arizona State University, Tempe, AZ, Canada

Wiebe Nijland Spatial Pattern Analysis and Research (SPAR) Laboratory, Department of Geography, University of Victoria, Victoria, BC, Canada

Tõnu Oja Department of Geography, University of Tartu, Tartu, Estonia

Ajith H. Perera Ontario Forest Research Institute, Ministry of Natural Resources and Forestry, Sault Ste.Marie, ON, Canada

Tarmo K. Remmel Department of Geography, York University, Toronto, ON, Canada

F. Jack Triepke USDA Forest Service, Albuquerque, NM, USA

Evelyn Uuemaa Department of Geography, University of Tartu, Tartu, Estonia

Zuyuan Wang Swiss Federal Institute for Forest, Snow and Landscape Research, Birmensdorf, Switzerland

Lars T. Waser Swiss Federal Institute for Forest, Snow and Landscape Research, Birmensdorf, Switzerland

Michael A. Wulder Canadian Forest Service, Natural Resources Canada, Pacific Forestry Centre, Victoria, BC, Canada

Abbreviations

0D	Zero-dimensional
1D	One-dimensional
2D	Two-dimensional
3D	Three-dimensional
ALS	Airborne laser scanning
AVHRR	The advanced very high resolution radiometer
BA	Basal area
BRT	Boosted regression trees
CART	Classification and regression trees
CHM	Canopy height model
CIR	Color-infrared
DBH	Diameter at breast height
DEM	Digital elevation model
DSM	Digital surface model
DTM	Digital terrain model
DMSP	Defense meteorological satellite program
FD	Fractal dimension
GIS	Geographical information systems
GLAS	Geoscience laser altimeter system
GLM	Generalized linear model
GPS	Global positioning system
GSD	Ground sample distance
ICESat	Ice, cloud and land elevation satellite
IGBP	International geosphere biosphere program
IMU	Inertial measurement unit
IP	Interpretation plot
ITC	Individual tree crowns
k-NN	*k*-nearest neighbor
LiDAR	Light detection and ranging
L_t	The travel time of a light pulse
MAUP	Modifiable areal unit problem

MMU	Minimum mapping unit
MODIS	Moderate resolution imaging spectroradiometer
MVCD	Minimum vegetation clearance distance
NAD 27	North American Datum of 1927
NAD 83	North American Datum of 1983
NDVI	Normalized difference vegetation index
NFI	National forest inventory
p	Number of points projected into a horizontal area (i.e., the point density)
R	Range for LiDAR
REDD	Reducing emissions from deforestation and degradation
RF	Random forest
RMSE	Root-mean-square error
SAR	Synthetic aperture radar
SLS	Spaceborne laser scanning
t_{L}	The total travel time of a single energy pulse
t_{rise}	Rise time of an energy pulse
TLS	Terrestrial laser scanning
TOF	Trees outside forest
UAV	Unmanned aerial vehicle
VHM	Vegetation height model
VHR	Very high resolution
WGS 84	World Geodetic System datum of 1984

Mapping Forest Landscapes: Overview and a Primer

Tarmo K. Remmel and Ajith H. Perera

Abstract In this chapter, we offer a primer that defines key terminology and that positions forest landscape mapping within a broad context that encompasses all readers of this book. We present the fundamentals of cartography, and emphasize the importance of data representations, map projections, scales, and data collection options and principles. We then formalize the term forest landscape, and clarify how to understand this term in the context of this book by considering the numerous characteristics that could be mapped and the influence of scale on this mapping. Next, we focus on mapping forest landscapes from the perspectives of both regions and boundaries between them by considering the fuzziness of both; we then extend these concepts beyond two dimensions. The chapter concludes with a long discussion of map utility and how maps can be interpreted, mined for information, and how scale affects these types of interpretations. We stress that data with higher spatial resolution is not always better and that multiple-scale and cross-scale analyses may yield more meaningful information than interpreting a landscape at only a single scale. We conclude with a summary and discussion of the assessment of accuracy, error sources, and overall map validation. Throughout, we draw attention to chapters in this book that advance the discussion of the topics introduced in this chapter.

Abbreviations

2D	Two-dimensional
3D	Three-dimensional
AVHRR	The advanced very high resolution radiometer
DBH	Diameter at breast height

T.K. Remmel (✉)
Department of Geography, York University,
4700 Keele Street, Toronto, ON, Canada, M3J 1P3
e-mail: remmelt@yorku.ca

A.H. Perera
Ontario Forest Research Institute, Ministry of Natural Resources and Forestry,
1235 Queen Street East, Sault Ste. Marie, ON, Canada, P6A 2E5
e-mail: ajith.perera@ontario.ca

T.K. Remmel, A.H. Perera (eds.), *Mapping Forest Landscape Patterns*,
DOI 10.1007/978-1-4939-7331-6_1

DEM	Digital elevation model
DMSP	Defense meteorological satellite program
GIS	Geographical information systems
LiDAR	Light detection and ranging
MAUP	Modifiable areal unit problem
MMU	Minimum mapping unit
MODIS	The moderate resolution imaging spectroradiometer
NAD 27	North American Datum of 1927
NAD 83	North American Datum of 1983
NDVI	The normalized-difference vegetation index
RMSE	Root-mean-square error
UAV	Unmanned aerial vehicle
WGS 84	World Geodetic System datum of 1984

Mapping Forest Landscapes: An Introduction

What is Mapping?

Historians and archaeologists continue to debate where and when the world's first map was produced. There are compelling arguments in favor of a recently discovered 14,000-year-old engraved landscape depiction that was discovered in a Spanish cave; however, it is broadly accepted that a 6200 BC wall painting that depicts buildings and an erupting volcano from Çatalhöyük in Turkey (near ancient Babylon) is the world's first map. Though neither can be considered a true map, constructed to exacting standards stemming from modern cartographic principles, both mark a possible prehistoric origin to the visual (geographic) representation of spatial entities and phenomena.

The union of geodesy (the mathematics and science of measurement of the Earth) with sampling, geographic measurement, and description has evolved into the discipline of cartography and has further led to the refinement of what are commonly referred to as thematic maps—cartographic representations of related features within specific thematic domains such as land use types. These representations of real landscapes are built upon the premise of accurately situating attributes and measured values of an area on a flat surface (the map), where distances scale proportionally to the corresponding real-world distances to permit measurements from the map itself. It is the fundamental concept of the thematic map that forms the basis of modern navigation, land registry and ownership, land cover mapping, and ecological modeling, and this concept touches virtually every aspect of geoinformatics. In this book we specifically address thematic maps of forested landscapes, but the ideas presented here extend well beyond the context of forestry.

Forest landscape mapping represents the confluence of many areas of scientific expertise: mathematics, geography, remote sensing, geographical information systems (GIS), spatial statistics, forest and landscape ecology, management of forests and other resources, and even economics. Given this variety of specialties that have become involved with forest landscape mapping, we feel that it's important to explain some basic concepts associated with this topic, thereby forming a context for the discussions that appear in the rest of this book. Therefore, our goal in this chapter is to provide an overview—to define what we mean by mapping, to elucidate what forest landscapes are and why we map them, and, finally, to provide context for the considerations and challenges involved in mapping forest landscapes. It is designed to cover broad considerations that will introduce the perspective of geographers to ecologists, and vice-versa.

Thematic Mapping Basics

Mapping can be defined as the depiction and attribution of components belonging to natural and cultural environments using visual (graphic) representations. Accurate representation of spatial relationships in two-dimensional maps requires precise positional measurements in horizontal, vertical, or both dimensions, together with appropriate use of color, symbols, and cartographic scale to construct truthful and effective representations of real environments. Though some aspects of thematic mapping extend to include mental abstractions that do not occupy physical spaces, the goal in this book is to focus on the mapping of physical environments while avoiding the more abstract mental aspects of mapping as well as the cultural and often anthropogenic elements, except where they intersect with forested landscapes. Because *abstraction* is a key point in this book, a working definition is important. Here, we define abstraction as the process of creating a visual image that does not exist in the real world, but that bears an obvious resemblance to the real world, though with some of the complexity eliminated to preserve only the most important points. For example, an aerial photograph is relatively concrete, but a line drawing based on that photograph is highly abstract.

Forested landscapes are traditionally mapped to delineate land ownership, typically to support deeds and title records and to support taxation. However, legal mapping (precise surveying) aside, modern forest mapping serves many more and wider reaching purposes. From forest management planning to disturbance tracking, mapping allows foresters and managers to organize their harvesting and regeneration strategies under the constraints imposed by natural environmental conditions. Ecologists might be concerned about the diversity of species (flora or fauna) or the ranges of mammals or birds. Forest scientists are curious about biomass accumulation in terms of fuel loading and wildfire risk, and climate change researchers may wish to quantify the carbon sequestered in forests and the underlying soils or its periodic, episodic, or catastrophic release. Park rangers interested in establishing new hiking trails and viewpoints are concerned with topographic and esthetic components, and most stakeholders are at some point interested in identifying and measuring changes between consecutive periods.

Traditional thematic maps can be constructed from data obtained through field observation and measurement, surveying, photographic and photogrammetric analyses, or interpretation of satellite images. However, as the past decades have demonstrated, spatial models are becoming equally likely sources of thematic maps, and the products from traditionally developed maps might become model parameters that are ultimately used to produce new maps.

A map will never be able to record the full complexity of a landscape; among other problems, it would need to be the same size as the landscape to do so. Thus, it is impossible to consider depicting a complete enumeration of all phenomena, interactions, and processes that act on an area of interest across the full spectrum of spatial scales. Forested landscapes can be complex and heterogeneous environments composed of diverse biotic and abiotic elements and the associated biological, chemical, physical, and energy exchanges and interactions. Capturing the true complexity of a forested landscape, including every element and entity from the submicroscopic level through connections with global atmospheric circulation and solar insolation, would be a daunting—indeed, impossible—task. An appropriate level of representation must therefore be chosen based on the needs of the map's users, and the cartographic construction will then rely on a series of generalizations that prevent the map from being cluttered with extraneous information that would make it difficult to comprehend and from distracting the user's attention from the most important points. Spatial objects must therefore be simplified with respect to their shape (e.g., edge complexity), generalized in terms of their thematic detail (e.g., the number of levels of a variable), and designed based on the minimum feature size that must be represented. Thoughtful design, combined with clear use of colors, symbols, spacing, and fonts, and the choice of an appropriate density of landscape elements, will permit the human mind to perceive the inherent complexities of the landscape without being distracted by the inclusion of peripheral information. Indeed, information designer Richard Saul Wurman considers "map" to be an acronym for "mankind's ability to perceive."

Forest landscapes are mapped for many reasons, and each creates requirements for the final map representation, its precision, and its accuracy. Similarly, the content desired in each map will dictate how landscapes are measured, represented, and ultimately portrayed as maps. In this section, we provide a background to the processes that cartographers use to convert real-world forest landscapes into maps that abstract the things or processes they are studying. These creative processes are all constrained by the considerations and caveats detailed in the following sections that simultaneously constrain and emphasize the need for such map products.

For any mapping exercise, the initial requirement is a clear articulation of the map's purpose and definition of the related requirements for the final product, as these will guide the data acquisition, representation, and cartographic processes that will be implemented. In most instances, forest landscapes will be mapped as one or more thematic maps to characterize their composition. Such maps delineate areas that are more similar to each other than they are to a different group of areas. Whether these thematic classifications characterize tree species, age classes, wood volume, height, or other measureable metrics will be determined by the initial step in which the cartographer articulates the map's purpose. Once the purpose is clear, the specific variables that must be measured can be identified and their measurement can proceed using traditional and proven methods.

Thematic maps can be as varied as the purposes that they are designed for. Depending on the extent of the area for which mapping is conducted, the level of detail required, and the spatial scale of the phenomena of interest, the outcome of the mapping exercise will differ. For forested landscapes, mapping generally revolves around the mapping of forest composition and factors (e.g., functional or utilitarian) that derive from the forest's resources, disturbances (e.g., natural or anthropogenic), and processes acting on the landscape. The emphasis is often on land cover mapping (composition), paying particular attention to tree genera and species assemblages (where possible) to produce forest inventory maps that can be updated regularly to support forest management planning. Such mapping also reflects the structural characteristics of landscapes (i.e., the assemblage of shapes into patterns and horizontal or vertical structures).

Factors that Influence Mapped Patterns

The decision to add or delete even a single feature will alter the observed spatial pattern in a map. Beyond the selection of which features to include in a map, several additional decisions will influence the design of the map and the patterns that it portrays. The choice of a measurement will alter the number of levels at which an attribute can be drawn (colors, shapes, symbol sizes) and determine whether the presentation will be qualitative or quantitative. Furthermore, the type of data representation will impose certain restrictions or allow certain freedoms on the representation of the geographic space. Within these constraints, the degree of feature generalization, the choice of geometric projection, and (ultimately) the scale of mapping will all contribute to defining the final landscape pattern that is displayed. The design decisions made within these constraints will influence symbolization, attribute definition, and observed patterns of depicted fragmentation, patch sizes, topographic variation, or complexity of nested geographic features, but in each case, small changes to decisions about data representation and map construction can have a substantial influence on the resulting spatial patterns. In the following sections, we will discuss each of these influences independently.

Attributes linked to geographic locations can be qualitative (generally observed and described) or quantitative (generally measured and numeric). Stevens (1946) defined specific levels of measurement that describe the types of attributes that can be associated with mapped locations and that depend on the application domain for which the map is being constructed: nominal, ordinal, interval, and ratio. Although these four categories cover most attribute types that may be desirable to map, Chrisman (1998) extended this list by six to include the categories of log interval, extensive ratio, cyclical ratio, derived ratio, counts, and absolute scales, all of which are particularly applicable in cartography. These categories are defined as follows:

Nominal categories permit the assignment of unordered labels to geographic elements (e.g., land use) that have no implied value beyond their identification of membership in a category. Colors, codes, and identification values for counting objects are all nominal; for example, a serial code for each tree growing in a plantation does not permit ranking of one tree over another.

Ordinal categories are ordered, but there is no specific or consistent numerical step between the classes (e.g., small, medium, large). There is an implied relative ranking, but there is no way to assess the absolute difference between categories (e.g., large is not necessarily twice as big as small).

Interval measurements express a meaningful quantitative difference between values, but there is no defined zero or origin (e.g., degree Celsius). It is possible to say that the difference between 10 °C and 20 °C is identical to the difference between 30 °C and 40 °C, but unlike 0 K, 0 °C does not represent the absence of heat.

Ratio measurements have an explicitly defined zero (or origin), such as the measurement of length, velocity, or mass (where 0 cm, 0 km/h, and 0 kg, respectively, all express the absence of what they explicitly measure).

Log intervals are similar to an interval, but using a logarithmic scale (e.g., Richter earthquake intensities). As a result, the magnitude of the difference between adjacent categories will change.

Extensive ratios are similar to a ratio, but the signifiers (e.g., symbols such as dots in a map) have a size that is proportional to their numerical value.

Cyclical ratios are similar to a ratio, but the values have a maximum and a minimum, and when the maximum value is exceeded, values start over again at the minimum (e.g., the 360° in a circle).

Derived ratios are similar to ratios, but are expressed with respect to a specific range of values. For example, in a choropleth map, the values of a variable can be expressed in shades of a color, with the intensity of the color proportional to the magnitude of the value. For example, if green is used to represent vegetation cover, 0% vegetation cover would be represented by white (0% green) and 100% vegetation cover would be represented by the darkest shade of green (100% green).

Counts are simple total numbers, such as the number of trees in a sampling plot or the number of insects on a tree.

Absolute scales are fixed and cannot be rescaled. For example, probabilities are an absolute scale that ranges from 0 to 1.

As the measurement level or framework changes, so does the ability to represent different types of data and perform certain types of analyses of the data. Chrisman (1998) proposed extensions to these classic measurement levels to facilitate the handling of attributes related to, for example, probabilities, angles, nonlinear intervals, dates, and times, all of which are relatively common and important measurements related to geographic data and mathematical modeling. Each measurement level has implications for how measurements can be presented by varied use of color and symbology.

Table 1 Sinton's (1978) scheme for distinguishing between vector and raster representations of data

	Vector	Raster
Time	Fixed	Fixed
Space	Measured	Controlled
Attribute	Controlled	Measured

Although the term "map" conjures images of analog depictions on paper, the modern approach to cartography is firmly rooted in the digital era. Though maps may appear to be analog constructions, their origins lie in the need to visualize digital information (nowadays, often very large databases). In the era of geographic information systems (GIS), such databases belong to one of the two broad families of geographic data: vector representation (in which lines and objects are defined as single mathematical entities) and raster representation (in which lines and objects are defined by groups of independent areas or points). Depending on whether we *fix*, *control*, or *measure* (geographic) *space*, *attributes*, or *time*, we can conceptualize the type of representation as either *vector* data (points, lines, polygons) or *raster* data (regular arrays of cells, generally squares). Table 1 summarizes some of the consequences of these decisions that were defined by Sinton (1978). Sinton noted that raster data representation is used when, at a fixed point in time, a regular tessellation of polygons (i.e., a grid) is used to control the spatial fabric within which a variable of interest is measured at the location of each cell within the tessellation. In contrast, vector data representation is used when, at a fixed point in time, the variable to be measured is controlled and then identified or delineated within the spatial fabric by defining a shape (e.g., a point, line, or polygon) that can range from a single local point to an area that encompasses a region (analogous to extending outward from a single cell to a group of cells in a raster representation). In vector representations, spatial complexity can be inherently higher, as spatial entities are represented by combinations of geographic primitives (points, lines, or area features); for example, polygons are constructed by connecting line segments to enclose an area, and locations of directional change are identified by points. In vector representations, the positions of these objects are recorded as spatial coordinates that do not necessarily align to the imposed size and spacing of the regular cells in a raster representation, and in a GIS database, each primitive is linked with corresponding attributes stored in data tables. Changes to the representation affect the graphic presentation, the level of detail or generalization of features, and the overall complexity of the shapes.

As the level of vector representation increases from points (0D) to lines (1D) and polygons (2D), the complexity of the topological and hierarchical data structures required to store this data in a digital database increases. (Three-dimensional (3D) representations are also possible, though they are a more recent innovation. We will discuss them later in this chapter.) Since polygons are enclosed by lines that comprise segments bounded by points that are either nodes (the endpoints of a line) or vertices (points of inflection along a line), the number and types of data that need to be stored with higher level representations are much greater than at lower levels. For this reason, the storage overhead for highly complex objects can be substantial.

If the operational or representational cartographic scale is small, then this stored complexity may never actually be seen, but it will nonetheless be present and will increase processing, handling, and storage requirements.

Several algorithms and filtering methods have been developed to reduce the complexity of linear features in digital geographic databases. The simplest is likely the notion of *weed tolerance*, which represents the minimum distance permitted between any two consecutive vertices along a line during line simplification. One or more vertices positioned closer together than this distance are "weeded out" during digitizing or data entry to simplify the representation. This differs from *grain tolerance*, which represents the same concept but is implemented during the process of digitizing a new linear feature. These approaches are especially useful when using tools that create a piecewise representation of lines (splining input tools) or that automatically identify vertices (automated vertex placement tools) based on the digitizer's settings or movement of the input device by its operator. The general implementation of weed tolerances is to filter data to retain points with a minimum spacing along a line.

Often, line generalization is performed on existing data to meet the needs of a specific application. In this case, algorithms are applied to systematically remove vertices to simplify the structure of a line while maintaining its general form. A well-known approach is that of Douglas and Peucker (1973), which recursively finds points to maintain and delete in complex lines, thereby simplifying the line. (Note that this is different from *smoothing*, since the abrupt directional changes that would be removed by smoothing are preserved.) Related processes exist for handling raster data, in which lines (represented by a series of adjacent cells that have a common value) can be *skeletonized* (*thinned*), such that the line is never more than a single-cell wide.

Although linear geometry, and hence the geometry of many spatial features, receives most attention when considering generalization in the context of changing mapping scales, attribute generalization can also be a substantial component of map simplification. Imagine, for example, an attribute table for a given geographic dataset of points; the rows are the observations (the number of features in the dataset or on the map) and the columns are attributes for which values or codes will be stored in the cells of the table. Generalization and simplification could reduce the number of features (rows) based on a query (e.g., to remove features that do not meet a specified threshold or that do not attain some other conditional requirement). However, it is also possible to remove certain attributes (columns) and thereby decrease the storage requirement and information content of the database. These are simplistic generalizations. It is also possible to constrain the number of levels in nominal or ordinal attributes, decrease the numeric precision of quantitative variables, limit the character width of textual fields, or force lower levels of measurement (e.g., nominal rather than ordinal). Although eliminating variables serves to greatly reduce data overhead, manipulating storage characteristics has the advantage of maintaining some proportion of the initial attribute information.

Generalization does not need to be permanent and can instead be an on-the-fly complexity reduction for mapping purposes while preserving the full information content in the database. It is possible to select feature subsets by means of queries that constrain the results to specific spatial characteristics or attributes, to aggregate features by using a common symbol, or to represent spatial clusters of features using a single symbol or marker.

Map Projections

Map projections are the mathematical models used to represent the curved and irregular 3D surfaces of the Earth on flat 2D maps. The choice of a projection is often taken for granted by map users, but selection of an appropriate projection is actually a critical step in the map-making process and something that receives considerable attention by cartographers. Since no projection can fully preserve all of the characteristics of the area being mapped, it is vital for cartographers to choose projections that represent an acceptable trade-off between competing properties such as size accuracy and shape accuracy in the representation, as these trade-offs will influence how the map can ultimately be used. Four broad types of mapping projection exist: mapping that preserves the area of objects (equal-area or equivalent projections), the shapes of objects (conformal or orthomorphic projections), the distances between points (equidistant projections), or the directions from a central point to all other locations (azimuthal projections). No single map can belong to all of these projection types simultaneously, and the preserved properties are generally not global, but rather are best suited to small through intermediate areal extents.

In an effort to reduce the amount of deformation when representing the true character of the Earth's surface on a map, an appropriate projection must be selected in which standard lines (or points) can be positioned at optimal locations so that they minimize the distortion of a target criterion such as area, shape, distance, or direction. Since scale distortion is theoretically eliminated along these lines or between these points, having them within the area of interest can greatly improve the map's ability to preserve landscape characteristics and improve measurement accuracy. Many projections can be produced geometrically by selecting a developable surface (a cylinder, cone, or plane) with an appropriate scale factor and orientation, onto which the Earth's surface will be projected. The developable surfaces can then be flattened (unrolled) without additional deformation (e.g., tearing or stretching). These developable surfaces yield the azimuthal, cylindrical, conic, and arbitrary families of map projections (Fig. 1). When these developable surfaces are positioned such that the plane touches either of the poles, the cylinder touches at the equator, or the cone touches a line of latitude, the aspect is said to be *normal*. If the position is rotated 90°, the aspect is considered *transverse*, and if the positioning is at any other angle, it is *oblique*. If the scale factor is such that the developable surface is the same size as the Earth, the projection is considered to be a *tangent case*, and when the developable surface is smaller than the Earth, it is considered to be a *secant case*. In the secant case, there are two standard lines for cylindrical and conical projections, and the standard point becomes a standard circle (or ellipse) for azimuthal projections.

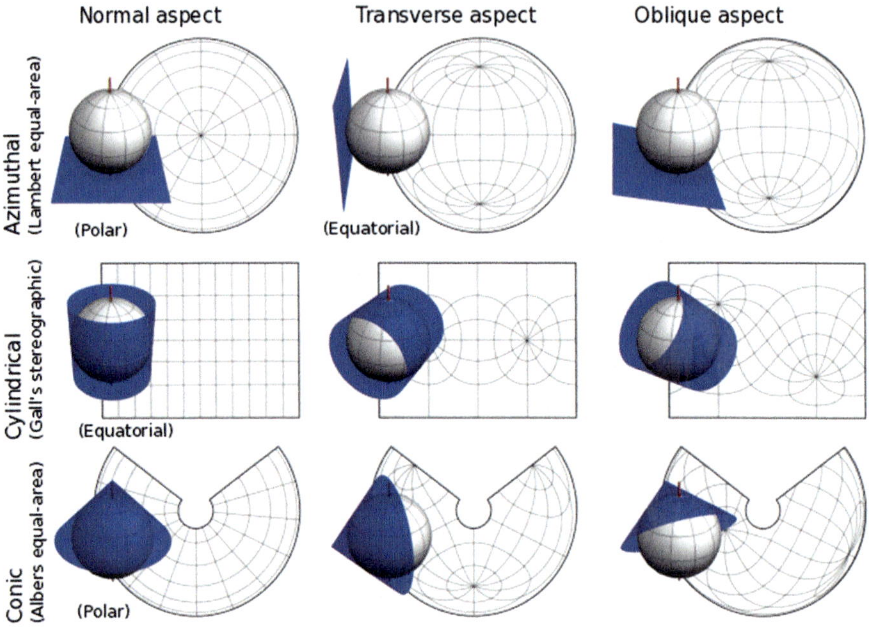

Fig. 1 Illustrations of the basic map projection aspects (normal, transverse, and oblique) and three major forms (azimuthal, cylindrical, conic), and the developable surfaces each can produce for their primary orientations (Reproduced with permission from the designer, Carlos Furuti, Progonos Consultoria)

In any mapping exercise, the cartographer must choose a projection that will minimize the error in the most important attribute of the data (e.g., area vs. size) based on the primary purpose of the map. This requires the selection of standard points or lines to preserve surface characteristics of interest while minimizing deformations and to select an appropriate datum that will ensure that a proper model of the Earth's shape is implemented in the positioning of the developable surface relative to the Earth's surface. A datum can be considered as a shift of the developable surface to make it correspond more closely to the Earth's true surface within a specific region; numerous datums exist and most have been devised for use in particular situations. If data from different projections or datums must be combined or overlaid, they must first be projected to a common form to avoid misalignment of the data due to incongruencies between Earth models. In North America, the North American Datum of 1927 (NAD 27) and the North American Datum of 1983 (NAD 83) are commonly used, but positions could differ horizontally by up to 200 m between these datums. However, with the arrival of GPS navigation, the World Geodetic System of 1984 (WGS 84) has surfaced as the most accurate version for worldwide use; for all but the most detailed survey work, it can be considered equivalent to NAD 83 in North America.

The important message regarding projections and datums is that the goal is to minimize deformations but also select a mapping form that will support the analyses and calculations required to produce the final product. For example, if area calculations are going to be the primary use of the map, then a map projection that preserves shapes or distances will be less appropriate than one that preserves sizes.

The term *scale* is widely used in the literature and is in regular use by practitioners from a multitude of fields, yet its meaning can be confusing if it is not appropriately contextualized (Dungan et al. 2002). The ecological definition for scale is such that a large scale refers to a large area or extent and a low degree of spatial detail; conversely, a small scale refers to a small area or extent and a high degree of spatial detail. Although this is logical, cartographers use an equally logical but diametrically opposed definition that stems from how they express the fractions that describe the relative number of units in the real world accounted for by the same unit on a map. For example, with a map scale of 1:10,000, one unit measured on the map equates to 10,000 of the same units measured in the real world (e.g., 1 mm represents 10 m). This unitless notion of scale provides us with the means to use maps to depict and measure the real world that they represent. However, ecologically small areas will have ratios that are much larger than ecologically large areas for the same map size. For example, a 1:100 scale would cover a small area of forest in high detail from the ecologist's perspective, but from the cartographer's perspective, this would be a larger scale (1/100 = 0.010) than a 1:1000 map (1/1000 = 0.001). It is these opposing meanings that can lead to confusion.

Thus, whenever the term *scale* is used, authors should clearly contextualize its use by stating whether the meaning is ecological or cartographic to avoid confusion. Such confusion becomes highly probable when ecologists, cartographers, and geographers work together and use their own familiar terminology. To further avoid ambiguity, it may be better to use the phrases *large area* or *small area* for ecological purposes, and use *scale* only for cartographic purposes. When it's necessary to use the term *scale* to describe both extent and level of detail, it is advisable to use concepts such as the spatial resolution (e.g., the size of an image pixel), or the minimum mapping unit (MMU), which represents the smallest discernable spatial entity in a database or on a map.

Data Sources Used in Mapping

The data used to construct maps can be obtained from several sources. These are not necessarily fixed, but rather are reflections of specific goals, map extents, cartographic scales, desired spatial resolution, and temporal scale at which processes controlling the observed pattern operate (Fig. 2), which are the reasons for creating the map in the first place. Field sampling, in situ sensors, aerial photography, and satellite imaging are common data sources, but even microscopic investigations (e.g., involving fungi, insects, and genetics) may be necessary in some cases. The following sections describe various components of the data collection spectrum that feed into the cartographic process.

Ever since it became necessary to map vegetation (particularly in forest landscapes), primarily driven by the vegetation's perceived value and the resulting need to delineate ownership, field-based mapping has formed a critical component in the

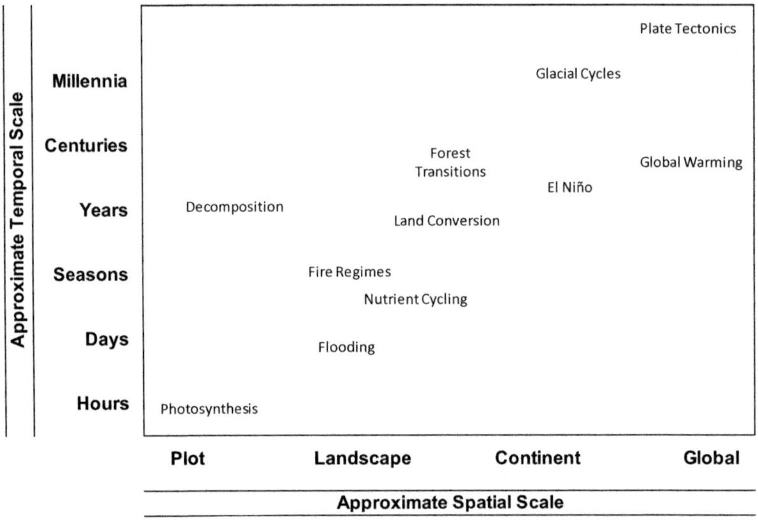

Fig. 2 Temporal and spatial scales for phenomena are related and reflect the optimal scales at which processes and their effects are best observed and mapped

construction of such maps. In situ observations by field technicians or sensors are governed by classical methods of mapping vegetation in the field, as established by Küchler (1955), and multiple variations are possible given the breadth of the sampling designs that can be implemented (e.g., complete, random, transect-based, clustered, based on convenience, purpose driven); some examples are provided in Fig. 3. These methods rely on the selection of an appropriate sampling design based on the data being collected and on the investigator having sufficient skill at identifying species, landforms, and terrain conditions that they can accurately convert field data into cartographic products with minimal error, uncertainty, and bias. Field survey data are often augmented by in situ sensor measurements (e.g., weather stations) to collect spatially distributed data for specific applications and to construct highly specialized maps. In traditional surveys, all objects (e.g., trees) within a defined area are included in the sample. However, the high cost of data collection in remote forest areas often results in the use of variable-area sampling, in which sampled data ultimately form the mapped features or form the basis for interpolation. Such variable-area sampling is common in forest mapping when variables such as basal area are estimated by using a wedge prism. In this approach, the degree to which the prism visually offsets tree stems is used to decide whether specific trees should be included or excluded from the sample as the prism is rotated about the center of the field plot. This approach provides a rapid assessment of basal area by sampling only trees of a required size class that are within an allowable distance from the plot's centroid.

Though field surveys and mapping are standard practices, particularly for validation of other mapping products such as satellite remote-sensing data, the data is simultaneously localized, focused on details near ground level, and biased toward a

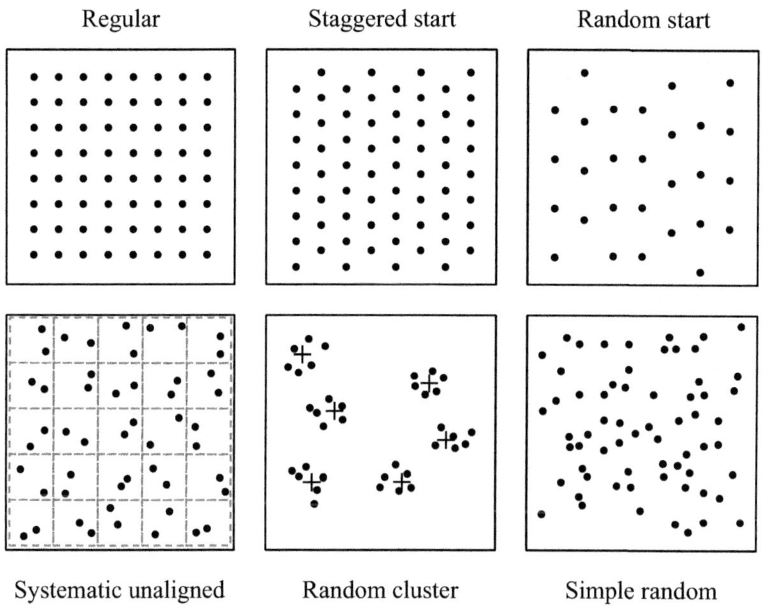

Fig. 3 Some standard field sampling designs for surveys under field conditions

generally horizontal view of the landscape from predominantly below canopy height. This limitation makes broader conceptualization and characterization of spatial relationships difficult. Thus, the past 150–200 years have seen numerous efforts to gain a different perspective to improve mapping efforts.

In an effort to gain a better vantage point and to gain more spatial context for their surroundings, people have continually climbed higher so they could look down on the landscape around them. The development of technologies involving various forms of flight has permitted many improvements in cartography as people attached photographic devices to kites, birds, hot air balloons, rockets, and more recently airplanes, satellites, and unmanned aerial vehicles (UAVs, or "drones") to acquire oblique and ultimately downward-looking views from above areas of interest (Fig. 4). The advances made possible by acquiring nadir (vertically downward-looking) and stereoscopic (overlapping) pairs of landscape photographs (or images) led to the development of photogrammetry (Kavanagh 2003), which is the science of making physical measurements of length (and hence area) and height from such visual products. The visible scene that forms an image can be described in terms of its geometry, which depends on the relationship between the ground and image planes, but also depends on the light or other forms of electromagnetic radiation reflected from the ground plane and focused through a camera lens to expose film or sensors on the image plane within the camera. The distance from the center of the lens to the image plane is called the *focal length* and the vertical axis connecting the centers of the image plane and ground plane through the center of the lens is called the *optical axis*. The centers

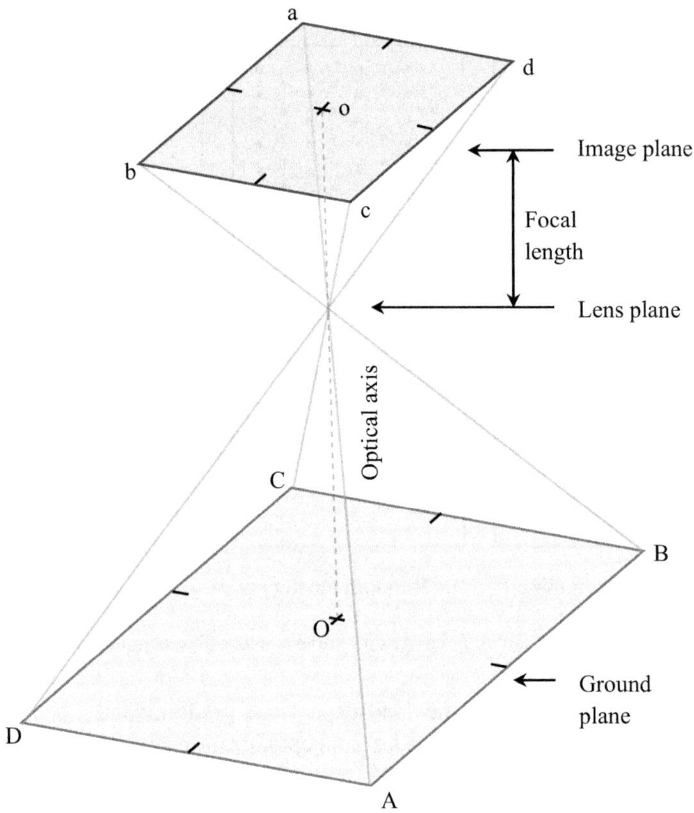

Fig. 4 The basic structural geometry of an aerial photograph image

of the ground and image planes (O and o, respectively) are referred to as nadir points. To aid in geometric measurements, halfway marks (*fiducial marks*) are added along the image plane to allow rapid location of the scene's center, and if overlapping scenes are available, these marks also aid in locating the aircraft's flight line during image acquisition.

Aerial imaging has been a dominant means of data acquisition for cartographic purposes for more than a century. Advances have come in terms of improved sensitivity and range of the imaging medium (e.g., films, digital sensors), stability of the imaging platform (e.g., gimbals), and improvements in imaging technology (e.g., lenses, acquisition rate, a change from analog to digital). Advances in film (and, more recently, in digital imaging arrays) span the gamut from standard panchromatic images (only visible colors, typically as grayscale images) to infrared, color, and color-infrared capabilities, while camera and imaging systems have undergone considerable miniaturization coupled with substantial improvements in photographic and imaging quality and resolution (spatial, spectral, and radiometric).

Rapid improvements in spatial resolution and in the acquisition of multi-spectral data have revolutionized mapping and brought aspects of a once complex science to regular citizens through the Internet and mobile services. However, issues related to cloud cover, shadows (due to tall objects or topographic relief), geometric distortions (e.g., radial displacement, the Earth's curvature, sensor lens distortion), and the obstacle created by the forest canopy continue to complicate data acquisition when mapping is done from satellites and airborne platforms. The technological advances experienced thus far have solved some mapping problems but introduced others; these remain areas of active research.

In an ongoing quest to acquire better data for mapping, the development and availability of imagery with high spatial, radiometric, spectral, and temporal resolution have opened up the floodgates for data procurement. Satellite platforms provide the ability to record multispectral or radar data over large areas, with regular revisit times, variable spatial resolutions, and options to observe landscapes in wavelengths beyond what the human eye can see. These platforms typically provide nadir views of landscapes (although pointable optics that permit oblique viewing are available on satellites such as the Satellite Pour l'Observation de la Terre [SPOT]) and permit the classification of data to extract a range of data products about the environment and to produce maps that support a range of research and management goals.

For Earth observation, images are collected simultaneously in multiple spectral bands, each of which represents a different slice of the electromagnetic spectrum; bands from the ultraviolet through to the infrared (200 to 3000 nm), as well as the thermal infrared (3000 nm to 1 mm), are the most common because of their usefulness in vegetation studies. Depending on the individual sensor, the spectral data that forms the images can come from any number of spectral bands, with an additional panchromatic band that covers the full visible electromagnetic spectrum (between wavelengths of approximately 400 and 700 nm) being common. Since vegetation that contains chlorophyll strongly reflects electromagnetic radiation in the near-infrared region of the spectrum (700 to 1100 nm), but strongly absorbs radiation in the adjacent red region (600 to 700 nm), these regions are typically well measured for Earth observation. The slope formed by connecting green-vegetation reflectance measurements in the adjacent red and near-infrared spectral regions (i.e., the *red-edge*), and its shift toward shorter wavelengths, can be used in diagnoses of vegetation health, moisture stress, biomass, and presence or absence of a species, among other uses.

Figure 5 displays a forest landscape from northwestern Ontario, Canada, as a false-color infrared composite multispectral image, with 3.2 m spatial resolution. In remote sensing, *false color* means that the cartographer replaces the actual colors (many of which cannot be seen by the human eye) with other colors that represent properties of the image, such as reflectance in the near-infrared band. In Fig. 5, the near-infrared, red, and green spectral bands are displayed as red, green, and blue, respectively. Numerous indices, such as the normalized-difference vegetation index (NDVI), and transformed indicators, such as the tasseled cap indicator, have been generated to exploit information about forest conditions from spectral data (Schroeder et al. 2011). Thermal remote sensing has been used for detecting wildfire hotspots (Fraser et al. 2000) and drought conditions (Coates et al. 2015).

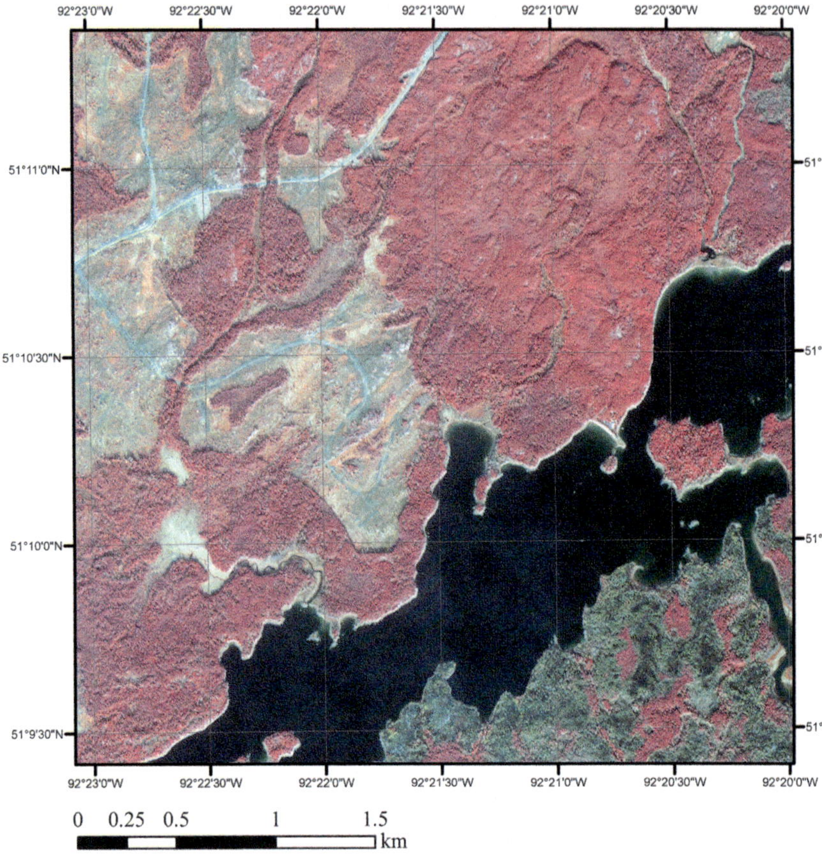

Fig. 5 A false-color infrared image from the Ikonos optical satellite for a forest landscape in northwestern Ontario, Canada. Reddish pixels represent healthy forest growth, *black pixels* represent water, *cyan pixels* represent gravel roads, and *greenish-grey pixels* represent fire-disturbed forest

Numerous satellite platforms exist, providing Earth observation at a multitude of spatial, temporal, spectral, and radiometric resolutions. Some common satellite programs and sensors include Landsat, SPOT, the Advanced Very High Resolution Radiometer (AVHRR), Ikonos, WorldView, Sentinel, and the Moderate Resolution Imaging Spectroradiometer (MODIS), each of which was selected for specific niche application areas and spatial extents. Great gains have been made in mapping fire disturbances and in managing carbon stocks using AVHRR and MODIS data (Sukhinin et al. 2004; Ruiz et al. 2012), in hotspot mapping during forest fires using the higher thermal contrasts provided by imaging at night with data from the Defense Meteorological Satellite Program (DMSP) satellites (Badarinath et al. 2011), and in assessing post-disturbance vegetation recovery by means of time-series mapping (Schroeder et al. 2011).

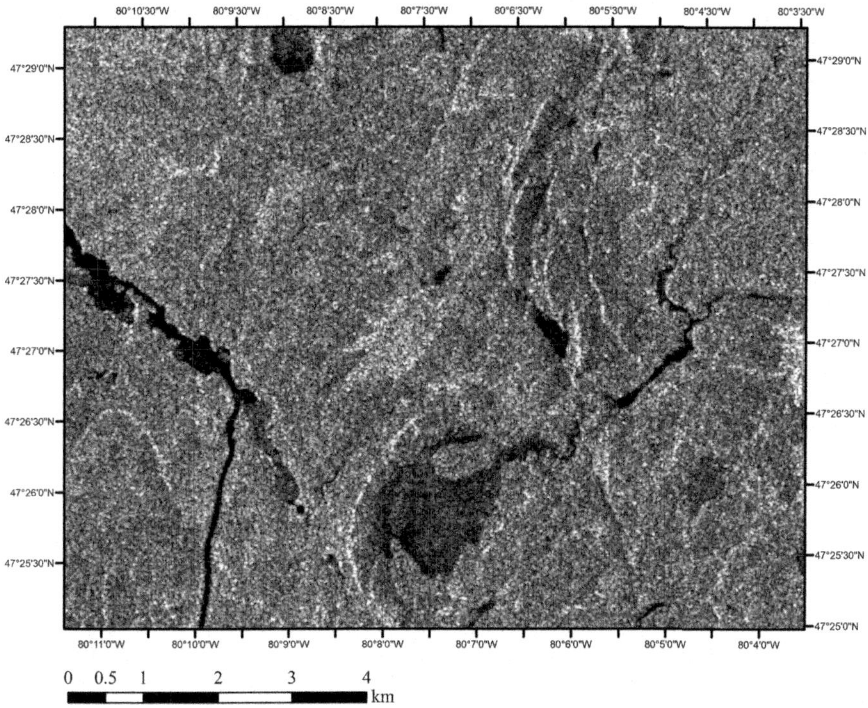

Fig. 6 Radarsat 1 image from northern Ontario, Canada (synthetic aperture radar, SAR WIDE 1 beam mode)

The utilization of satellites circumvents the need for a plane, pilot, and camera system, but introduces considerations regarding the fixed times when new images are acquired (which depend on the satellite's orbit), atmospheric conditions at the time of image acquisition, and potentially high cost. One unfortunate reality of near-optical satellite data is that many images are degraded by interference from cloud cover and haze, and received signals are also affected by atmospheric attenuation, which reduces the signal strength. Another problem is that most satellite sensors are passive, and detect reflected light rather than generating their own signals; as a result, most can only provide images during the day. One solution is to rely on radar imagery (Fig. 6), since the longer wavelengths (on the order of centimeters) penetrate haze, clouds, and light precipitation. Moreover, because radar is an active sensor (i.e., it generates its own signal and reflection), images can be obtained at night. Figure 6 shows a scene from northern Ontario, Canada, at 12.5 m spatial resolution. Radar can provide information that near-optical wavelengths cannot, primarily regarding physical properties of vegetation such as the moisture content based on an analysis of the vegetation's dielectric properties, texture, and scattering of the radar signal. Some radar can even penetrate the ground or snow to reveal buried structures. Radar data has been used to map fire scars (Bourgeau-Chavez et al. 2002), but not at national or global scales. One drawback of radar is that the image data are much more difficult to interpret due to geometric distortions (e.g., layover, radar shadows, foreshortening), speckles, and extremely bright spots due to corner reflectors. *Layover* represents the case in which tall landscape features appear to lean toward the direction of the radar source due to the tops of those features being closer to the sensor than their base, *radar shadows* represent areas with no reflection because the signal was blocked by an impenetrable object, and *foreshortening* represents the extreme distortions formed by very tall features such that the bottom of the feature becomes unseen due to the degree of layover. *Speckles* represent noise (distorted data) with a granular visual texture that results from interference among reflections from complex terrain. *Bright spots* are generated by vertical features such as tree trunks or cliffs that strongly reflect radar signals due to their orientation relative to the signal. Both radar and near-optical remote sensing images can be distorted when the image covers a wide swath due to variations in topography and the general curvature of the Earth.

What is a Forest Landscape?

In forestry, most maps are created to describe a forest landscape. Intuitively, the term "forest landscape" invokes a visual image of a vast expanse of land that contains trees, similar to a vista gained from a high vantage point (Fig. 7). Although this perception is not incorrect, it is insufficiently precise because it is a subjective

Fig. 7 An intuitive understanding of the term *forest landscape* implies the kind of large visual image that can be obtained from a high vantage point

interpretation (i.e., it depends on the individual and circumstances) that includes ambiguous (i.e., implicit and unclear) descriptors, and lacks a scientific foundation (i.e., is not based on explicit knowledge or logic, and hence is not repeatable). Asking the question "what is a forest landscape?" and pursuing an objective and an unambiguous answer founded in science are not merely an academic exercise. Though the answer may vary depending on the specific circumstances, the question must be explicitly posed and answered before each mapping exercise, since it clearly defines the goals and constraints. Mappers should not assume that this question has been asked and answered; on the contrary, it's necessary to develop an explicit and objective description, even if this is only a preliminary working definition, before beginning any mapping exercise. As we shall see later, this step is important because it defines the scope, scale, and methods that are most appropriate for a specific mapping goal. In our attempt to articulate what we mean by a forest landscape in the context of mapping, a brief foray into basic ecological concepts will be helpful because it explains the organization of organisms and their environment within a landscape. To do so, we must first decompose the term "forest landscape" and examine its two major components separately.

What is a Forest?

The common use of the term *forest* includes three levels of ecological organization (Fig. 8). First, a forest could mean a single-species population of trees that form a spatial cluster of individual stems. This connotation is sometimes accurate, as in the case of an even-aged, mono-specific tree plantation. The definition and spatial delineation of the forest is relatively easy and precise in this case. Second, forest may be used to describe a community, which in ecology means a collection of populations of different species that share the same space; this is perhaps the most common usage of the term. In this case, the forest may include several populations—each comprising different tree species as well as other plant life forms such as shrubs, herbs, and grasses, all of varying ages. Such component populations would share the same space and coexist, spatially interspersed within the community. For example, *forest stand* is a common forestry term that refers to a spatially integrated mixed-species plant community. The definition and spatial delineation of such a forest are more difficult than identifying and demarcating the population of a single tree species. Even though the term *stand* is used widely, in practice the term denotes a highly subjective, ambiguous, and implicit spatial entity; given this ambiguity, its boundaries are difficult to define. Third, forest can be an even more integrative and holistic concept that refers to a spatial unit that contains many terrestrial communities, both plant and animal, which are sometimes intermingled with aquatic communities such as lakes and rivers. In ecology, this means *an ecosystem*, where many communities coexist in close spatial proximity and interact with each other and with their physical environment at multiple scales.

Therefore, a tree-based (forest) ecosystem may contain many other plant, animal, and aquatic communities that are spatially interspersed, that interact with each other, and that typically share a common climate. Given the level of abstraction

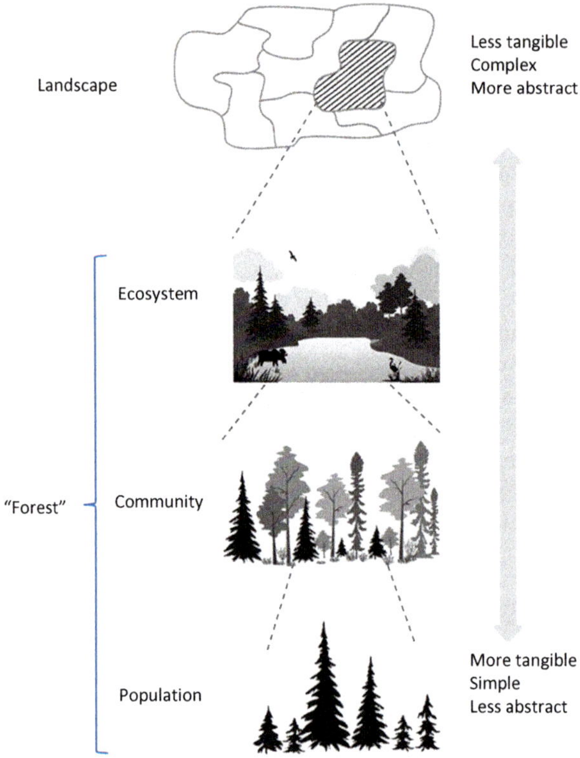

Fig. 8 Levels of ecological organization and the associated degrees of complexity and abstraction along the spectrum from populations of trees to a forest landscape. Lower levels are constituents of higher levels and are successively nested within the levels above

involved in conceiving an ecosystem and the high degree of internal complexity, an unequivocal definition or spatial delineation is difficult to achieve. As illustrated in Fig. 8, the degree of abstraction changes among the different levels of ecological organization. Here, we have tried to simplify and generalize ecological concepts that have been discussed and debated for more than 75 years. For those who want further information on the concepts that underlie ecological organization, most basic texts on ecology provide the necessary information. In particular, we recommend Allen and Hoekstra (1992), who elegantly elucidate the principles of ecological organization in relevance to scale.

What is a Landscape?

In ecological theory, a landscape represents the level of ecological organization above an ecosystem (Fig. 8). That is, it comprises an assembly of different ecosystems that are spatially interspersed and that interact with each other. Therefore, by definition, it is composed of a series of nested sublevels: individual organisms are nested within populations of their species, these combinations assemble to form communities, communities combine to create ecosystems, and ecosystems combine to create a landscape. This does not mean that all landscapes are large. If the populations of individual organisms are relatively small, then their associated communities, ecosystems, and landscapes will also be relatively small. For example, an arthropod landscape could be a single fallen tree, even though that tree would be a single organism from the perspective of trees. If the populations of one or more organisms are large, then all levels of the hierarchy will occupy relatively large spatial extents. This is the most common use of the term landscape, which implies a large geographical expanse primarily because it is based on a human perspective. In fact, most early discussions of landscape ecology address the structure, function, and utility of vast extents of land exclusively as they relate to people (e.g., Naveh and Lieberman 1984; Forman and Godron 1986). Only recently has research on landscape ecology become more comprehensive and holistic (e.g., Allen and Hoekstra 1992; Turner and Gardner 2015).

Forest landscapes, which are composed of populations of trees and other organisms that combine to form forests and other ecosystems, will be large in spatial extent, and are likely to include many different types and sizes of ecosystems that are spatially interspersed, but connected so that they interact with each other. The proportion of the forest component may vary from very high (almost completely forested landscapes) to very low (sparsely forested landscapes). Examples of the former include the vast expanses of temperate boreal landscapes, where relatively few tree species compose the populations, communities, and ecosystems (Fig. 9). Even though this landscape is predominantly forested, spatial interspersion of other ecosystems—terrestrial (e.g., shrubs), semiterrestrial (e.g., wetlands), and aquatic (e.g., lakes)—may occur to varying degrees, increasing the overall spatial heterogeneity. At the other extreme, there are landscapes where non-forest ecosystems dominate the forest ecosystems. These may be designated as landscapes of their dominant ecosystem type; for example, agricultural landscapes are dominated by agricultural ecosystems (Fig. 10a) and savannah landscapes are dominated by grasslands (Fig. 10b), though they may also contain patches of forest. Most of the world's forest landscapes fall between these extremes, and are interspersed with several other ecosystem types to form a spatial mosaic (Fig. 11). In landscape ecology, a landscape's less dominant component ecosystems are called *patches*, and the background (typically the most dominant and contiguous ecosystem type) is called the *matrix* (Forman and Godron 1986). In the case of forest landscapes, forest ecosystems can be the matrix (Fig. 9) or the patches (Fig. 10).

Given this complexity, how can we define a forest landscape in a reasonably objective way? Many definitions are found in the literature; Turner and Gardner (2015) provide a list. However, all of the definitions emphasize the spatial patterns, heterogeneity, and a focal factor—for example, a species or an ecological function. This also suggests that the definition of a landscape depends on the context in which the landscape is being considered (e.g., from a forestry perspective). Building on these definitions, we can adopt a broad and inclusive definition

Fig. 9 Contiguous boreal forest landscapes composed of (**a**) different populations and communities (revealed by different leaf colors), and (**b**) a terrestrial wetland surrounding aquatic ecosystems

Fig. 10 (**a**) An agricultural landscape where the forest component is sparse and patchy, and (**b**) a savannah landscape where the forest component is sparse and sporadic

Fig. 11 Most forested landscapes are likely to form a mosaic that includes non-forest components (other ecosystems). For example, in this image there are spatially interspersed natural forests, forest plantations, farmland, grassland, and human settlements

for forest landscapes: large geographical units that contain a mosaic of forest cover types, often interspersed with non-forest cover types, including those of anthropogenic origin (modified from Perera et al. 2000). Of course, the geographical extent of a forest landscape and its spatial delineation will require specific and explicit definitions for each research question and analytical method and for each management goal.

What are the Characteristics of Forest Landscapes?

As depicted in Fig. 8, the lower levels of the ecological hierarchy (e.g., individual trees and their populations) are tangible and spatially distinct structural entities. In contrast, the higher levels relevant to our topic—forest ecosystems and landscapes—are functional constructs that are less distinct. In fact, the level of abstraction required to perceive these entities increases continuously from a group of individual trees (tangible; simple to describe, recognize, and delineate) to forest landscapes (less tangible and more abstract; difficult to describe, recognize, and delineate). Consequently, there is a high degree of ambiguity in recognizing, delineating, and characterizing forest landscapes. Furthermore, the components of forest landscapes (i.e., embedded constituents) are successively nested within the higher levels (i.e., trees compose populations, populations compose communities, communities compose ecosystems, and ecosystems compose landscapes), leading to a multiple-scale structure. This also results in a high degree of internal heterogeneity (compositional diversity and spatial variability of components) within a forest landscape. All constituents of forest landscapes are dynamic—their structure and functions change over time. Such changes are relatively rapid at lower levels (e.g., months and years for trees) but much slower at higher levels (e.g., decades and centuries for forest landscapes). Thus, both the structures and the functions within a forest landscape occur at multiple scales. In summary, the nature of forest landscapes is complex: they are abstract entities that occupy large extents, are spatially heterogeneous, contain fuzzy components, and change slowly.

Given this description, what traits can be used to characterize a forest landscape? In adherence to the principles of ecological organization and parsimony (i.e., preferring the simplest description that can explain all the key details), forest landscapes should be characterized by their immediate subcomponents, namely the ecosystems; these, in turn, can be characterized by their subcomponents (communities) and so on, moving down the hierarchy. In practice, however, the traits typically used to characterize forest landscapes are a cross-scale mixture of elements from populations, communities, and ecosystems. This array of traits can be very long, so we have listed only a few here for illustrative purposes, grouped as descriptors of various aspects: composition, structure, function, processes, patterns, and utility (Table 2).

Table 2 Examples of traits used to characterize aspects of forest landscapes at different scales

Aspect	Forest landscape trait	Comments and sample references
Composition	Life form	Composition is the most common characterization of forest landscapes at local, regional, and global scales (e.g., Reed et al. 1994; Martin et al. 1998; Potapov et al. 2008)
	Cover type	
	Phenological type	
	Leaf type	
	Genus or species mixtures	
Structure	Tree age	These descriptors are borrowed from tree physiology and forest ecology, and are typically used at local extents, but are also used occasionally at regional and global scales (e.g., Nelson et al. 1988; Martin and Aber 1997; Treitz and Howarth 1999; Asner et al. 2012)
	Density of tree stems	
	Basal area of stems	
	Canopy height	
	Canopy closure	
	Canopy volume	
	Foliage density	
	Chlorophyll content	
	Nutrient content	
	Biomass	
	Diameter at breast height (DBH)	
Utility	Timber volume	These descriptors are commonly used at local and regional scales to characterize forest landscapes (e.g., Riitters et al. 1997; Falkowski et al. 2005; Hill et al. 2014)
	Fuel mass	
	Erosion control	
	Conservation	
	Natural capital	
	Wildlife habitat	

(continued)

Table 2 (continued)

Aspect	Forest landscape trait	Comments and sample references
Function	Photosynthetic rates	These are conventional descriptors of forest ecology that are used as ecosystem-level characteristics. They are increasingly being used at regional and global scales (e.g., Cook et al. 1989; Running et al. 1989; Kimball et al. 2000)
	Carbon sequestration	
	Nutrient cycling	
	Hydrological cycling	
	Habitat provision	
Processes	Wildfire	These descriptors are commonly used at regional and global scales to characterize forest landscapes (e.g., Macomber and Woodcock 1994; Wang et al. 2002; Wulder et al. 2004a, b)
	Flood	
	Drought	
	Windthrow	
	Insect epidemics	
	Disease incidence	
	Timber harvest	
	Land conversion	
	Road network growth	
	Regeneration	
Patterns	Patch geometry	These descriptors have recently been used to characterize forest landscapes, mostly at regional scales but rarely at global extents (e.g., Turner et al. 1989; McGarigal and Marks 1995; Haines-Young and Chopping 1996; Vogt et al. 2009)
	Patch boundary	
	Matrix geometry	
	Complexity	
	Inter-patch distances	
	Boundary complexity	

Compositionally, forest landscapes are typically characterized by various discrete groupings of their most obvious basic elements—the plant species. Commonly used species-based traits are life forms (e.g., trees, shrubs, herbs, grass), cover type (the dominant species in mixed-species communities), phenology (e.g., differences in the ability of deciduous and evergreen trees to retain their leaves throughout the year), and leaf type (e.g., needle leaves versus broad leaves). Similarly, various structural aspects of forest constituents could also be used to characterize forest landscapes: discrete metrics such as the density of groupings of stems (e.g., dense versus sparse forest) and continuous metrics such as the basal area of stems, degree of canopy closure, canopy volume, foliage density, chlorophyll content, nutrient content, and biomass. Functional derivatives are also used, though perhaps less commonly, to characterize forest landscapes. Examples include photosynthetic rates, biomass, carbon sequestration capacity, growth rates, and even nutrient and hydrological cycling. Forest landscapes are also characterized by their internal processes; these include abiotic disturbances (e.g., wildfire, flood, drought, windthrow), biotic disturbances (e.g., pests, diseases), and anthropogenic activities (e.g., land conversion, road construction, mining). Spatial patterns within forest landscapes are also increasingly used to characterize them, using various measures of the geometrical properties of the constituents. Such landscape metrics address the characteristics of patches (e.g., extent, shape, number), of the relationships among patches (e.g., edges, boundaries, ecotones), and of the whole landscape (e.g., fragmentation, contagion, interspersion, pattern). Perhaps the most common characterization of forest landscapes is utilitarian, based on how humans perceive forest landscapes for management purposes. The derivatives therefore include timber volume (for forest harvesting), fuel mass (for wildfire management), and wildlife habitat (for conservation or hunting). All these characteristics are frequently used in the literature as attributes of forest landscapes that affect their spatial portrayal as maps.

The spatial and temporal scales of observation are fundamental to characterizing forest landscapes. An explicit description of the scale of forest landscape components (i.e., ecosystem, community, or population) is essential, but it's also necessary to deliberately determine the appropriate spatial resolution (mostly, but not exclusively, for compositional and structural characteristics) and temporal interval (mostly, but not exclusively, for functional and process characteristics). Ideally, the spatial and temporal scales of characterization must be based on the ecological scale of the traits being measured. However, in practice, researchers commonly characterize forest landscapes at the highest possible spatial resolution and shortest possible temporal interval based on the belief that the most precise information is the most informative and accurate. This is not necessarily the correct approach. Since we will address the topic of appropriate scale in detail later in this chapter (in section "Map Scale"), we will state for the moment only that the spatial resolution and temporal interval must be suitable for the scale of the goal.

Another consideration in characterizing forest landscapes is how to determine their extent and boundaries. From the theoretical perspective of system function, the extent of a forest landscape would be delimited by the relative strength of the interactions among its subsystems; that is, the ecosystems within a given forest landscape interact more strongly with each other than they do with ecosystems outside the landscape. From a compositional perspective, the internal heterogeneity of ecosystem constituents within a given forest landscape will be different than the ecosystem heterogeneity outside its boundaries. However, the literature suggests that both the theoretical and compositional approaches are impractical; the extent and boundaries of forest landscapes (when studied or managed) remain case dependent, and rely on features that may be geographical (e.g., topography), administrative (e.g., forest management units), or arbitrary (e.g., satellite image extent). Therefore, it is of paramount importance to develop explicit working definitions, extents, and boundaries before mapping forest landscapes.

In conjunction with land cover and species mapping, foresters and ecologists are interested in mapping additional biophysical attributes and parameters (e.g., biomass, carbon, diameter at breast height (DBH), stand age, and height) that define the state of the forests that occupy a landscape (Hall et al. 1997). Table 2 provides a representative list of forest aspects and traits that are generally mapped, as well as some key references that describe each group of traits. Together, these attributes let managers develop harvesting and regeneration plans, design road layouts, minimize water crossings, and avoid low and wet areas or wetlands. Similarly, road construction can be optimized by designing road layouts that account for proximity to sources of aggregate (e.g., gravel) suitable for road surfaces and relatively flat topography to reduce construction costs; thus, mapping of forest landscapes is often coupled with detailed topographical and hydrological mapping within a multicriterion framework (Tampekis et al. 2015). The multitude of variables that can be measured may be separated into different maps, or aggregated as indicators within a single map composed of multiple GIS layers. Sometimes the mapped products are the outcomes of predictive or deterministic models, and other times they are interpretations, interpolations, or derivatives of these models (Saarinen et al. 2016).

For disturbance mapping, it is possible to consider different stages of the landscape pattern produced by the disturbance regime. Fires can be assessed prior to burning by estimating the probability of ignition and burning based on fuel distribution maps (Arroyo et al. 2008), and wide-area and long-term trends can be mapped and analyzed in terms of disturbance regimes (Morgan et al. 2001). Processes or events can be mapped in real time, as in the example of detecting and mapping active fires or hotspots (Giglio et al. 2009), or can be mapped after the fire has been extinguished by detecting the footprints (Remmel and Perera 2001) and determining the severity of the burn (Mitri and Gitas 2008). Similar conditions and effects can be assessed for other dominant forms of forest disturbance, such as wind damage (Fransson et al. 2010), insect infestations (White et al. 2004), disease outbreaks (Chen et al. 2015), drought (Goerner et al. 2009), or flooding (Voormansik et al. 2014). In each unique case, the purpose of the mapping will be slightly different, as will the variables measured and the types of maps produced. Whereas hotspots or fire instances might be represented by a series of points, disturbed areas are likely to be delineated as a polygon or a related set of polygons, and blowdown maps may be delineated as linear features oriented with respect to the prevailing wind direction. Regardless of the subject, each thematic map will contain the geographic positions of the entities considered by the analysis along with descriptors of those entities, recorded at one or more of the measurement scales discussed earlier.

Considerations in Forest Landscape Mapping

Describing Spatial Patterns

Areas within a landscape express some element of similarity or connectivity, whether this is in terms of an attribute, process, function, or service provided by the landscape; these related areas may form a region with characteristics that are distinct from those of other regions. Depending on the premise that connects the areas, the definitions of the regions may change from being formal (based on internal characteristics) to functional (based on function), but either approach identifies areas that are more homogeneous internally than they are with respect to their surroundings. The regional approach to thematic mapping involves the conceptual approach of moving from inward to outward; in this approach, regions are

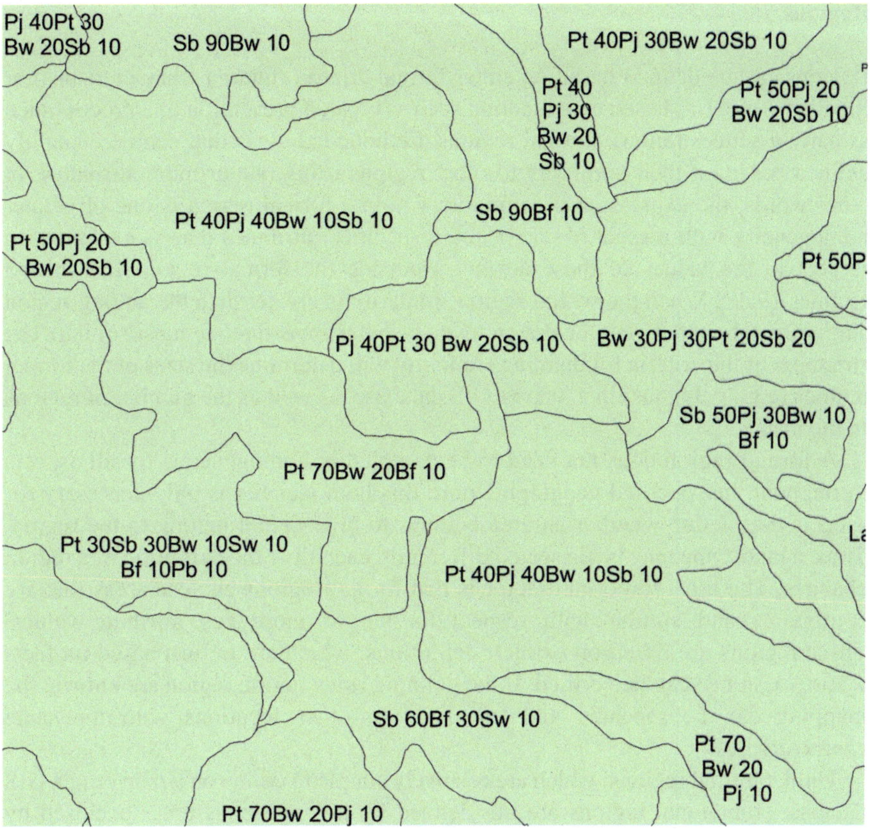

Fig. 12 Example of a Canadian forest resource inventory map of polygons that depict different stands and species compositions

seeded (given an initial position) at identified locations and set to grow outwards from those seeds based on algorithms that accumulate areas based on specific criteria (e.g., similarity of attributes, textures, brightness values, reflectance).

Since area data (e.g., polygons or clusters of raster cells) that partition landscapes into two-dimensional entities dominate the representation of forested landscapes at most operational scales (Fig. 12), we focus here on two broad approaches for conceptualizing and constructing the corresponding spatial units. The first is to focus on the areas themselves by building polygons, contiguous clusters of cells, or patches outward from specified centroids (the geometric center or the center of mass) using "greedy" algorithms (i.e., algorithms that seek local optima at each stage of the analysis) to build homogeneous landscape units. The second approach focuses on discovering boundaries by measuring and isolating high-contrast locations, and then drawing the boundary segments that separate these locations; eventually areas will become enclosed and will delineate the most homogeneous patches that are separated by these boundaries.

Regions

Regions can be defined based on either formal criteria (internal characteristics) or functional criteria (based on a function such as erosion prevention or a process such as carbon sequestration). Formal regions are bounded areas that can be explicitly defined based on their similarity to other regions using one or more attributes. In other words, the expression of similarity within a formal region is one of greater homogeneity with respect to one or more specified attributes than is evident with respect to the values of those attributes outside the formal region (Brown and Holmes 1971). When the within-region similarity is greater than the among-region similarity, a boundary can be drawn to enclose the areas that are most similar. The strictness of the criteria for defining similarity will determine the sizes of the formal regions (e.g., polygons) in a map or GIS database, as well as the number of regions in the map.

A formal region does not need to be completely homogeneous for all aspects throughout the specified geographic unit; this homogeneity is only necessary for aspects that define whether an area belongs to or does not belong to the region. Thus, a landscape may be dissected differently each time the definition of a region changes. The most important factor is that formal regions enclose areas that are contiguous and similar with respect to one or more key attribute values. These regions are based on explicit definitions, which are in turn based on facts whose existence can be verified. If the defining rules for the region are known, the mapping can be repeated and translated to other locations with the same characteristics.

Unlike formal regions, which are relatively simple to define on a map or in a GIS database, functional regions are not defined by homogeneous areas enclosed by clearly marked boundaries. Instead, functional regions are defined by relationships that connect multiple entities. Nodal regions are a special case in which there is only one central location (the node) to which connections are made (Brown and Holmes 1971). The connections can be obvious, as in the case of transportation networks (e.g., roads connecting harvesting sites to mills), but the connections can also be process based (e.g., wind distribution vectors, animal migrations, nutrient cycling pathways). Functional regions are associated with multiple locations based on the connections or processes that link them across landscapes, and the locations of the constituent areas do not need to be contiguous or even proximal.

Whereas formal regions are useful in defining areas to represent (for example) inventory data that is spatially clustered or homogeneous, functional regions are important for defining and representing operative connections that are important in process-based modeling applications. Consider the dispersal of seeds from a tree (a point source) into the surrounding landscape. In a simple gravity-only model, the dispersal area may mimic the polygonal area of the tree crown projected on the ground. However, in reality, wind may disperse the seeds farther and in the prevailing wind directions, and animals or birds may carry these seeds even farther, over considerable distances. The interactions between dispersal mechanisms, the physical controls on the processes that govern the functioning of the dispersal mechanisms, and the characteristics of the point source all combine to form a complex functional region that may not be optimally represented by the polygons or contiguous sets of raster cells of a model based on formal regions. In strong contrast, functional regions are characterized by connectors (often represented by arrows or links on maps) between a central location and one or more distant locations. For functional regions, the linkages among locations are due to interconnectivity through a process rather than the simple physical adjacency or proximity that is easily represented by a single area.

Formal and functional representations can be combined to take advantage of their offsetting strengths and weaknesses. In an insect infestation, when the spread vector (the travel distance and direction) of the infesting species is relatively short, infested areas that develop during a single outbreak are likely to be closely clustered in space and time, with only short periods of low colonization activity or small dispersal gaps within the landscape. In such scenarios, the multiple infested formal regions can be linked in a map by labeling them with a common attribute code to indicate that they all belong to the same functional outbreak region. Similarly, a wildfire event may leave unburned residual patches in the landscape within the predominantly burned area or may burn smaller patches outside of the burned area; these multiple landscape components can have vastly differing land cover composition and other characteristics yet still be connected functionally to the same event. The linking of formal regions by means of one or more functional relationships permits the study and mapping of complex landscape processes and interactions that would otherwise be impossible to discern. In forested landscapes, it is particularly pertinent to consider functional connectors in the context of connectivity, as in the case of migratory routes for animals, with habitat patches linked by corridors or travel routes, such as avian pathways. The connectivity may also be related to physical processes, such as moisture and nutrient movements within complex forest environments.

Attribute Fuzziness

Cartographically speaking, an ideal world would be composed of a limited number of features evenly distributed within a geographic space. Each feature would be easily identifiable and describable, with clear and distinct boundaries that never be overlapped with those of other features and aligned perfectly with the sampling frame or raster grid onto which these features were mapped. Similarly, the features would all exist and interact at the same scale. Unfortunately, the world is not cartographically ideal any more than all measured quantities are integers; just as some properties or processes must be represented by decimal values rather than integers, reality cannot always be divided into discrete chunks. Even a relatively uniform and homogeneous forest stand is inherently complex in its horizontal, vertical, temporal, and attribute dimensions. Any decision to map these environments will be affected by decisions surrounding the cartographic scale, observation time, and choice of representation (e.g., vector, raster, point cloud).

Imagine a 10 m × 10 m square in a forested environment. The number of ways in which this 100 m^2 patch of land could be described is staggering. Which attributes should be included? Should the attributes be qualitative or quantitative? How many levels should be defined for any of the attributes? Now imagine the compounded effect if this area were 30 m × 30 m or even larger, such as a landscape. These are the realities of data collected by imaging with medium spatial resolution or even field sampling. At some scale and extent, even conceptually simple attributes such as composition become mixed within the MMU. Thus, it is possible to imagine that data presented at the scale of the MMU is not compositionally pure, but rather represents a mixture of two or more entities. Consider, for example, a pixel in a satellite image that falls at the boundary between water and land and therefore includes both surface types. In that context, a better representation would perhaps be to incorporate this fuzziness into the attribute definition. The notion of fuzzy sets (Zadeh 1965) indicates that membership in a specific MMU is actually a set of probabilities for each attribute or attribute value that sum to 1 (i.e., that account for all possibilities) and that represent the relative abundance or likelihood of certain entities existing within the specified spatial extent.

Fuzziness describes a situation based on probability rather than discrete values; probabilities range from 0 to 1, whereas discrete values can only be 0 or 1. The fuzziness in the membership of a pixel or other spatial unit can be used to produce a range of output maps (or inputs to spatial models) that can be used to explore potential output domains. Since this approach is highly flexible, a clear specification of the attribute definitions is essential to ensure logical consistency in the stored data. It is also imperative that each definition relate to a specific scale, ranging from local to regional or even global. This means that if local data is to be aggregated or connected with datasets that cover a larger area (e.g., at a regional scale), the definitions must be broad enough to incorporate subtle changes in nuances of the definition that result from the changed scale.

In addition, when it's necessary to slice continuous variables into categories, what rules will be followed to make this division a reasonably objective process? There are many possible methods (Slocum et al. 2009), ranging from simple (e.g., dividing the range into equal intervals) to complex (e.g., identifying natural breaks in the distribution), each influencing the outcome (Csillag et al. 2008). Finally, the units of measurement used to quantify an attribute (e.g., density, vegetation cover, proportions) all have underlying spatial contexts. When the spatial context changes, the values stored in the attributes may no longer be representative; thus, procedures need to be defined for how to adjust the values in such instances. This may lead to additional problems that must be solved, such as rounding errors and classification adjustments; each solution is likely to incorporate some degree of fuzziness into the attribute definitions. A final consideration is how to deal with values that occur at boundaries. For example, should a tree that is exactly 10 m tall be grouped with the size class from 9 to 10 m or the class from 10 to 11 m? This leads us to consider the concept of boundaries in more depth, both in terms of attributes and spatial geometry.

Focus on Boundaries

The opposite approach to identifying regions (polygons, areas, patches) is to shift the focus from spatial homogeneity within an area to the task of identifying highly contrasting areas that reveal the possibility of boundaries (contrast interfaces) between areas. Line segments can be used to represent where such boundaries exist and can then be connected to form lines that eventually envelop areas that can be considered to be distinct regions based on substantial dissimilarity from their surrounding regions or based on a lack of internal contrast within the region. Since boundaries can persist in landscapes for long periods (Jordan et al. 2008), many methods exist for identifying boundaries, including the identification of locations where a property changes rapidly (i.e., wombling; Philibert et al. 2008) and assigning membership probabilities to continuous boundary functions (Mark and Csillag 1989). Contrast enhancement, which uses the available data to improve the distribution of image shades or colors to create a better visual image, can be performed either algorithmically or by selecting imagery from appropriate dates; for example, mapping coniferous versus deciduous tree stands can be improved by using winter scenes because more snow adheres to conifers than to deciduous species, thereby improving light versus dark contrast (Peterson et al. 2004), and this difference in properties has proven useful for identifying boundaries. The following sections explore these methods for dealing with the problem of boundary fuzziness and the impacts of spatial scale.

Membership Function

The boundary between two adjacent patches is rarely an absolute and infinitely fine line scribed between two obviously different regions. Although such an ideal description may be true for legal boundaries and is approached by some built or highly contrasting surface features (e.g., walls), the transition between one patch and an adjacent one in most natural environments is much more gradual, and there is often a degree of overlap between adjacent patches. (Consider, for example, how the crowns of adjacent trees interpenetrate.) This gradual transition between patches will vary based on the combinations of all the influencing factors that have been discussed thus far in this chapter, but given the additional complexity of defining regions, these complications are often disregarded or considered negligible.

However, some researchers have taken up the challenge of describing gradual transitions and explaining them by means of various functional forms (Mark and Csillag 1989) or by treating transitional regions between patches separately to mimic ecotones, where land cover types mix (Zhang and Suart 2001). The questions then arise as to what form the patch-membership function should take, how wide the transition zone should be, and whether the transition zone should have sub-transition zones. This quickly becomes a question of scale choice.

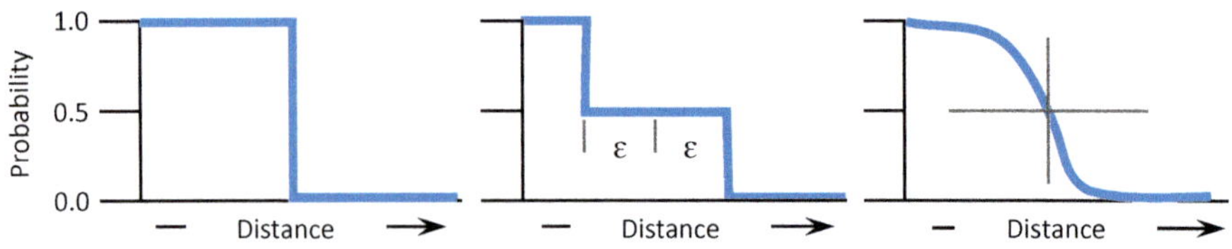

Fig. 13 Typical boundary functions between two areas that can be used to define membership or the probability of membership in a given class. In the second panel, ε represents the width of the error band at the boundary of each class. In the third panel, the crosshairs represent the midway point between the two classes. Modified from Mark and Csillag (1989)

Figure 13 illustrates three conceptual approaches to the handling of a boundary and the resulting membership function. If the vertical axis represents the probability of membership in a class or an area (or some other indicators of belonging), the first panel indicates a binary case, in which something belongs to one patch or the other and the boundary between them is crisp and unmistakably defined. The second panel indicates a case in which some distance on either side of the boundary is allocated to one of the two error bands, with one band associated with each focal patch. A location that falls within either of the error bands suggests that the location belongs to an ecotone (i.e., a transitional space between two classes), rather than belonging unmistakably to either of the classes. Variations of this form include variability in the width of the error band relative to some characteristic (e.g., direction, local patch width, land cover-type contrast). In the final panel, membership is described by a smooth function that relates distance along the curve to some degree of membership. This functional form could be linear, Gaussian, stepped, or any other defensible functional form. Once one of these three functional forms (or another form that we haven't discussed) is selected, then the patch boundary can be mapped, and any point along the gradient can be classified as a member or nonmember according to its probability of belonging to a given patch. Conversely, the boundary can be considered as being fuzzy, and the function can be used to determine the probability of membership in a given patch at a specific location along the horizontal distance between the centroids of the patches. Representative examples of these extreme cases are presented for the example of a managed boreal forest site in northern Ontario, Canada (Fig. 14).

Boundary Location Fuzziness

Once the rules for demarcating the boundaries that delineate regions have been decided, and the resulting maps have been constructed, how confident can we be in the boundary locations? Whether the boundaries in question were identified by image classification, digitizing, GPS data acquisition, or conversion of perimeters sketched on a manually annotated map to digital images, both the physical position and the uncertainty of the boundary's position must be scrutinized to determine their accuracy. In some cases, boundaries are constructed to have a width that describes the probability gradient for membership in the adjacent patches. The width could also be like the error band (ε in Fig. 13) that dictates the uncertainty

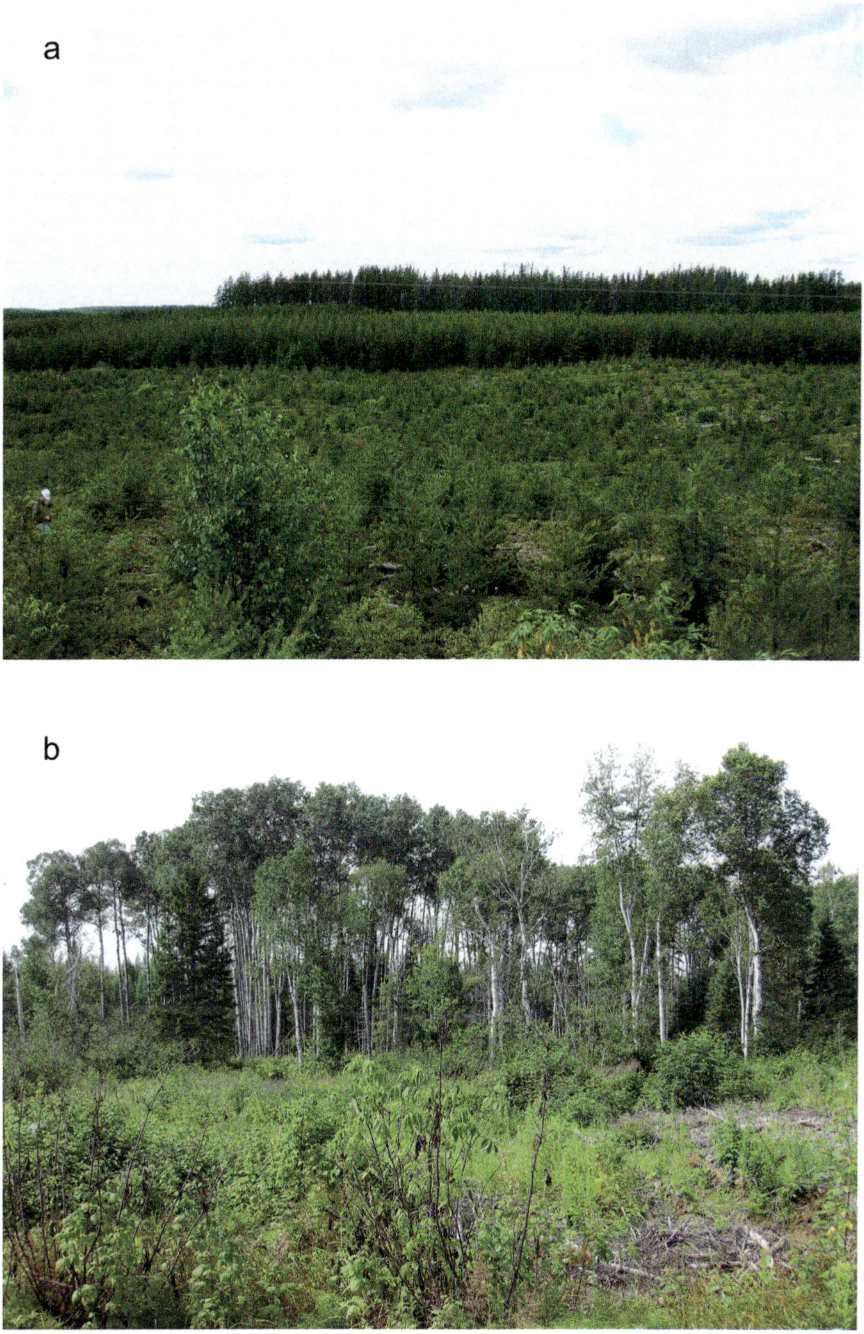

Fig. 14 Images of forest stand boundaries (ecotones) for a managed boreal forest site in northern Ontario, Canada: (**a**) stepped, (**b**) gradual, and (**c**) abrupt. Figure 13 illustrates the membership functions for these boundary types. Photographs by Tarmo Remmel

Fig. 14 (continued)

in its horizontal positioning (Abeyta and Franklin 1998; Kronenfeld 2011). Much work has been done to characterize the vagueness and uncertainty of forest stand boundaries (Edwards and Lowell 1996; Brown 1998; Jordan et al. 2005).

In reality, data analysis typically involves data with multiple spatial resolutions and integrated queries across multiple scales and levels of generalization. In this context, fuzziness of the boundary location becomes a central issue. As boundary lines are simplified, weed tolerances are altered such that the minimum vertex spacing changes, so line complexities will adjust to reflect those changes. These alterations will influence the geometry, curvature, and spatial footprint of the boundaries and therefore introduce some level of fuzziness into their definitions. Whether those locations are represented in vector or raster forms, the positional uncertainty may represent the scaling uncertainty rather than the inability to precisely select a boundary location. In raster cases, a coarsening of the spatial resolution will increase the mixtures within cells (i.e., a line may not fall completely within one cell of the grid, and may instead fall partially within two adjacent cells) and widen the boundaries. These actions alter both the accuracy and the precision of the boundary position.

Boundaries as Transitional Regions

Depending on the specific landscape, the scale of data collection and mapping, and the definition of the attribute concerned, the boundary enclosing a region may itself be region. Consider, for example, a riparian zone that separates a river from the surrounding forest. In ecological parlance, transitional zones that are wide enough to possess their own unique conditions are called ecotones, and can be described as being different from the two regions they separate. They are uniquely defined by characteristics related to habitat type, light penetration, airflow, moisture regimes, or other factors. Ecotones often provide niche habitats for species, or define zones in which predation rates differ from those in core patch areas. In such cases, the traditional definition of a sharp boundary (or even a fuzzy boundary) may not suffice. Instead, the boundary becomes its own region and may necessitate its own delineated patch structure. The structures of ecotones have been studied to identify optimal ways they can be characterized and delineated in terms of the uncertainty and abruptness of the change across these wide transitional zones (Bowersox and Brown 2001; Arnot et al. 2004). However, in addition to the issue of defining the boundary, the ecotone's internal structure must also be considered; ecotones may possess patches and other complex structures, depending on the observational scale used.

Beyond 2D Data

Although the 2D plane is the traditional domain for mapping, it is becoming increasingly important and fruitful to map additional dimensions. Many environments (at high spatial, temporal, radiometric, and thematic resolution) have important 3D characteristics (i.e., topography), and because of the importance of temporal changes, a fourth dimension (time) will be important in many contexts. Forests are a great example of multidimensional entities; in addition to their 2D horizontal extent, they grow vertically and change over time. The inclusion of additional dimensions has important implications for data acquisition, storage, handling, cartographic representation, and analysis. In the following sections, we explore some of the interesting new elements that can be measured and analyzed by extending our traditional 2D perspective to include additional dimensions of forested landscapes, including vertical extensions and time. These additional dimensions have implications for data handling, storage, and processing time, but in exchange for these costs will permit many new and interesting interpretations and analytical possibilities.

Vertical Structural Dimension

The measurement of vertical landscape elements in forested landscapes is difficult in the field. Although it is possible to directly measure, triangulate, or estimate tree heights, these values become increasingly difficult to attain as the height of trees or the stem density increases. For this reason, photogrammetric approaches that rely on the overlap of successive air photos and the parallax of human sight have been important throughout the past century for mapping and measuring properties of landscapes beyond simply the horizontal plane. Such measurements have been especially useful for measuring tree heights, describing topography, separating understory from overstory vegetation, estimating DBH and biomass, and calculating several additional biophysical metrics.

Recently, the availability of light detection and ranging (LiDAR) technology has spurred the automated processing of data from wide areas to extract these types of values (Lim et al. 2003a, b; Coops et al. 2007). LiDAR provides both terrestrial and airborne capabilities for directly collecting 3D data to characterize topography and land-surface features (Shan and Toth 2009). Chapter "Airborne LiDAR Applications in Forest Landscapes" by Ko and Remmel discusses this subject in detail. Opportunities for simultaneously producing digital elevation models (DEMs) with high spatial resolution, computing the heights of trees, and assessing biomass and several other biophysical parameters became a reality. Although there are limitations to this technology due to shadowing (blocking of some features by other features), signal attenuation, and errors associated with parameter estimations (Lim et al. 2003a, b), multiple passes over an area can provide scans from multiple directions and produce datasets with very high densities of LiDAR points. Since the accuracy of this data capture is on the order of centimeters to millimeters, the scientific community is still only discovering the possibilities that this technology may provide.

LiDAR requires that the laser source be equipped with an inertial navigation system and global positioning system receiver if mobile, or that it be accurately located if stationary, so that the measured ranges can be post-processed by means of inversion to produce a set of (*x*, *y*, *z*) coordinate triplets that identify points within the cloud of LiDAR points that represent important objects such as tree crowns and the ground surface (Fig. 15). Because each light pulse emitted by a LiDAR system can generate more than one return signal, most systems can process these multiple returns; these are called discrete-return LiDAR systems. Other systems (described later in this section) can capture and process the whole waveform of the return signal, and thereby obtain even more data on the spatial structure. Both technologies allow characterization of multiple levels of a forest canopy. The first returns occur closest to the detector, generally at the top of the canopy, whereas the last returns are farthest from the detector and generally represent the ground or points near the ground.

Fig. 15 A false-color set of 3D LiDAR points in an (*x*, *y*, *z*) coordinate space that show variations in the 3D structure of a forest landscape. Here, the lighter shading indicates higher absolute elevation of the LiDAR return point (e.g., the top of the tree canopy)

With increasing extension of traditional planar 2D surveys into 3D, spurred by technological advances such as LiDAR, new data structures are being developed to accommodate these new data demands. For example, LiDAR points are identified by 2D coordinates in traditional (x, y) coordinate space, but also include a z-coordinate that captures the vertical position with respect to a defined vertical datum. These points can be rendered in 3D, and due to the large number of locations in a typical dataset appear like a cloud of points; hence, they are referred to as a LiDAR *point cloud*. Unlike remote sensing systems in which geographic space is fully partitioned into cells for which measurements are made, LiDAR datasets are primarily empty space that separates points that represent locations where laser pulses interacted with something solid that reflected some of the energy back toward the detector.

Depending on the specific LiDAR system used, and budget constraints that limit how much information can be purchased from a surveyor, each point within the cloud can have a number of attributes attached to it. In addition to the 3D position of each point and the date and time when it was recorded, the dataset can include attributes such as the scan angle or the intensity of the return pulse. Classification or feature identification and extraction algorithms are still being produced that can handle the multidimensionality of the non-connected points in the point cloud, and this is proving to be a difficult but rewarding and interesting task. Further increasing the complexity of the analysis, and therefore requiring high computing power and data storage even for simple 3D positions, multispectral LiDAR technologies have been developed that use lasers with different wavelengths to acquire even more detailed data about the environment.

Emerging LiDAR technologies are permitting the storage of full-waveform results. Thus, whereas discrete-return LiDAR may provide several returns for each outgoing pulse, full-waveform LiDAR returns a continuous wave function for the returned energy. Full-waveform LiDAR can potentially provide much greater detail along near-vertical transects through a forested canopy, but at the cost of much greater data storage requirements and longer processing times. Full-waveform methods are being tested for detailed mapping of single trees, genera, or even species-level classifications, as well as for understory detection and removal to allow the calculation of ground-surface elevations. Much of this work is still in its infancy, but major contributions are expected.

Mapping Dynamic Processes

All landscapes change; some simply change more quickly than others, or more frequently at certain scales. The notion of dynamic processes is thus context and scale specific, but can be taken to mean that there is an expectation that a landscape will often have changed on a shorter interval than can generally be measured, imaged, or mapped. Hence, mapping dynamic landscapes can be thought of in the context of change detection or the quantification of differences that occur between successive points in time. Rapid forest disturbances (hours to days, weeks, and months) create the dynamism of forested landscapes (e.g., wildfire, blowdown, insect infestations), with alteration of the composition, configuration, and morphology of the landscape. In addition, landscapes are dynamic over long periods (from years to decades and centuries), allowing less frequent surveys to capture changes due to tree growth and mortality and the resulting effects on biomass accumulation and other processes such as carbon sequestration.

To adequately map and study dynamic processes, the critical first step is to identify the time period that would best capture the essence of that specific process (this is the temporal equivalnet of the MMU in spatial data). This means that the measurements must be sufficiently frequent to capture changes in the state of the landscape that can then be interpreted to infer how a process creates specific landscape patterns. The linkage between patterns and processes can then be studied. As with any map-based comparison, standard issues related to scale, spatial resolution, extent, and composition of the overall landscape mosaic need to be considered in the study design, and appropriate hypotheses must be articulated to define the data collection needs.

Change detection and updating of maps are critically important for maintaining representations of the landscape that are up to date and complete. Temporal considerations constrain the methodology for answering questions related to change during a specified period, for assessing the effect of a specific disturbance (e.g., fires, insect infestations, windstorms, disease outbreaks, harvesting), or for identifying where specific events or processes occurred. Change detection assumes that the initial conditions are known or can be inferred, and that a change or difference between this state and another one can be described some time after the initial state was described. This typically leads to one-directional analyses (the change in a given area during a given time), but may also incorporate probabilistic techniques or transition matrices to describe the changes between states over time. The benefit of knowing past conditions when classifying satellite imagery or other sources of landscape data is that labeling accuracy will be higher than if prior knowledge is unavailable, since certain conditions are more or less likely under those initial conditions. Thus, prior knowledge supports decision making and should produce better final results.

Utility of Forest Landscape Maps

Maps serve many purposes. Historically, maps were used primarily for navigation and for identifying locations of markets for trade or the acquisition of goods and resources (e.g., villages, hunting grounds) and for facilitating the transport and relocation of goods (e.g., land and water routes). Traditionally, maps were static and showed the locations of things at the time the maps were created; these maps facilitated movement among known locations because the spatial scale was consistent across the map, thereby allowing distance (and time) calculations. Though many maps still serve these requirements, the mapping of complex and dynamic landscapes such as forests can serve many additional purposes.

Forest landscapes are often subject to intensive management, whether for conservation, extraction of timber, provision of ecosystem services, or preservation or development of certain characteristics (e.g., biodiversity) in a region. Since effective management requires baseline data and ongoing updates to provide information on the landscape's current state, maps capable of accurately characterizing all forms of forest conditions are essential for effectively managing forests. Such maps record topography (e.g., land cover and elevation), constructed features, and detailed resource layers (e.g., biomass, species, height, age) that can be used independently or in conjunction with other mapped information. These maps provide the starting points for future decision-making capacity or to monitor changes to baseline conditions.

Change detection in forest landscapes includes both anthropogenic changes (e.g., road network expansion, mapping of harvest sites) and natural events (e.g., forest fires, wind, and insect damage). Incorporation of time as a dimension permits the detection and mapping of explicit landscape changes (first-order effects) and drawing of inferences about processes and their properties (second-order effects). The temporal and spatial scales of the data acquisition and mapping, along with the map's extent, will govern the suite of discernable forest landscape characteristics that can be mapped and the changes that can be detected. It is particularly important to understand that although static displays at a single point in time provide evidence of the current state of the landscape, change detection allows inferences to be drawn about the processes responsible for that state. This improves our understanding of how and why a landscape is changing.

Increasingly, many maps are not the desired end product, but rather serve as inputs for other products, such as simulation models and decision-support systems. Thus, forest landscape maps can be the results or outputs of simulation models, in which case the mapped information is not what was originally detected or measured, but rather the derivative of one or more modeling decisions or interpretations.

A relatively recent problem that arises from the distribution of large volumes of data over the Internet relates to how to determine the origin, accuracy, and utility of maps. Many maps result from overlaying or combining data from a range of sources, some of which may have been produced by inappropriate means or may not be sufficiently similar to permit easy or reliable combination of the disparate datasets. The utility of maps that are produced by black-box models, based on undeclared assumptions or on data from unclear or unknown sources, should be questioned and the conclusions drawn from such maps should be used with heightened caution.

Map Representations

By definition, geographic data includes one or more entities (items, objects, or things) that are described by both their location (coordinates) and other useful attributes. In a digital environment such as a GIS, attribute tables contain descriptive

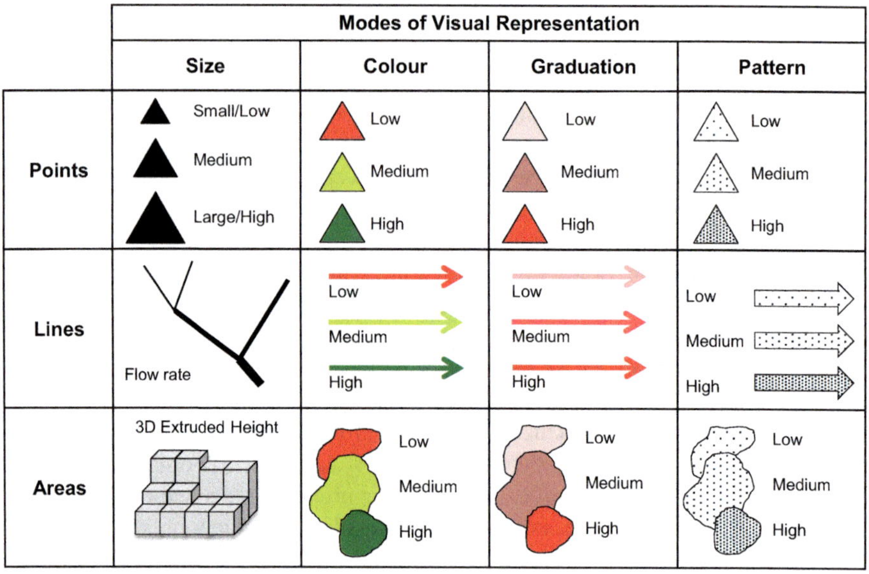

Fig. 16 Visualization of quantitative and qualitative data based on alterations of size, color value, color graduation, or symbol patterns to convey elements of data representation among the geographic primitives (points, lines, and areas)

data on each attribute (e.g., biomass), including its locations in the map. Assigning colors and symbols to different types of spatial entities or to various levels of their attributes, and then displaying these in a map, links the descriptors to their locations. In this way, it is possible to convey vast amounts of information through the resulting maps. The visual effect of colors and symbols in a map produces the spatial patterns that can be characterized, measured, and compared.

Options for coloring or labeling a map (Brewer et al. 1997; Slocum et al. 2009) depend on both the nature of the entities that must be represented and the type of attribute that is to be mapped for each entity (Fig. 16). First, markers or symbols can be chosen to represent specific types of point data, and their size or color can be modified to represent values of certain attributes; for example, circles could be used to represent deciduous trees and triangles could be used to represent conifers. Similarly, linear features can be labeled using lines with different widths, colors, and patterns (e.g., solid versus dashed) to communicate attributes attached to those features such as their boundary. Area data can also be displayed by adopting the techniques used for points and lines; for example, polygon boundaries and the fill color or opacity can be modified to communicate attributes of the polygons. Second, the symbol, line pattern, or color can be adjusted to communicate the attribute values directly. If the data is nominal or ordinal, then unique colors or symbols can be assigned to each unique value. As the measurement framework becomes more complex, quantitative data can be grouped or divided in various ways to produce categories that can be represented by monochromatic scales (e.g., using shades of black in 20% intervals to distinguish five categories) or multicolored scales (e.g., red to indicate danger, yellow to indicate uncertainty, and green to represent safety). If quantitative values are not divided into discrete categories, a continuous color ramp can be applied to the data, with the percentage of the full color value proportional to the value of the attribute; this provides a smooth, continuous range of attribute values. These approaches can be used with both raster and vector data.

Morphological Interpretations

Moving beyond the extensive body of traditional landscape pattern metrics (Baker and Cai 1992; McGarigal and Marks 1995; Riitters et al. 1995; Uuemaa et al. 2011), researchers are developing a body of pattern analysis methods that consider morphological landscape characteristics (Soille and Vogt 2009). Both branches of pattern characterization provide numerous options for characterizing and quantifying various aspects of spatial patterns, but *morphological* methods (also called *morphometric* methods) focus on quantifying forms or shapes (i.e., morphologies). These methods seek to partition binary landscapes that present the presence and absence of a category (e.g., forest) into morphological elements that describe the fundamental forms that cells or clusters of cells represent from a structural perspective. For example, basic morphological landscape pattern elements can include cores, edges, and islets (among others). Each image cell will be coded as having membership in only one mutually exclusive morphological element class (e.g., a cell cannot be considered an edge and a core simultaneously). These classifications can have deep ecological meaning and are often used to further partition the analysis of land cover that comprises individual morphological element classes.

Traditional maps portray spatial entities in either raster or vector representations by labeling the entities with either colors or symbols that represent the associated attribute values or levels within a hierarchy or sequence. The combination of these labels and their arrangement in space can reveal emergent properties of spatial patterns. Landscape pattern metrics permit the measurement (quantification) and summarization of these patterns either globally (for an entire landscape) or at the class or patch levels. These summaries result both from the attributes being mapped and from their geometry, and they are strongly influenced by spatial resolution, geographic extent, and composition and configuration of the entities.

Morphological metrics let researchers quantitatively describe and compare landscapes, though the precautions raised by Remmel and Csillag (2003) for landscape metrics extend to morphological elements too, since the distributions of these metrics are unknown and it is difficult to produce null models against which comparisons could be made. Thus, the challenge of comparing maps based on pattern is not as simple as it might appear at first glance. However, despite these precautions, morphological metrics provide a means for mapping shape-based classifications of landscape elements, thereby producing spatial summaries of landscape components that visualize the relationships among their position, connectivity, and overall shape. Since the results can be directly mapped, the results allow spatial inquiries, such as asking whether elements with certain morphologies are clustered.

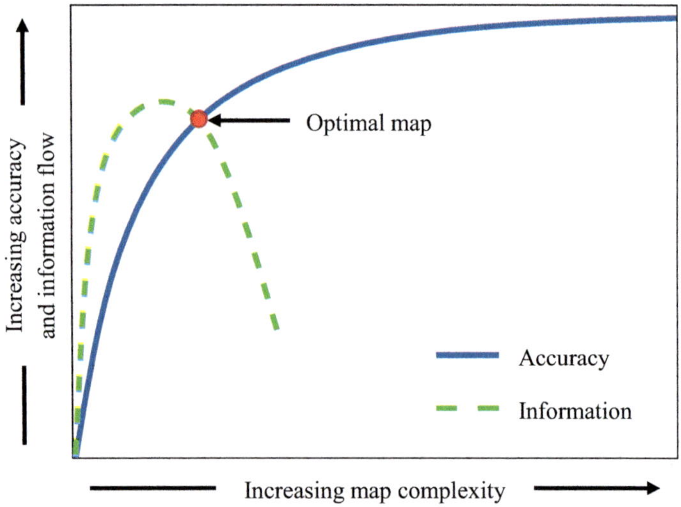

Fig. 17 Optimal maps exist in which the maximum map information flow and accuracy intersect with the curve for complexity. Modified from Jenks and Caspall (1971)

Map Scale

The concept of scale is central to mapping, and particularly so for mapping forest landscapes, which exist as amalgams of entities, flows, and interconnected processes that exist and interoperate at multiple extents and scales. The construction of maps requires a consideration of scale at all stages, from survey planning and obtaining data to analysis, interpolation, and interpretation, and finally to combining data from multiple surveys or maps and using the mapped and georeferenced information to support research or forest management. Jenks and Caspall (1971) proposed a theoretical model in which an optimal map exists for any purpose and in which both the information and accuracy are maximized; to do so, they proposed a graph with axes for map complexity and for accuracy and information flow, where information flow refers to the ability to grasp the concepts presented in a map (Fig. 17); when maps are overly complex, their interpretation often becomes more difficult and thus the information flow to the reader eventually starts to decrease. Although it's not yet clear how to formalize these functions to permit identification of the characteristics of this optimal map, the idea is clear: produce a map that combines the greatest flow of accurate information with the minimum complexity.

There is no easy answer to the question of how to choose the most appropriate scale for analysis. However, there are some useful rules of thumb. First, and most important, the scale must be such that it lets you see the features you are attempting to study. That is, the MMU must reveal the distribution and patterns of the features being studied. This means that the smallest feature of interest should be represented by at least one cell in a grid or one visible shape in a vector display, with a sufficiently high resolution that it will be visible as a core area surrounded by different features and with boundaries between them. At the same time, the internal heterogeneity should not be so large that its identity as a distinct feature is concealed and the analysis becomes unnecessarily complicated. Similarly, the cells and shapes should not be so large that they oversimplify the surface through smoothing. Woodcock and Strahler (1987) clearly summarize the scales at which objects are best observed based on measures of local variance along a continuum of spatial resolution.

Scale influences the amount of detail that is visible. This is most easily conceptualized by thinking about spatial resolution in terms of the pixels that comprise a satellite image: images with fine spatial resolution show more detail than those with coarse spatial resolution. Hence, spatial resolution is often a good proxy for scale. This concept extends to include time, since the interval between successive measurements (i.e., the observation frequency) and the duration of each observation will both influence the types of changes or dynamic processes that can be observed, measured, and mapped. Scale also depends on the extent of the area that must be mapped to reveal the conditions, patterns, and processes being studied and the methods that can acquire the necessary data most easily at that spatial extent (Fig. 18).

Given these definitions, mapping of naturally complex forest landscapes by detecting and measuring phenomena, patterns, and processes is inherently a scale-dependent task. Scale dependence in ecological patterns and processes has been an important theme for several decades (e.g., Turner et al. 1989; Wiens 1989; King 1991; Levin 1992; Peterson and Parker 1998). Advances are expected in this domain as increasing familiarity is achieved with specific environments and data sources, but determination of the optimal scale will always remain context specific. Difficulties remain in determining a single scale for mapping and analysis because many phenomena and processes operate at different scales or over a range of scales. Moreover, two processes that interact may operate at different scales. One key breakthrough that mitigates the difficulty of scale dependency is the use of GIS software and other computer tools, which allow researchers and managers to interactively change the scale until it is optimal for a given purpose and to reveal important differences among scales.

As spatial resolution decreases (becomes coarser), narrow patches and ecotones may become too small to be represented and may disappear completely from datasets and maps. Area and perimeter (hence, shape) computations will be affected by this disappearance, as will simple frequency counting to determine whether certain classes (particularly rare classes) exist within a landscape. The resulting changes in patch shapes will also influence computed spatial relationships such as nearest-neighbor calculations, boundary widths, contrasts, or abruptness of a boundary. McIntire and Fortin (2006) showed that boundary widths (and especially the gradient steepness and boundary heterogeneity) change substantially in response to changes in spatial resolution in landscapes affected by mountain pine beetle infestations and wildfires. This consideration of spatial resolution is important for the analysis of satellite spectral data, since as spatial resolution decreases, the amount

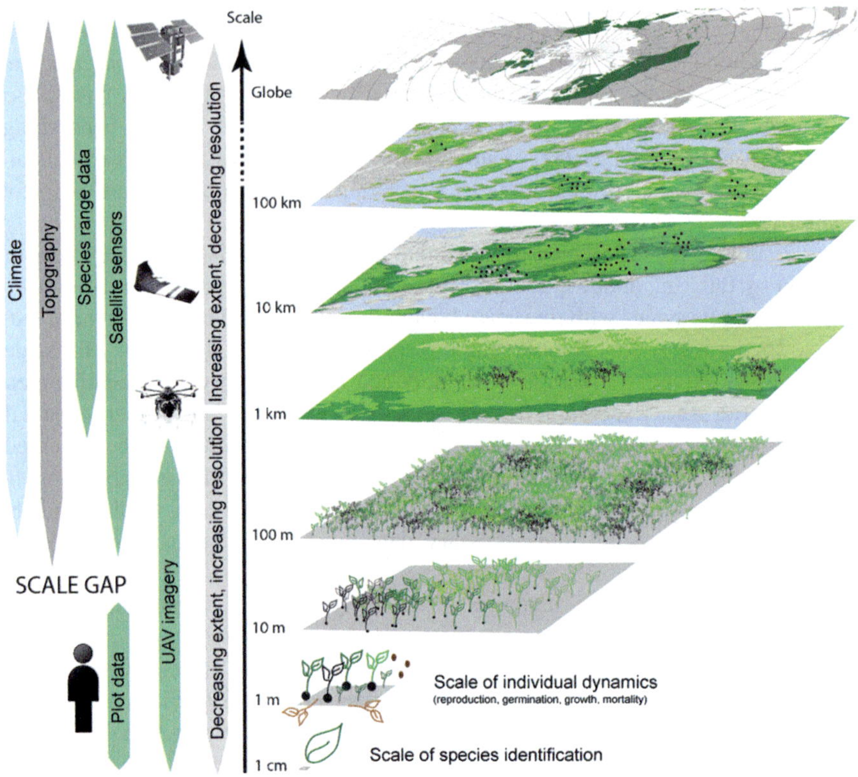

Fig. 18 Illustrations of the relationships among scale, resolution (level of observable detail), and the patterns revealed at each scale, along with the typical methods of data acquisition. Modified from *http://conf2016.uas4rs.org.au/speaker/urs-treier-university-of-aarhus-denmark/. Used with permission*

of mixing of spectral signatures within cells increases and classification accuracy decreases, and although the decrease may be small (Boudewyn et al. 2000), its magnitude will depend on the classes, their configuration, the landscape characteristics, and other factors. Therefore, researchers (e.g., Remmel and Csillag 2006; Gaucherel 2007) have attempted to analyze maps and patterns at multiple spatial scales, since single-scale analyses do not always provide the complete picture.

Spatial Resolution

It's important to account for the impact of spatial scale on how landscapes are represented and the level of information a map can retain. Here, we examine this problem from a patch (or area) perspective by considering both the geometric and attribute elements of a landscape and the impacts upon them as scale changes.

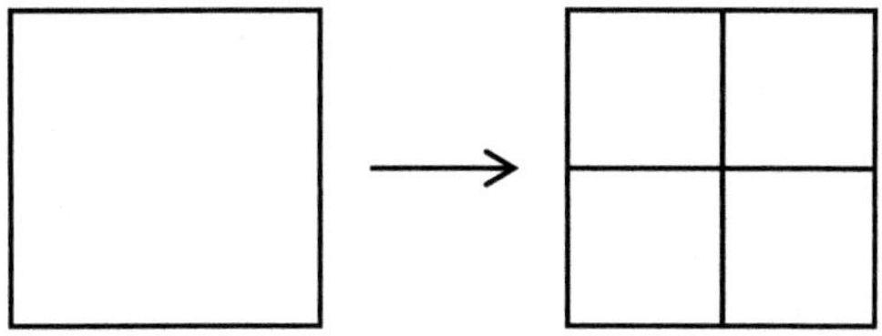

Fig. 19 A doubling of the spatial resolution equates to a quadrupling of the number of raster cells that must be analyzed

In raster environments, the impact of spatial resolution can be significant, since halving the pixel size (i.e., doubling of the spatial resolution) quadruples the number of cells that must be analyzed, which increases both the storage requirements and the processing time (Fig. 19). Although this change has operational considerations (i.e., defines how data must be collected), a new type of data implication arises: the modifiable areal unit problem (MAUP). In this problem, aggregation of area-based measures of spatial data introduces various forms of statistical bias that depend on the scale of the aggregation. The biases can affect the means or variances, depending on how the spatial aggregation or zoning is conducted (Jelinski and Wu 1996; Dark and Bram 2007). The characterization of forest stands is strongly influenced by MAUP. For example, if forests are classified to a dominant species at some fine mapping extent, the aggregation of these spatial units to coarser ones without starting from the original raw data can introduce bias so that the aggregated maps no longer reflect the true species distribution on the ground. This concept is similar to the popular vote and aggregated Electoral College results being in disagreement with each other, as occurred in the US presidential election of 2016 (Collin 2016). Although the MAUP is most often seen as a problem, Openshaw (1977) suggests that the effect can be highly useful once it is understood and controlled for. Menon (2012) presented an example in which different aggregations (zoning) of underlying counties (spatial partitions) were applied to compute a specific index and control its value.

As spatial resolutions decrease, the area represented by each raster cell increases. In many cases, this means that the attribute definitions must be revised (or carefully reinterpreted) to account for the fact that they represent a combination of larger areas that leads to less local detail and complexity. The use of coarser spatial resolution also creates an averaging effect that weakens the ability to detect areas with specific spatial characteristics because the attribute represents the average of two or more areas with different spatial characteristics. (In remote sensing, this is called the *mixed-pixel effect.*) One important implication is that extreme values may disappear or become less frequent, which is problematic when those extreme values are the important characteristics being studied. This effect is clearly seen when a DEM is resampled. DEMs are typically raster grids in which each cell contains a value that represents its height above some reference datum. Because of the large number of cells in a grid that covers a large area, it's often necessary to choose a coarser resolution to allow mapping, modeling, or assessment of large areas in a reasonable amount of time. Unfortunately, the landscape becomes increasingly smooth as the spatial resolution decreases. This can both decrease slopes in areas with a steep slope and lead to the loss of peaks and valleys. This will affect interpretations of erosion potential, the development of drainage networks, and the angle of incoming solar radiation. This can have consequences for both environmental research and forestry operations such as planning and building roads. A smoother DEM may also contradict a tree canopy height model created from LiDAR data with high spatial resolution, leading to a skewed interpretation of tree heights. In contrast, the use of finer spatial resolution does not necessarily improve the level of thematic detail (e.g., mapping the extents of specific forest stand types), particularly when the edges of these objects account for a small proportion of their total area.

Changes to spatial resolution have also been shown to have potentially significant additional consequences, including alterations of patch shape, size, and edge complexity, all of which affect the values of metrics based on these factors. Thus, comparing maps created at different spatial resolutions can introduce bias. In vector environments, the cartographic scale, MMU, and grain or weed tolerances will all affect the level of observable detail that can be presented. Researchers and managers often use line generalization functions to simplify linear features, but as is the case in a raster environment, altering the number of vertices that define a line will affect estimates of its length and complexity, and the area of any associated polygons. When spatial resolution becomes overly coarse for the landscape being observed, smaller or narrower patches will begin to disappear from the final map, since the dominance of neighboring patches will weigh more heavily in the assignment of attributes to cells or delineated polygons. This alters the observed spatial pattern, since it decreases the number of patches that comprise the landscape, thereby decreasing the landscape's perceived complexity and potentially resulting in an overly simplistic understanding.

Increased Spatial Resolution Versus Marginal Information Gain

In a world of ever-increasing desire for increased spatial resolution (e.g., high-definition television, digital cameras with more megapixels), it's not surprising that the same desire for detail exists with satellite image acquisition. Although imagery with high spatial resolution can provide more or better visual context, thereby benefiting the human element of identification and classification work, it often complicates automation of mapping and spatial analysis because of the larger amounts of information that must be processed. In addition, there is the question of diminishing returns: at some point, increasing the resolution does not provide enough new or useful information (the marginal information gain per unit of increased resolution) to justify the cost of acquiring and processing the data. In addition, the amount of information obtained may become overwhelming, preventing humans from seeing important overall patterns, as in the phrase "failing to see the forest because of the trees."

One advantage of using data with coarser spatial resolution is that complex environmental states are generalized, smoothed, and averaged, leading to decreased spatial heterogeneity and allowing us to see patterns that would be difficult to detect amidst a sea of details; in effect, it becomes easier to see the forest when one is not distracted by all the individual trees. Calculations and data processing also become simpler and faster. One powerful advantage of tools such as GIS software is the ability to interactively adjust the resolution, thereby allowing researchers to adjust the scale until they can focus on the most important details. Notwithstanding the perceived benefits and esthetic qualities of images with high spatial resolution, researchers and managers must always ask whether the marginal information gain obtained from increased resolution justifies the costs (i.e., increased processing time and complexity). They must carefully consider the trade-off between image extent, spatial resolution, and information content.

Multiple-Scale and Cross-Scale Analyses

With the growing availability of spatial data at multiple spatial resolutions and growing interest in characterizing landscapes or processes across multiple scales, it has become necessary to develop new methods capable of performing these types of analyses and to refine existing methods (Johnson et al. 2001; Mysorewala et al. 2009). For example, multiple-resolution assessments of species diversity have been achieved by using quadtrees (Csillag et al. 1992), which partition a space into quadrants only if a specified threshold for diversity is reached. The result is a surface tiling with tile size that varies across the study area. Similar approaches have been used to segment land-cover maps to improve summary statistics by reducing the error terms generated by modeling (de Bruin et al. 2004) and to compare raster-format categorical maps (Remmel and Csillag 2006). Continued development of multiple-resolution analyses, including wavelet convolutions (Bradshaw and Spies 1992) and fusion of LiDAR and image data (Dalponte et al. 2008), will broaden our understanding of the spatial processes that act on forested landscapes and permit the construction of increasingly informative maps. Major gains will come from streamlining the fusion and simultaneous analysis of data from multiple sources and obtained at multiple scales.

When comparing forest maps at different levels of thematic resolution or performing comparisons between data products with different thematic definitions, the results can differ greatly among the algorithms used to perform the aggregation (Remmel et al. 2005). Aggregation of thematic data with different resolutions has been shown to significantly affect the characterization of spatial patterns, particularly when landscape metrics (see Chapter "Mapping the Abstractions of Forest Landscape Patterns" by Uuemaa et al.) are used to prepare quantitative summaries and abstractions (Buyantuyev and Wu 2007). When disparate data products are combined into a common analytical framework, it becomes imperative to align the thematic resolutions and definitions. For more information, see Chapter "Towards Automated Forest Mapping" by Waser et al. This will require substantial effort to ensure semantic consistency and to identify thematic ranges that are comparable across products. We predict that as methods develop and access to data products increases, standards will be developed for thematic classification and description, because these issues will become as important as geometric accuracy has been in the past.

Error Assessment and Validation

Sources of Errors

When maps are validated, their positional and geometric errors are often assessed, but thematic labels and overall spatial patterns should also be assessed for completeness and accuracy. Positional accuracy can be assessed in either relative or absolute terms, depending on the context and the desired use of the maps. When maps will not be used in conjunction with other spatially explicit data, then the absolute location does not matter. When the absolute locations of features (e.g., a specific latitude and longitude) are not required, then a less restrictive relative assessment of positioning accuracy can be applied. Such assessments examine the relative size and positioning of geographical features (e.g., adjacency, proximity, area, perimeter) without requiring that these features be tied to real-world locations.

However, when map overlays or change detection must be performed, then the positional accuracy must be assessed to ensure that the coordinates are equivalent between maps. Often, the positional error permitted on maps will be given as a horizontal or vertical distance error (e.g., ±2 m), which often depends on the cartographic scale and can be summarized using metrics such as the root-mean-square error (RMSE). RMSE summarizes the average separation between the actual (A_i) and observed (O_i) map coordinates of point i between a sample (N) of control locations across the map (Eq. 1):

$$\text{RMSE} = \sqrt{\frac{\sum_{i=1}^{N}\left(A_i - O_i\right)^2}{N}} \tag{1}$$

RMSE can be further partitioned to describe the error in specific directions (e.g., in the x and y horizontal directions) or for categorical data (Ding et al. 2014). RMSE values can be mapped as vectors that display the magnitude of the difference (which is proportional to the scaled length of the vector) and the associated offset direction. When displayed graphically at each control point A_i, trends and variability in positional accuracy can be visualized. RMSE generally summarizes how well a dataset matches the validation data, or how well two superimposed maps align at a series of control points.

Although RMSE values are often calculated as an overall value for an entire map or some region of the map, this fails to account for the fact that errors have many sources, and should therefore be calculated separately for each source of error. To find ways to improve accuracy by decreasing RMSE and other error metrics, it's necessary to partition this error among its many components. Such errors arise from the projection, datum, spatial resolution, geometric registration, and precision chosen by the researcher. Furthermore, Dauwalter and Rahel (2011) demonstrated that the patch sizes that comprise a landscape, rather than their shape or complexity, influence positional accuracy. Ongoing research will be needed to establish acceptable practices for assessing positional accuracy (both globally and locally) and for presenting this information in statistical and graphical forms. A greater focus needs to be placed on partitioning spatial accuracy into error, uncertainty, and bias components for all input data simultaneously, resulting in a multidimensional assessment.

When the statistical distribution of a dataset (e.g., normal vs. Gaussian) is known, standard statistical techniques can be used to convert the RMSE into a physical distance at a given scale (USGS 1999). For example, the 95% confidence interval for normally distributed data would indicate that errors should not be greater than 1.64 times the RMSE. The United States Geological Survey, a major producer of maps, adheres to a series of heuristic targets for horizontal, vertical, and thematic accuracy. The targets depend on the cartographic scale of the original map (USGS 1999).

Digital maps can now be easily rescaled in GIS software and produced at virtually any cartographic scale that is desired, but it is crucial to understand that both the new scale and the map's accuracy relate directly to the original scale at which data were measured and converted into maps. Errors also appear when information is converted to a coarser resolution, which results in the averaging problem described earlier in this chapter, or interpolated to a finer resolution, which requires assumptions about the missing data (Jelinski and Wu 1996). Errors also appear as maps progress through sequences of aggregation, overlaying, or other processing stages because the errors from each process can combine, interact, and propagate in derivative products (Wu 1999). In ecology, this is termed *aggregation error* or *transmutation error* (O'Neil and King 1998), or *extrapolation error* (Miller et al. 2004). The existence of such errors and the need to deal with them during cartographic analysis emphasize the importance of knowing the origin of maps and the need to obtain complete and informative data about the origin and characteristics of the maps (i.e., metadata).

Map Validation

Maps promise that specific entities will be present at specific georeferenced points. Accordingly, a mapped point is *accurate* if that particular entity is encountered at that specific location. Maps can be considered *valid* if all points or a statistically representative sample of points has been verified, and *validity* means that the map can be used for the intended purpose. The process of verification is scale specific; accuracy must be assessed at the same spatial and temporal resolutions that were used to generate the mapped information and that will be used to create the map.

Missing data is a familiar problem in the geoinformatics community. Whether the gap results from missing attribute information or the need to "clean" a dataset to remove erroneous data (e.g., cloud-contaminated pixels in satellite images), these gaps pose problems in subsequent analyses. The problem arises at all scales and for all platforms. For example, satellite data will continue to be plagued by atmospheric conditions that block transmission or reception of a signal. Similarly, ground-based data collection will be subject to human error, such as missed survey dates due to illness, bad weather, or an inability to survey during the whole-field season because of budget constraints. Data collected by automated sensors is sometimes lost due to power failures, running out of storage, vandalism, or damage by animals. For these and other reasons, incomplete data are an expectation rather than a rare occurrence. One solution to this problem is to use multiple data sources to fill gaps in a data series or in coverage. This approach requires the merger of data with multiple spatial and thematic resolutions, leading to the data fusion problems described earlier in this chapter.

When gaps in data are detected, it is possible to simply mask out those areas from further analysis, but this has the drawback of leaving literal holes in the map. To avoid these holes, researchers can instead use various forms of simulation (e.g., Stamatellos and Panourgias 2005) or interpolation (e.g., Hessl et al. 2007) to predict the real value of the missing data. Because it is not always possible to obtain data on a study area before and after some change (e.g., in response to a disturbance such as forest harvesting), researchers often use a carefully selected area as a proxy for the initial conditions. This *space-for-time substitution* (Ullman 1974) has been highly effective, but relies on the assumption that the proxy is valid, which may not be correct if undetected conditions cause the proxy to differ from what it is intended to represent in some important way. Furthermore, data representation conversions (e.g., from vector to raster) or type conversions (e.g., from point to area) create additional concerns that warrant special attention. For example, converting point data to a raster format will introduce errors related to the area and perimeter of the resulting data. Each time that data are modified from one form to another, there are implications for the resulting spatial patterns, and these may affect the interpretation of the results.

As we noted earlier, the quantification and comparison of spatial patterns focus on the composition, configuration, and morphology of the landscape as the dominant descriptive metrics. Although tools for computing these metrics are widely available, concerns regarding their validity for comparing landscape patterns have been raised and explored (Remmel and Csillag 2003). These approaches are reasonably effective for obtaining diagnostic indicators of possible landscape pattern changes, but pose numerous complications when statistical comparisons are necessary. A comprehensive review of scale effects on landscape indices (Šímová and Gdulová 2012) emphasized that the most useful metrics are those that are simple to compute and easy to interpret, and that the metrics are most useful when grain sizes and image extents can be kept consistent.

The greatest gains in quantifying and comparing spatial patterns will be made through the development of methods or metrics that simultaneously capture composition, configuration, and morphological attribute values at multiple scales (e.g., Remmel and Csillag 2006), and that will permit valid statistical comparisons of these metrics. Great advances have been made in recent years, but the next phase will involve connecting the various disparate approaches so they can be integrated within GIS software. The assessment of patterns will need to follow these developments in analysis and mapping at multiple scales. MacDougall (1975) focused on the accuracy of map overlays, and found that inaccuracy can result from the comparison of low-quality maps and that some boundary errors can be minimized by patch enlargement or reduction operations. Chrisman (1987) reassessed MacDougall's work, and showed that overlaid maps produced both large polygons that agreed between the maps and small areas that disagreed between maps, which he referred to as "slivers" (impurities). More importantly, he discussed the importance of the disagreements, and noted that some categorical or spatial disagreements were not important for certain applications.

For this research to move forward, it will be important for researchers to use terminology consistently. Although the terms *accuracy*, *ground-truthing*, and *validation* are often used interchangeably, they have significantly different meanings. *Accuracy* refers to the correctness of a mapped result. *Ground-truthing* is one way that accuracy is determined: researchers obtain field data at the mapped sites and compare it with the data that appears in the map. This approach assumes that it is possible to determine the truth about what exists at a specific location at any scale, mixture of land cover types, and landscape complexity, or position (i.e., theoretical perfection) and that this information can then be used to validate the results of a classification or mapping process at the same location. In contrast, *validation* refers to achieving a result that is not perfect, but rather is "good enough" for the purpose of the cartographic exercise.

There is always a possibility that some gap in our understanding of ecosystems can lead to discrepancies of various magnitudes; even something as simple as vegetation cover or crown closure can have definitions that differ among studies, and measuring such values in the field can be difficult and will produce results with different biases when different measurement methods have been used. Hence, there is a general preference toward emphasizing validation over accuracy. This philosophy aims to align what is possible through cartographic analysis with what we can actually achieve in the real world. For example, the explicit definitions of land cover categories are commonly designed such that the inherent variability can be accounted for by means of a validation effort such as ground-truthing. *Accuracy* is then used to refer to one or more metrics that quantify how well the data presented in a map correspond to the actual landscape conditions in the field.

Though many measures of accuracy have been proposed and used, most rely in some form on the initial construction of a *confusion matrix* (sometimes referred to as an *error matrix*) that compares the mapped data with validation data obtained by means of ground-truthing (Congalton 1991). The mapped data generally form the rows of this matrix and the validated categories generally form the columns (Fig. 20). Each cell of the confusion matrix records the frequency (f) of the agreement between the row and column (mapped and validated) data for i categorical labels. The values in each row and column are summed to produce marginal distributions for the rows and columns. The sum of the marginal distributions for all rows in the confusion matrix (which equals the sum for all columns) represents the number of sample locations where validation data was collected. There has been considerable debate over the number of samples required to validate a dataset, but researchers have generally accepted a worst-case conservative sample size (Eq. 2), which is well documented by Congalton and Green (1999):

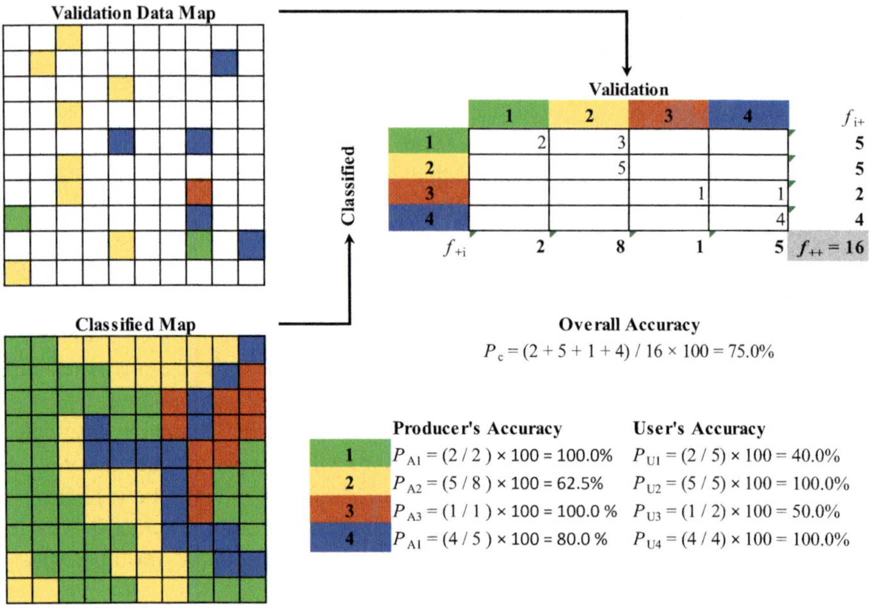

Fig. 20 An example of a confusion matrix constructed for mapped and reference data with four mapping categories. See the text for an explanation of producer and user accuracies and the frequency (f) terms

$$n = B\prod_{i}\left(1-\prod_{i}\right)/\,b_i^2 \tag{2}$$

where n represents the conservative estimate for the number of required samples; B is determined from a chi square distribution with 1 degree of freedom and $1 - \alpha/k$, where α represents the probability of a type I error and we assume a classification scheme with $k = 4$ classes; $\prod_i$ is the proportion of the map covered by class i (e.g., 30% in the example in Fig. 20); and b_i is the desired precision (5% or 0.05 given a desired 95% confidence interval). Thus, for this example, B would be $\chi^2_{(1,0.9875)}$ = 6.239. Thus, Eq. 2 would yield $n = 524$ samples, or approximately 131 samples per land cover category.

The confusion matrix can be described using various summary statistics, of which the most popular ones are the overall accuracy (the proportion of correctly classified values, P_c), the user's accuracy (the proportion of errors of commission, P_{Ui}), and the producer's accuracy (the proportion of errors of omission, P_{Ai}). Here, the overall accuracy is a global measure of the number of agreements between mapped and validated samples (e.g., pixels) across all land cover categories ($P_c = \sum_{k}^{i=1} f_{ii} / f_{++}$). The user's accuracy represents (individually for each category) the number of correctly labeled locations on the map (f_{ii}) divided by the total number of samples mapped into that same category (f_{++}): $P_{Ui} = f_{ii}/f_{i+}$. The producer's accuracy represents (individually for each category) the number of correctly labeled samples (f_{ii}) divided by the total number of samples identified in that category via field validation (f_{+i}): $P_{Ai} = f_{ii}/f_{+i}$.

Alternative summary statistics exist that compensate for the probability of correct labeling due to random chance between mapped locations and field-validated data using the kappa coefficient of agreement (κ; Foody 1992) and the construction of confidence intervals for summary metrics (Congalton and Green 1999; Csillag et al. 2006). The kappa coefficient of agreement for a thematic map is calculated as follows:

$$\hat{\kappa} = \left(p_{\mathrm{o}} - p_{\mathrm{c}}\right) / \left(1 - p_{\mathrm{c}}\right) \tag{3}$$

where p_{o} is the proportion of correctly classified cases and p_{c} is the proportion of correctly classified cases that would be expected based on random chance. However, because p_{c} relies heavily on a priori knowledge of the marginal distributions of the error matrix, its use has been questioned (Foody 2007). All of these approaches should be supported by a sufficiently large sample size to permit a statistically significant comparison (Chuvieco 2016). Stehman (2004) cautions against the normalization of the confusion matrix (i.e., mathematical transformation to produce a matrix that sums to unity), stating the risk that it will "cloak the assessment in a veil of analytic mystique that hinders understanding and proper interpretation."

Foody (2002) provides an extensive review of accuracy assessment methods and challenges to these methods, and underscores the importance of accuracy assessment, particularly when mapping large areas or monitoring changing landscapes. Mitchell et al. (2008) investigated how to mine the classification cluster statistics in remote sensing data to produce ratios of the probability of the most probable land cover class to the probability of the second most probable class for all pixels and develop statistical measures of reliability for each pixel location. The state of the art in accuracy assessment has not changed greatly since this period, when de facto standards for accuracy assessment were forged. However, Remmel (2009) extended the use of the confusion matrix to measure the uncertainty of landscape configuration and provide a deeper interpretation of the spatial distribution of accuracy. Their approach extracted information on spatial variability from an otherwise nonspatial confusion matrix. Olofsson et al. (2014) agreed with Foody's (2002) review and continued to outline best practices for determining the optimal sample size and developing confidence intervals specifically suitable for use in area estimation and landscape change.

Summary

Mapping has a history that dates back several thousand years. During this span, the field has undergone substantial improvement by refining both the art and the science of cartography to produce current standards for mapping excellence. The focus of this chapter has been to introduce the underlying concepts that govern mapping in general, but with a view toward applying these general rules to mapping of forest landscapes. At the core of all mapping is the creation of an abstraction of reality, and abstraction represents acceptance of the notion that the seemingly infinite complexity of the real world must be simplified in some way to create summaries that can be presented in map form and comprehended by users of the map. Decisions about the best way to generalize information are influenced by, and often limited by, how we choose to represent real-world phenomena (e.g., raster cells, vector primitives, tessellations). They are also influenced and limited by the types of attributes, topological relationships, and geometries that are possible or that we choose to account for. The spatial and temporal resolutions of a map (which are proxies for its scale) and of the underlying data acquisition further constrain the options available for mapping, and must be considered in all decisions surrounding plans for data acquisition, the actual surveys or measurements that provide that data, cartographic design, analysis, and accuracy assessment.

These considerations are complicated by the need to choose formal definitions of forests, landscapes, and characteristics that we generally map within the context of forest landscape mapping. We drew attention to approaches that focus on mapping of areas (formal and functional regions) and the identification of boundaries that separate a map into groups of areas with similar properties that are distinct from groups of areas with different properties. Both area-based and boundary-based approaches can be further extended to account for fuzzy logic, in which an area's membership in a given class is not crisp, but rather is described by a probability distribution function or by boundaries that represent transition zones (e.g., ecotones). Either of these extensions, alone or in combination, can lead to very complex maps.

The extension of maps from traditional flat 2D formats to 3D or even higher dimensions can express the dynamics, change, and vertical complexity of landscapes. This extension is facilitated by recent technological developments in computing, computer graphics, and data acquisition capabilities (e.g., LiDAR, satellite images). Maps are no longer produced simply to demarcate the locations of things, resources, or phenomena, but rather to understand broader spatial patterns and associations, and to make inferences about ecological processes. Maps, though often the final desired product, are increasingly serving as intermediate or initial data layers that record spatially explicit information that will be used as input for models, analytical engines, or visualizations that illustrate the stages dynamic systems pass through over time or in space.

Forest landscape maps are used extensively in conservation and forest management efforts, and are perceived as scientific documents. In this context, robust and statistically defensible techniques should be used to design the maps and assess their accuracy, and validation tools have been developed to accomplish this. The most popular tools generally involve cross-tabulations of observed versus expected attribute values (e.g., confusion matrices). Summary statistics are commonly derived to quantify the accuracy and validity of maps that are produced at specific scales and that are well documented in the associated metadata.

Techniques for georeferencing data (e.g., GPS receivers, ground-control points for satellite images), for performing statistical and spatial analyses of data, and for displaying and visualizing data have significantly improved in the last few decades. These improvements have accompanied the rapidly growing demand for spatially and temporally explicit information about forest landscapes. This demand has been motivated by the growing recognition of the need for sustainable development and by global concerns (e.g., climate change), regional concerns (e.g., wildfires, biodiversity), and local concerns (e.g., ecosystem services). These concerns arise from the need to balance conservation and extraction of natural resources, and the need to mitigate the consequences of this balance. The interaction between requirements dictated by user demands and the availability of new data collection technologies and analytical methods is fortunate, because it benefits all stakeholders, but it requires a sound understanding of mapping techniques and constraints (by the users) and of forest landscape ecology (by the mappers). This chapter was written to help provide this understanding.

References

Abeyta AM, Franklin J (1998) The accuracy of vegetation stand boundaries derived from image segmentation in a desert environment. Photogramm Eng Remote Sens 64(1):59–66

Allen TFH, Hoekstra TW (1992) Toward a unified ecology. Columbia University Press, New York, NY

Arnot C, Fisher PF, Wadsworth R, Wellens J (2004) Landscape metrics with ecotones: pattern under uncertainty. Landsc Ecol 19(2):181–195

Arroyo LA, Pascual C, Manzanera JA (2008) Fire models and methods to map fuel types: the role of remote sensing. For Ecol Manag 256:1239–1252

Asner GP, Mascaro J, Muller-Landau HC, Vielledent G, Vaudry R, Rasamoelina M, Hall JS, van Breugel M (2012) A universal airborne LiDAR approach for tropical forest carbon mapping. Oecologia 168:1147–1160

Badarinath KVS, Sharma AR, Kharol SK (2011) Forest fire monitoring and burnt area mapping using satellite data: a study over the forest region of Kerala state, India. Int J Remote Sens 32(1):85–102

Baker WL, Cai YM (1992) The r.le-programs for multiscale analysis of landscape structure using the GRASS geographical information system. Landsc Ecol 7(4):291–302

Boudewyn P, Seeman D, Wulder M, Magnussen S (2000) The effects of polygon boundary pixels on image classification accuracy. In: Proceedings of the 22nd Symposium of the Canadian Remote Sensing Society, Victoria, British Columbia, 20–25 August 2000. pp 637–643

Bourgeau-Chavez LL, Kasischke ES, Brunzell S, Mudd JP (2002) Mapping fire scars in global boreal forests using imaging radar data. Int J Remote Sens 23(20):4211–4234

Bowersox MA, Brown DG (2001) Measuring the abruptness of patchy ecotones—a simulation-based comparison of landscape pattern statistics. Plant Ecol 156(1):89–103

Bradshaw GA, Spies TA (1992) Characterizing canopy gap structure in forests using wavelet analysis. J Ecol 80(2):205–215

Brewer CA, MacEachren AM, Pickle LW, Hermann D (1997) Mapping mortality: evaluating color schemes for choropleth maps. Ann Assoc Am Geogr 87(3):411–438

Brown DG (1998) Classification and boundary vagueness in mapping presettlement forest types. Int J Geogr Inf Sci 12(2):105–129

Brown LA, Holmes J (1971) The delimitation of functional regions, nodal regions, and hierarchies by functional distance approaches. J Reg Sci 11(1):57–72
Buyantuyev A, Wu J (2007) Effects of thematic resolution on landscape pattern analysis. Landsc Ecol 22:7–13
Chen G, Metz MR, Rizzo DM, Meentemeyer RK (2015) Mapping burn severity in a disease-impacted forest landscape using Landsat and MASTER imagery. Int J Appl Earth Obs Geoinf 40:91–99
Chrisman NR (1987) The accuracy of map overlays: a reassessment. Landsc Urban Plan 14:427–439
Chrisman NR (1998) Rethinking levels of measurement for cartography. Cartogr Geogr Inf Sci 25(4):231–242
Chuvieco E (2016) Fundamentals of satellite remote sensing, 2nd edn. CRC Press, New York, NY
Coates AR, Dennison PE, Roberts DA, Roth KL (2015) Monitoring the impacts of severe drought on Southern California chaparral species using hyperspectral and thermal infrared imagery. Remote Sens 7:14276–14291
Collin K (2016, November 17) The electoral college badly distorts the vote. And it's going to get worse. Citizens in the least populous states have about 2.5 times more voting power than those in the most populous states. *The Washington Post*. washingtonpost.com
Congalton RG (1991) A review of assessing the accuracy of classifications of remotely sensed data. Remote Sens Environ 37(1):35–46
Congalton RG, Green K (1999) Assessing the accuracy of remotely sensed data: principles and practices. Lewis Publishers, Boca Raton, 137 p
Cook EA, Iverson LR, Graham RL (1989) Estimating forest productivity with thematic mapper and biogeographical data. Remote Sens Environ 28:131–141
Coops NC, Hilker T, Wulder MA, St-Onge B, Newnham G, Siggins A, Trofymow JAT (2007) Estimating canopy structure of Douglas-fir forest stands from discrete-return LiDAR. Trees 21(3):295–310
Csillag F, Kummert Á, Kertéz M (1992) Resolution, accuracy and attributes: approaches for environmental geographical information systems. Comput Environ Urban Syst 16(4):289–297
Csillag F, Kabos S, Remmel TK (2006) Confidence in coincidence. Int J Remote Sens 27(5–6):1269–1276
Csillag F, Kabos S, Remmel TK (2008) A spatial clustering perspective on autocorrelation and regionalization. Environ Ecol Stat 15:385–401
Dalponte M, Bruzzone L, Gianelle D (2008) Fusion of hyperspectral and LIDAR remote sensing data for classification of complex forest areas. IEEE Trans Geosci Remote Sens 46(5):1416–1427
Dark SJ, Bram D (2007) The modifiable areal unit problem (MAUP) in physical geography. Prog Phys Geogr 31(5):471–479
Dauwalter DC, Rahel FJ (2011) Patch size and shape influence the accuracy of mapping small habitat patches with a global positioning system. Environ Monit Assess 179:123–135
De Bruin S, De Wit AJW, Van Oort PAJ (2004) Using quadtree segmentation to support error modelling in categorical raster data. Int J Geogr Inf Sci 18(2):151–168
Ding YL, Ge Y, Hu MG, Wang JF, Wang JH, Zheng XM, Zhao K (2014) Comparison of spatial sampling strategies for ground sampling and validation of MODIS LAI products. Int J Remote Sens 35(20):7230–7244
Douglas D, Peucker T (1973) Algorithms for the reduction of the number of points required to represent a digitized line or its caricature. Can Cartogr 10(2):112–122
Dungan JL, Perry JN, Dale MRT, Legendre P, Citron-Pousty S, Fortin M-J, Jakomulska A, Miriti M, Rosenberg MS (2002) A balanced view of scale in spatial statistical analysis. Ecography 25(5):626–640
Edwards G, Lowell KE (1996) Modeling uncertainty in photointerpreted boundaries. Photogramm Eng Remote Sens 62(4):337–391
Falkowski MJ, Gessler PE, Morgan P, Hudak AT, Smith AMS (2005) Characterizing and mapping forest fire fuels using ASTER imagery and gradient modelling. For Ecol Manag 217:129–146

Foody GM (1992) On the compensation for chance agreement in image classification accuracy assessment. Photogramm Eng Remote Sens 58(10):1459–1460
Foody GM (2002) Status of land cover classification accuracy assessment. Remote Sens Environ 80(1):185–201
Foody GM (2007) The evaluation and comparison of thematic maps derived from remote sensing. In: Caetano M, Painho M (eds) Proceedings of the 7th international symposium on spatial accuracy assessment in natural resources and environmental sciences, 5 to 7 July 2006. Lisboa. Instituto Geográfico Português, Lisbon, pp 18–31
Forman RTT, Godron M (1986) Landscape ecology. Wiley, New York, NY
Fransson JES, Pantze A, Eriksson LEB, Soja MJ, Santoro M (2010) Mapping of wind-thrown forests using satellite SAR images. In: IEEE International Geoscience & Remote Sensing Symposium, June 25–30, Honolulu, Hawaii, USA. IEEE, New York, NY
Fraser RH, Li Z, Cihlar J (2000) Hotspot and NDVI differencing synergy (HANDS): a new technique for burned area mapping over boreal forest. Remote Sens Environ 74:362–376
Gaucherel C (2007) Multiscale heterogeneity map and associated scaling profile for landscape analysis. Landsc Urban Plan 82(3):95–102
Giglio L, Loboda T, Roy DP, Quayle B, Justice CO (2009) An active-fire based burned area mapping algorithm for the MODIS sensor. Remote Sens Environ 113(2):408–420
Goerner A, Reichstein M, Rambal S (2009) Tracking seasonal drought effects on ecosystem light use efficiency with satellite-based PRI in a Mediterranean forest. Remote Sens Environ 113(5):1101–1111
Haines-Young R, Chopping M (1996) Quantifying landscape structure: a review of landscape indices and their application to forested landscapes. Prog Phys Geogr 20(4):418–445
Hall FG, Knapp DE, Huemmrich KF (1997) Physically based classification and satellite mapping of biophysical characteristics in the southern boreal forest. J Geophys Res-Atmos 102(D24):29567–29580
Hessl A, Miller J, Kernan J, Keenum D, McKenzie D (2007) Mapping paleo-fire boundaries from binary point data: comparing interpolation methods. Prof Geogr 59(1):87–104
Hill A, Breschan J, Mandallaz D (2014) Accuracy assessment of timber volume maps using forest inventory data and LiDAR canopy height models. Forests 5:2253–2275
Jelinski DE, Wu J (1996) The modifiable areal unit problem and implications for landscape ecology. Landsc Ecol 11(3):129–140
Jenks GF, Caspall FC (1971) Error on choroplethic maps: definition, measurement, reduction. Ann Assoc Am Geogr 61(2):217–244
Johnson GD, Myers W, Patil GP, Taillie C (2001) Fragmentation profiles for real and simulated landscapes. Environ Ecol Stat 8:5–20
Jordan GJ, Fortin M-J, Lertzman KP (2005) Assessing spatial uncertainty associated with forest fire boundary delineation. Landsc Ecol 20:719–731
Jordan GJ, Fortin M-J, Lertzman KP (2008) Spatial pattern and persistence of historical fire boundaries in southern interior British Columbia. Environ Ecol Stat 15:523–535
Kavanagh BF (2003) Geomatics. Prentice Hall, Upper Saddle River, NJ
Kimball JS, Keyser AR, Running SW, Saatchi SS (2000) Regional assessment of boreal forest productivity using an ecological process model and remote sensing parameter maps. Tree Physiol 21:761–775
King AW (1991) Translating models across scales in the landscape. In: Turner MG, Gardner RH (eds) Quantitative methods in landscape ecology. Springer, New York, NY, pp 479–517
Kronenfeld BJ (2011) Beyond the epsilon band: polygonal modeling of gradation/uncertainty in area-class maps. Int J Geogr Inf Sci 25(11):1749–1771
Küchler AW (1955) Comprehensive method of mapping vegetation. Ann Assoc Am Geogr 45(4):404–415
Levin SA (1992) The problem of patterns and scale in ecology. Ecology 73:1943–1967
Lim K, Treitz P, Baldwin K, Morrison I, Green J (2003a) Lidar remote sensing of biophysical properties of tolerant northern hardwood forests. Can J Remote Sens 29(5):658–678

Lim K, Treitz P, Wulder M, St-Onge B, Flood M (2003b) LiDAR remote sensing of forest structure. Prog Phys Geogr 27(1):88–106
MacDougall EB (1975) The accuracy of map overlays. Landsc Urban Plan 2:23–30
Macomber SA, Woodcock CE (1994) Mapping and monitoring conifer mortality using remote sensing in the Lake Tahoe Basin. Remote Sens Environ 50:255–266
Mark DM, Csillag F (1989) The nature of boundaries on "area-class" maps. Cartographica 26(1):65–77
Martin ME, Aber JD (1997) High spectral resolution remote sensing of forest canopy lignin, nitrogen, and ecosystem processes. Ecol Appl 7(2):431–443
Martin ME, Newman SD, Aber JD, Congalton RG (1998) Determining forest species composition using high spectral resolution remote sensing data. Remote Sens Environ 65:249–254
McGarigal K, Marks BJ (1995) FRAGSTATS: spatial pattern analysis program for quantifying landscape structure. U.S. Department of Agriculture, Forest Service, Pacific Northwest Research Station, Portland, OR. Gen Tech Rep PNW-GTR-351
McIntire EJB, Fortin M-J (2006) Structure and function of wildfire and mountain pine beetle forest boundaries. Ecography 29:309–318
Menon C (2012) The bright side of MAUP: defining new measures of industrial agglomeration. Pap Reg Sci 91(1):3–29
Miller JR, Turner MG, Smithwick EAH, Dent CL, Stanley EH (2004) Spatial extrapolation: the science of predicting ecological patterns and processes. Bioscience 54(4):310–320
Mitchell SW, Remmel TK, Csillag F, Wulder MA (2008) Distance to second cluster as a measure of classification confidence. Remote Sens Environ 112(5):2615–2626
Mitri GH, Gitas IZ (2008) Mapping the severity of fire using object-based classification of IKONOS imagery. Int J Wildland Fire 17(3):431–442
Morgan P, Hardy CC, Swetnam TW, Rollins MG, Long DG (2001) Mapping fire regimes across time and space: understanding coarse and fine-scale fire patterns. Internat J Wildl Fire 10(3–4):329–342
Mysorewala MF, Popa DO, Lewis FL (2009) Multi-scale sampling with mobile agents for mapping of forest fires. J Intell Robot Syst 54:535–565
Naveh Z, Lieberman AS (1984) Landscape ecology: theory and application. Springer-Verlag, New York, NY
Nelson R, Krabill W, Tonelli J (1988) Estimating biomass and volume using airborne laser data. Remote Sens Environ 24:247–267
O'Neil RV, King AW (1998) Homage to St. Michael: or why are there so many books on scale? In: Peterson DL, Parker VT (eds) Ecological scale: theory and applications. Columbia University Press, New York, NY, pp 4–15
Olofsson P, Foody GM, Herold M, Stehman SV, Woodcock CE, Wulder MA (2014) Good practices for estimating area and assessing accuracy of land change. Remote Sens Environ 148:42–57
Openshaw S (1977) A geographical solution to scale and aggregation problems in region-building, partitioning and spatial modelling. Trans Inst Brit Geogr 2:459–472
Perera AH, Euler DE, Thompson ID (2000) Ecology of a managed terrestrial landscape: patterns and processes of forest landscapes in ontario. UBC Press, Vancouver, BC
Peterson DL, Parker VT (eds) (1998) Ecological scale: theory and applications. Columbia University Press, New York, NY
Peterson U, Püssa K, Liira J (2004) Issues related to delineation of forest boundaries on landsat thematic mapper winter images. Int J Remote Sens 25(24):5617–5628
Philibert MD, Fortin M-J, Csillag F (2008) Spatial structure effects on the detection of patches boundaries using local operators. Environ Ecol Stat 15(4):447–467
Potapov P, Yaroshenko A, Turubanova S, Dubinin M, Laestadius L, Thies C, Aksenov D, Egorov A, Yesipova Y, Glushkov I, Karpachevskiy M, Kostikova A, Manisha A, Tsbikova E, Zhuravleva I (2008) Mapping the world's intact forest landscapes by remote sensing. Ecol Soc 13(2):51–67
Reed BC, Brown JF, VanderZee D, Loveland TR, Merchant JW, Olsen DO (1994) Measuring phonological variability from satellite imagery. J Veg Sci 5:703–714

Remmel TK (2009) Investigating global and local categorical map configuration comparisons based on coincidence matrices. Geogr Anal 41(1):113–126
Remmel TK, Csillag F (2003) When are two landscape pattern indices significantly different? J Geogr Syst 5(4):331–351
Remmel TK, Csillag F (2006) Mutual information spectra for comparing categorical maps. Int J Remote Sens 27(7):1425–1452
Remmel TK, Perera AH (2001) Fire mapping in a northern boreal forest: assessing AVHRR/NDVI methods of change detection. For Ecol Manag 152(1–3):119–129
Remmel TK, Csillag F, Mitchell S, Wulder MA (2005) Integration of forest inventory and satellite imagery: a Canadian status assessment and research issues. For Ecol Manag 207(3):405–428
Riitters KH, O'Neill RV, Hunsaker CT, Wickham JD, Yankee DH, Timmins SP, Jones KB, Jackson BL (1995) A factor analysis of landscape pattern and structure metrics. Landsc Ecol 10(1):23–39
Riitters KH, O'Neill RV, Jones KB (1997) Assessing habitat suitability at multiple scales: a landscape-level approach. Biol Conserv 81:191–202
Ruiz JAM, Riaño D, Arbelo M, French NHF, Ustin S, Witing ML (2012) Burned area mapping time series in Canada (1984–1999) from NOAA-AVHRR LTDR: a comparison with other remote sensing products and fire perimeters. Remote Sens Environ 117:407–414
Running SW, Nemani RR, Peterson DL, Band LE, Potts DF (1989) Mapping regional forest evapotranspiration and photosynthesis by coupling satellite data with ecosystem simulation. Ecology 70(4):1090–1101
Saarinen N, Vastaranta M, Honkavaara E, Wulder MA, White JC, Litkey P, Holopainen M, Hyyppä J (2016) Using multi-source data to map and model the predisposition of forests to wind disturbance. Scan J For Res 31(1):66–79
Schroeder TA, Wulder MA, Healey SP, Moisen GG (2011) Mapping wildfire and clearcut harvest disturbances in boreal forests with Landsat time series data. Remote Sens Environ 115:1421–1433
Shan J, Toth CK (2009) Topographic laser ranging and scanning: principles and processing. Taylor and Francis, Boca Raton, FL
Šímová P, Gdulová K (2012) Landscape indices behavior: a review of scale effects. Appl Geogr 34:385–394
Sinton D (1978) The inherent structure of information as a constraint to analysis: mapped thematic data as a case study. In: Dutton G (ed) Proceedings of the first international advanced study symposium on topological data structures for geographic information systems. Harvard University, Laboratory for Computer Graphics and Spatial Analysis, vol 7. Harvard Papers on GIS, Cambridge, MA, pp 1–17
Slocum TA, McMaster RB, Kessler FC, Howard HH (2009) Thematic cartography and Geovisualization, 3rd edn. Pearson-Prentice Hall, Upper Saddle River, NJ
Soille P, Vogt P (2009) Morphological segmentation of binary patterns. Pattern Recogn Lett 30:456–459
Stamatellos G, Panourgias G (2005) Simulating spatial distributions of forest trees by using data from fixed area plots. Forestry 78(3):305–312
Stehman SV (2004) A critical evaluation of the normalized error matrix in map accuracy assessment. Photogramm Eng Remote Sens 70(6):743–751
Stevens SS (1946) On the theory of scales of measurement. Science 103(2684):677–680
Sukhinin AI, French NHF, Kasischke ES, Hewson JH, Soja AJ, Csiszar IA, Hyer EJ, Loboda T, Conrad SG, Romansko VI, Pavlichenko EA, Miskiv SI, Slinkina OA (2004) AVHRR-based mapping of fires in Russia: new products for fire management and carbon cycle studies. Remote Sens Environ 93:546–564
Tampekis S, Sakellariou S, Samara F, Sfougaris A, Jaeger D, Christopoulou O (2015) Mapping the optimal forest road network based on the multicriteria evaluation technique: the case study of Mediterranean Island of Thassos in Greece. Environ Monit Assess 187(11):687
Treitz PM, Howarth PJ (1999) Hyperspectral remote sensing for estimating biophysical parameters of forest ecosystems. Prog Phys Geogr 23(3):359–390

Turner MG, Gardner RH (2015) Landscape ecology in theory and practice: pattern and process, 2nd edn. Springer, New York, NY
Turner MG, Dale VH, Gardner RH (1989) Predicting across scales: theory development and testing. Landsc Ecol 3:245–252
Ullman EL (1974) Space and/or time: opportunity for substitution and prediction. Trans Inst Brit Geogr 63:125–139
USGS (1999) Map accuracy standards. U.S. Geological Survey, U.S. Department of the Interior. USGS Fact Sheet 171–99. https://pubs.er.usgs.gov/publication/fs17199
Uuemaa E, Roosaare J, Oja T, Mander Ü (2011) Analysing the spatial structure of the Estonian landscapes: which landscape metrics are the most suitable for comparing different landscapes. Estonian J Ecol 60(1):70–80
Vogt P, Ferrari JR, Lookingbill TR, Gardner RH, Riitters KH, Ostaowicz K (2009) Mapping functional connectivity. Ecol Indic 9:64–71
Voormansik K, Praks J, Antropov O, Jagomagi J, Zalite K (2014) Flood mapping with TerraSAR-X in forested regions in Estonia. IEEE J Select Topics Appl Earth Obs Remote Sens 7(2):562–577
Wang Y, Colby JD, Mulcahy KA (2002) An efficient method for mapping flood extent in a coastal floodplain using Landsat TM and DEM data. Int J Remote Sens 23(18):3681–3696
White JC, Wulder WA, Brooks D, Reich R, Wheate RD (2004) Mapping mountain pine beetle infestation with high spatial resolution satellite imagery. For Chron 80(6):743–745
Wiens JA (1989) Spatial scaling in ecology. Funct Ecol 3:385–397
Woodcock CE, Strahler AH (1987) The factor of scale in remote sensing. Remote Sens Environ 21:311–332
Wu J (1999) Hierarchy and scaling: extrapolating information along a scaling ladder. Can J Remote Sens 25(4):367–380
Wulder MA, Hall RJ, Coops N, Franklin SE (2004a) High spatial resolution remotely-sensed data for ecosystem characterization. Bioscience 54(6):511–521
Wulder MA, Kurz WA, Gillis M (2004b) National level forest monitoring and modeling in Canada. Prog Plan 61:365–381
Zadeh LA (1965) Fuzzy sets. Inf Control 8:338–353
Zhang J, Suart N (2001) Fuzzy methods for categorical mapping with image-based land cover data. Int J Geogr Inf Sci 15(2):175–195

Fuzzy Classification of Vegetation for Ecosystem Mapping

F. Jack Triepke

Abstract Vegetation classification and mapping are important tools for addressing natural resource management, ecosystem restoration, and other contemporary ecological issues. Though classical set theory is most often applied for mapping problems, natural landscapes are often expressed as fuzzy sets. Where contrast among map categories or geometric objects is often weak in ecological contexts fuzzy approaches offer the advantage of identifying and utilizing the degree of membership among multiple possibilities, enabling opportunities for alternative outputs and for the careful analysis of error structure. In this chapter, fuzzy systems are explored for purposes of describing ecological features, for interpretation and mapping of those features, and for analyzing the uncertainty of spatial information. Some ecological applications that lend themselves to fuzzy logic are discussed along with examples of the effective use of fuzzy techniques for mapping and analysis, with explanations of the advantages of fuzzy approaches over crisp methods. Finally, in a look to future, I discuss advanced classifier methods, some Web-based solutions, and the potential for applying fuzzy systems to interactively generate user-defined map products, neutral of á priori ecological classification, according to the precise needs of natural resource managers and researchers.

Introduction

The mapping of ecological features is fundamental to the management and study of ecosystems (Brewer et al. 2006). Maps provide readily accessible graphic representations of ecosystem features and, within the data domain, an efficient means of facilitating spatial analyses when multiple spatial layers are brought together to respond to a question or an issue. Ecological mapping is the process of delineating the geographic extent of chosen features of ecosystem structure and composition or the extent of other biophysical expressions such as potential vegetation and ecological systems (Daubenmire 1978; Comer et al. 2003). With some of these spatial data,

F.J. Triepke (✉)
USDA Forest Service, 333 Broadway Blvd SE, Albuquerque, NM 87102, USA
e-mail: jtriepke@fs.fed.us

T.K. Remmel, A.H. Perera (eds.), *Mapping Forest Landscape Patterns*,
DOI 10.1007/978-1-4939-7331-6_2

remote sensing may have a limited role in their development and makeup, though image analysis and interpretation techniques are no less valuable. The reader is referred to Chapter "Mapping Forest Landscapes: Overview and a Primer" for an overview of important concepts and components of vegetation mapping and for a discussion on the variables that affect such products.

Mapping requires that map units and the underlying classification be identified prior to map development (Running et al. 1994; Brohman and Bryant 2005; Brewer 2007), without which leads to map data which are inconsistent, vague, unsuited for comparison, and problematic for GIS analysis. While fuzzy mapping approaches can offer solutions to the indiscrete nature of ecological features on the landscape, it is nevertheless important to contemplate a scientific classification so that the resulting map units are consistent in logic, comprehensive relative to the thematic resolution, and mutually exclusive. Even continuous non-categorical mapping, such as tree cover or NDVI renderings (USGS 2014), reflects classifications of some type characterized in a metric (e.g., aerial extent) and a definition that describes the feature. As detailed in the following sections, the fuzzy nature of natural landscapes challenges the common conception of mutual exclusivity among map classes.

Classic spatial analysis and image classification rely on the strengths of the underlying ecological classification and map unit framework, and the assumptions that the project area is made up of various mutually exclusive and internally homogeneous units that can be discerned and subsequently modeled with training data (Lillesand and Kiefer 2000). To an extent this perspective is the same for the ecological classification, itself, since it relies on a degree of similarity within classes and maximal dissimilarity among classes (Daubenmire 1966; Hoppner et al. 1999). However, as is more often the case, even ecological classes express themselves across landscapes as continua within and among other units—fuzzy sets. Thus the character of ecological features across landscapes often has significant fuzziness within any arbitrary class that humans impose to challenge a key assumption of conventional image classification (Wood and Foody 1993). That is not to say that fuzzy sets do not have medoid properties, objects that represent a central concept to the map unit that can be useful for understanding and conveying the unit's essence and for generating training data. But these medoid representations may have limited utility in framing a model for a fuzzy set given the variability of the unit reflected over the landscape.

The fuzziness of a map theme may even be revealed at the scale of individual image objects or pixels. In the case of mixed pixels, ecological features are of finer grain than the spatial resolution of the spectral imagery used to interpret those features (Foody 1997; Zhang and Foody 1998; Campbell 2002). The idea of mixed pixels is that their spatial extent is not completely occupied by a single homogeneous map theme. The exact issue may stem from the inherent fuzziness in the ecological features, as elaborated above, or due to the size of objects being small in relation to the underlying variability of ecological features (Li et al. 2014). It may be not only erroneous to assign one category to a mixed pixel, but by its messiness a mixed pixel confounds our ability to interpret and assign even the most likely category. At the same time, the spatial resolution of the spectral imagery may do justice to some map themes and not others so that acquisition of the preferred image source is determined as the least objectionable alternative when combining sources is not operationally feasible. With road verges, for instance, an ordinary image source such as Landsat TM may be too coarse to precisely map the narrow strip of lawn adjacent to urban roadways, though road verges may be a map theme of lesser interest. The choice of spectral imagery may also be a matter of cost, so that an affordable alternative such as Landsat TM may indirectly influence how a map legend is composed. In the case of road verges, map developers may opt for a more general and practical map theme, such as "residential," so that Landsat TM is an option, particularly if accurately mapping road verges is a low-priority objective.

Situations of gradual or mixed membership can be accommodated with fuzzy techniques. The need for fuzzy classification and related spatial analysis approaches was recognized as least since the 1990s (e.g., Wang 1990), and has since been applied in various vegetation mapping, fire severity, soils and geology, and sociological mapping (Zhang and Foody 1998; Triepke et al. 2008; Remmel and Perera 2009). In this chapter the application of fuzzy systems is reviewed within two realms: in the characterization of ecological features, both in map unit concept and in output attribution of a geodatabase. The attribution can, in turn, be used for the assessment of uncertainty. Second, fuzzy systems are explored as classifiers of image objects, expressed as rules for the interpretation of predictor data. Finally, some discussion of recent and future technology and methods is included, but first, an overview of fuzzy logic is provided.

Overview of Fuzzy Systems

Fuzzy logic is an approximate form of reasoning used in set theory to represent knowledge (Cox 1992). Since fuzzy systems represent approximate reasoning, they can provide accurate levels of abstraction for many ecological circumstances that, by their nature, are ambiguities among arbitrary human categories. In this context, fuzzy logic provides an efficient means of representing features with more effective metaphors and fewer rules than classic Boolean approaches (Rickel et al. 1998). Traditional map themes, and their associated rule sets, assume discrete boundaries among themes. Fuzzy logic, however, accommodates the gradation and overlap among map themes that is common in natural systems so that any pixel or object can be assigned a value between 0 and 1 by the strength of its identity to a theme (Wang 1990); that is, an object can have partial membership to one or more themes (Zhang and Foody 1998). Membership to a theme at any locale represented spatially by pixels or objects can be represented as "no," "yes," or "somewhat," respectively, as 0, 1, or any value between 0 and 1 (Zadeh 1965; Klaua 1966; Goguen 1969). As with Boolean logic, fuzzy logic suggests the most probable identity for a given object. But unlike Boolean functions (i.e., crisp or hard functions), fuzzy approaches leave open the possibility of partial membership to more than one map theme, offering a distinct advantage over crisp methods for refining map unit concepts, generating spatial outputs, and assessing uncertainty (see section "Fuzzy Approaches for Identifying and Utilizing Uncertainty").

Fuzzy systems are fundamentally probabilistic (Kosco 1995) and membership values can be assigned to a map object depending on the probability of membership to each of the categories conceptualized in a map legend. Membership values can be determined such that all individual memberships for each category total to 1, with classification rules constructed accordingly (Zadeh 1965; Goguen 1969). Membership is alternatively determined through suitable algorithms including unitless distance such as Euclidean or Mahalanobis functions, with outputs scaled to a range from 0 to 1 to represent the degree of inclusion to a category for each object. In the case of Mahalanobis distance, membership is the distance between an object and the distribution D of actual values for a given category (Knick and Rotenberry 1998). Here, the distance is a measure of the number of standard deviations between the object and the mean of D, with distance increasing as the value of an object and the mean of D increase. As with other approaches, Mahalanobis distance can be computed for each map category so that each object is comprised of a set of membership values for all categories. Later in the chapter other classifier approaches are explored. First however, we will step back and look at rule operators as a means to understand connections between fuzzy membership values and the classification of ecological features.

Fuzzy Systems: Key Concepts for Mapping

While fuzzy systems can be used to address either thematic or spatial mapping problems, for simplicity the following descriptions are based on familiar circumstances of thematic mapping. Some geometric situations are summarized in sections "Spatial Uncertainty" and "Simultaneous Considerations of Thematic and Spatial Uncertainty." Fuzzy logic is used for both characterizing ecological features and for classifier approaches, including the combining of rules into decision trees for image classification. Fuzzy rules can also be used to determine fuzzy membership values for categorical data for purposes of image classification and spatial analysis. The following sections describe the role of fuzzy systems for characterizing ecological features and for classifying them through the interpretation of image data.

Map Unit Concepts

To develop a system of map themes, first an ecological classification is adopted or developed based on the business needs or the research criteria of a program or project. Map units are made up of one or more ecological classes, often an iterative process that balances the capacity of the classification with mapping objectives, technological constraints, and available resources devoted to the particular map project (Jennings et al. 2003; Brohman and Bryant 2005; Brewer et al. 2006). In the

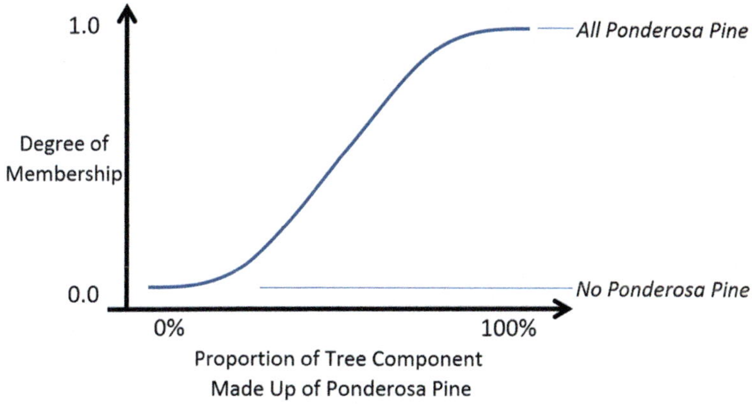

Fig. 1 Fuzzy operator used to show increasing membership to a "Ponderosa Pine" map unit as a logistic curve, based on the proportion of ponderosa pine in an image object

end, a one-to-one relationship between an ecological classification and the map units in a legend may not be possible or desired. Also, it may be necessary to normalize training samples so that there is a logical consistency between training data and the proposed map categories. While not discussed here, Wang (1990) and others have advocated attributing training data with fuzzy membership values between 0 and 1 for every map category, likewise extending the approach to accuracy samples. More recent work has shown that training maps with mixed objects, where membership to each map category is determined, can be an asset to improving map accuracy rather than an issue (Costa et al. 2017). In the end, map units are ideally underpinned by a scientific classification and characterized in fuzzy terms.

With a fuzzy mapping system, it is helpful to consider that that every pixel or object in the dataset is attributed with a membership value between 0 and 1 for every map theme. Now consider a fuzzy mapping system based on a single-species cover type concepts. Figure 1 shows a fuzzy membership function for a "Ponderosa Pine" vegetation type, where the strength of membership bears on the percentage of ponderosa pine (*Pinus ponderosa*) in the tree component of an object. With this function, membership values increase by the proportion of ponderosa pine, so that objects with tree components that are comprised only of ponderosa pine (100%) have fuzzy membership values of 1; conversely, objects with no ponderosa pine have membership values of 0. The relationship between membership and the classification can be sigmoid as shown in Fig. 2, or expressed by some other mathematical function. In the end every object of the map has a membership value for ponderosa pine between 0 and 1.

Now we will look at a more complex and realistic scenario with two map unit concepts in combination, and take an initial look at the characterization of uncertainty. Two common montane forest tree species of western North America—ponderosa pine and Douglas-fir (*Pseudotsuga menziesii*)—represent primary constituents of forests of the Cordillera in the United States and Canada (Morin

1993). These species commonly co-occur as frequent-fire components in relatively warm-dry forested settings. If a map legend includes these two conifer tree themes, and a given plant community has both species present, say, in a proportion of 60% and 40%, the community would be mapped as "Ponderosa Pine" assuming uniform classification rules. Likewise, if the respective percentages are 40% and 60%, the community would be mapped as Douglas-fir. If the proportions were equal, then a tie-breaker rule could be imposed, a mixed cover type could be introduced, or map objects could simply be attributed with fuzzy membership values for each map theme treated with no set-upon legend. In the latter case, classification schemes could be imposed á posteriori and ad hoc along with uncertainty characterizations.

Structure type concepts, such as tree size, likewise necessitate membership schemes and rule-making to determine what conditions point to the assignment of a given object to a particular map theme (Rickel et al. 1998). Where trees of more than one size class occupy the same object, as is often the case in North American forests, map unit concepts can be derived from a consistent means of assigning fuzzy membership. One possible scheme involves the expression of membership gradients within each unit of a category framework. In these situations, the overlap (fuzziness) among categories is often depicted in linear relationships between neighboring classes and by complete exclusivity in all other class-to-class relationships (Nauck and Kruse 1999). As shown in Fig. 2, an object can share properties of up to two categories, say seedling-sapling (<15 cm) and small-diameter trees (<15–25 cm); that is, the object representing a stand of trees has characteristics of both seedling-sapling trees and small trees, reflected in the canopy cover of each tree size class, and can be discerned by the comparative membership of the object to the two neighboring size classes. Of course the approach assumes consistent relationships among categories—e.g., that seedling-sapling trees do not co-occur with trees of size categories other than small. While this representation is unnecessary for univariate data such as the total tree cover, it can be useful for categorical data of

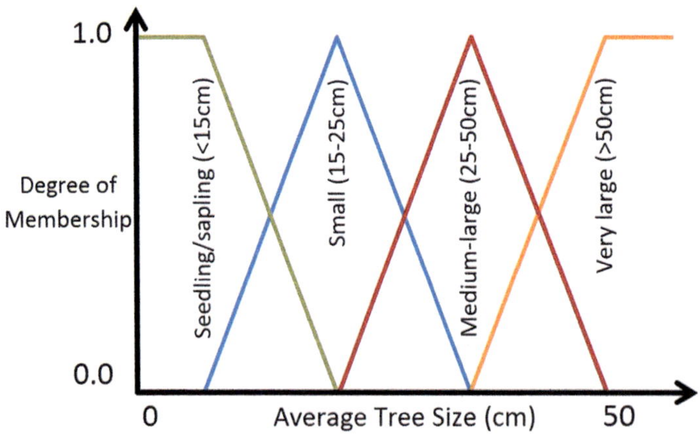

Fig. 2 Fuzzy operator for categorical data, showing hypothetical membership values for each class of tree size

multivariate conditions such as tree size or species. The approach assumes binary conditions since the object can have properties of only two categories, though the relationships could just as well be redrawn to show overlap in three or more categories granted certain additional assumptions.

Fuzzy approaches to building map rules and assessing uncertainty bring much-needed precision to these practices, and allow end users a greater range of options in spatial applications and the development and analysis of map data. Building on fuzzy map unit concepts, the following section provides a survey of common classifier methods that employ fuzzy approaches for the production of map data.

Fuzzy Classifiers

With concepts of fuzzy classifiers introduced in section "Mapping with Fuzzy Classifiers," what follows is a more detailed examination of these classifiers and the attribution of image objects. Here, fuzzy relationships between predictor data and ecological classes are identified and utilized as a means of building maps through image classification and often depicted in rule sets. Fuzzy classifiers do not bear on fuzzy map unit concepts (e.g., Fig. 1) nor do they necessarily result in fuzzy outputs, though there are advantages to both.

With fuzzy classifiers, rules are developed for each map unit or theme based on the fuzzy membership patterns among predictor variables including spectral information and biophysical data layers such as elevation. For example, in the southwestern USA steep slopes comprise a partial signature for fire-adapted shrubland types such as interior chaparral and mountain mahogany shrubland. In a rule-building process, each potential predictor can be represented by a membership function (Zadeh 1965). Figure 3 illustrates the relationship between one predictor variable, slope, and fuzzy membership to fire-adapted shrublands.

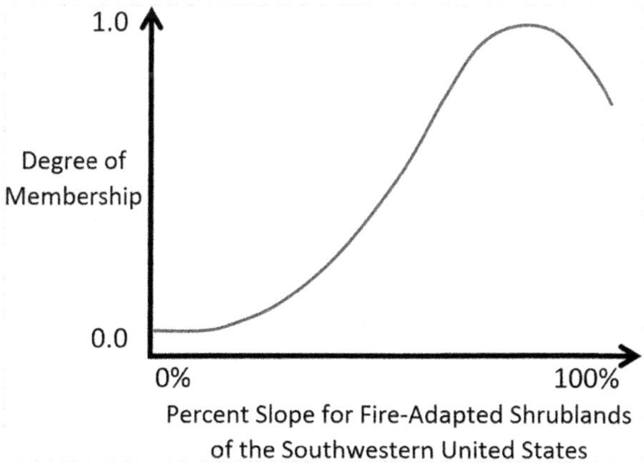

Fig. 3 Fuzzy operator showing the hypothetical relationship between fire-adapted shrublands and slope

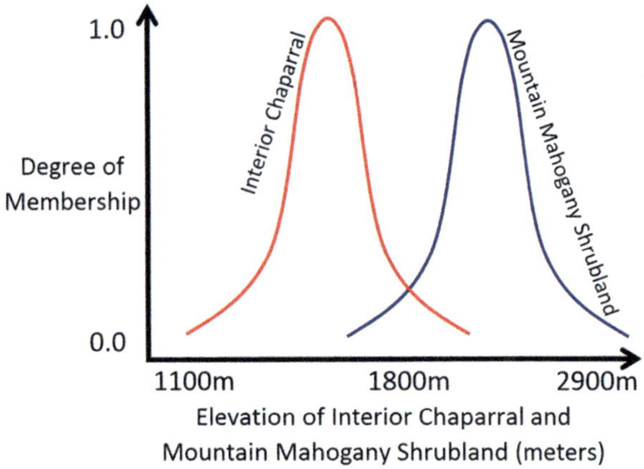

Fig. 4 Fuzzy operators showing the hypothetical relationships between elevation and the Interior Chaparral and Mountain Mahogany Shrubland types

Obviously fire-adapted shrublands are not the only cover type that can occur within the slope range represented by this function. Slope may be useful for differentiating these shrublands from other shrub types but other predictor variables, and consequently other rules and decision nodes, would be necessary to differentiate fire-adapted shrublands from forested and grassland cover types, and also to subdivide fire-adapted shrubland feature space into finer shrub units such as "Interior Chaparral" and "Mountain Mahogany Shrubland" (Brown et al. 1998).

Additional variables may be needed to more accurately model and separate themes depending on the degree of overlap among map themes for a particular model variable. For instance, the membership profiles for slope may be useful in helping to discern fire-adapted shrub systems from other shrub types in the Southwest, but additional differentia are needed to separate the two themes from one another. Figure 4 shows membership functions for elevation for the two shrubland systems that share slope affinities.

By combining variables in a decision tree, multiple membership functions can be brought together for particular classifier problems using hierarchical reasoning to combine membership functions into a rule set. Also, membership functions can be either fuzzy or Boolean and, in fact, can be combined with other algorithms such as nearest neighbor at specific decision nodes that makes up the rule set. Membership functions can be combined using classic mathematical operators such as if-then, greater than, less than, and, and or to build and join rules to determine a consequence theme (Mansoori et al. 2007). Fuzzy rule sets can be elaborate (Ishibuchi and Nakashima 2001) and require successive refinements to maximize precision and separation among map themes when developed manually. Small adjustments in membership can have considerable effects on the classifier outputs. The generation and combination of fuzzy rules can either be a manual exercise, as with eCognition software (Definiens 2003; Triepke et al. 2008) or an automated process (Nauck and Kruse 1998; de Oliveira 1999; Pajares et al. 2009), or both (e.g., Sameen and Pradhan 2017).

Finally, the map theme assignment that is ultimately given to a pixel or map feature in a decision process is the one of greatest overall membership among permutations of the decision network. In the process of assigning individual map themes, classifier outputs are defuzzified, replacing membership values with crisp outputs and map assignments (Rickel et al. 1998). By this process membership data can facilitate responses to a range of problems associated with ecological input data, ecological patterns, and demands of end users. The following sections expand on advantages of fuzzy classifiers over conventional crisp methods for addressing image classification and modeling dilemmas such as gradual membership among themes, modeling uncertainty, and possibility of multiple outcomes with changes in classification rules. Fuzzy logic in accuracy assessment is also discussed.

Fuzzy Representation with Continuous and Categorical Data

Spatial analysis with fuzzy classifiers can leverage both continuous and categorical predictor data. Continuous data in particular lend themselves to portrayal in probability surfaces and fuzzy sets. Not all continuous data may be representable in a fuzzy surface depending on whether each object in the dataset can be attributed by membership to the same theme. Typically there are also competing themes to a fuzzy set that are represented in membership values. Nevertheless many continuous data and legacy sources, such as solar insolation or wetness layers, can be depicted as fuzzy map surfaces.

The greater challenge is in representing categorical data with fuzzy surfaces, though it is possible for some datasets. A habitat model based on the amount of tree cover across the landscape can be informed by categorical data, even if continuous data are preferred, barring the need for more precise information. Categorical data can be made useful for map modeling and environmental analysis if suitable membership values can be determined for each category (Nauck and Kruse 1999). The example in Fig. 5 reflects a continuum for tree canopy cover where, conceptually, objects with no tree cover have a membership of 0 and communities with complete tree cover have a membership of 1. In this case, membership values have been determined based on the midpoint of the four a priori tree canopy cover classes—sparse (0–9.9%, midpoint 5%), open (10–29.9%, midpoint 20%), moderate (30–59.9%, midpoint 45%), and closed (60%+, midpoint 80%).

Such an approach to categorical data may be applicable for any number of datasets and ecological variables including species dominance. Again considering the abundance of ponderosa pine, the amount of ponderosa pine could be represented by cover percentages as with the example in Fig. 5 rather than a system of gradual membership suited for continuous data (Fig. 1). So a forested stand where ponderosa pine comprised 70% of the total tree cover would have a membership value for ponderosa pine of 0.80. With individual map themes, as with the amount of ponderosa pine cover or the amount of total tree cover, such a scheme is useful for exploiting categorical data within the context of a fuzzy approach.

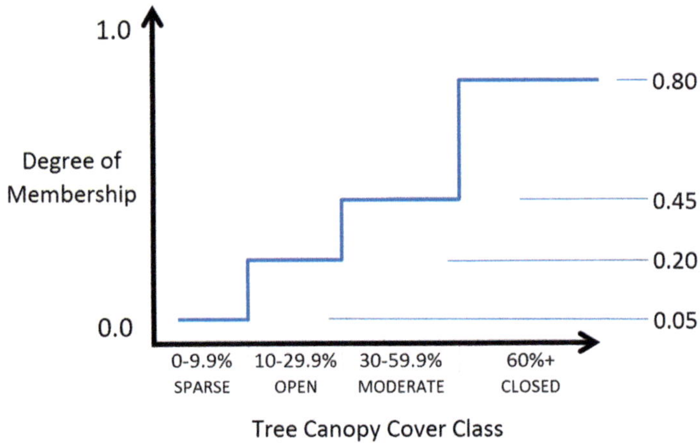

Fig. 5 Fuzzy operator for categorical data, showing hypothetical membership values for four classes of tree canopy cover

In mapping other plant composition or structural features of interest, it may be necessary to weight multiple map categories coincidentally, requiring rule-making to determine what conditions warrant the assignment of membership for a given map category to a given object. In utilizing tree cover type data, for instance, it is likely that the abundance of ponderosa pine would be factored in relation to the abundance of other tree species prompting rules that discern map units in feature space. Pixels of a mixed type, such as "Ponderosa Pine-Douglas-fir," compel the assignment of membership values for both tree species by some set of rules. In these cases, the underlying vegetation classification or map unit concepts are likely to play a role in building rules and determining standard membership values. In an upcoming example in section "Vertical Structure Mapping," rules were developed for producing membership values for forest canopy layering (storiedness) from categorical data. With ingenuity some categorical data can be made useful for map modeling purposes if the analyst can build a membership scheme that is appropriately suited to the predictor data (Rickel et al. 1998; Nauck and Kruse 1999).

Mapping with Fuzzy Classifiers

Pattern recognition and image classification of ecological features involve the search for signatures in spectral and biophysical datasets for purposes of rendering the features of interest in a two-dimensional model. While the term image classification most often refers to the classification of pixels or objects that have been generated from a remote sensing source (Lillesand and Kiefer 2000), here the term is inclusive to other ancillary predictor data including biophysical layers. Remmel and Perera in Chapter "Mapping Forest Landscapes: Overview and a Primer," provide useful overviews of mapping constructs, objectives, important data sources, and process of map development. Remmel further contrasts conventional functional forms behind image classification including binary solutions, quantification of uncertainty, and fuzzy membership. Here, only the background information that is immediately relevant to fuzzy classifier methods is given. Suffice to say that the major steps of image classification include the determination of a suitable map unit scheme, selection of training samples, preparation of the predictor data layers, selection of a classifier method, model development (training or feature extraction), post-processing of classifier results, accuracy assessment, and final processing of outputs into a geodataset (Lillesand and Kiefer 2000; Brohman and Bryant 2005; Brewer 2007; Lu and Weng 2007). Predictor data are those layers of complete coverage of the project area, sometimes referred to as census data, which are interpreted and used to model map themes. Remote sensing techniques were generated in earnest in the 1980s and included unsupervised technology such as *k*-means and ISODATA, and supervised classifiers including maximum likelihood, and various hybrid techniques (Li et al. 2014). While some of these approaches are still in use, other nonparametric classifiers have come into practice including neural networks and decision tree classifiers that include fuzzy rules and other knowledge-based classification that use both pixels and image objects as base model units (Lu and Weng 2007). Object-based techniques begin with the creation of a polygon configuration of image objects, each object a grouping of pixels with similar properties (e.g., spectral response) (Myint et al. 2011) to become the base units for an image classification exercise (see Chapter "Portraying Wildfires in Forest Landscapes as Complex Objects"). Identifying and refining suitable classifier procedures, commensurate with the input data and project objectives, are necessary for the successful development of a map layer of maximum accuracy.

This brief section focuses on the concepts of fuzzy systems in classifier methods for the development of map data. First, some clarification: fuzzy classifiers can be used to produce both crisp and fuzzy map data outputs. Also, fuzzy outputs can be post-processed to render crisp categorical mapping as described in section "Simultaneous Considerations of Thematic and Spatial Uncertainty." The application and value of fuzzy maps will be covered in this latter section. Second, interpolation and distance algorithms as well as conventional classifiers such as maximum likelihood estimation can be used to produce fuzzy maps that do not employ the fuzzy decision rules which typically comprise fuzzy classifiers. The focus of this section is on the former circumstances and the use of fuzzy classifiers to make maps, along with some advantages of fuzzy approaches over other classifier methods.

As framed at the beginning of section "Overview of Fuzzy Systems," fuzzy systems offer advantages in thinking about landscapes and responding to the heterogeneity and complexity of ecological features. In the context of classifiers, fuzzy systems offer a solution to the problem of imprecise predictor data, which Bonissone and others (Bonissone et al. 2010) summarize in this way: first, some imprecision can safely be ignored as in the case of ambiguities that are completely encompassed by relatively general map themes. Second, data which are significantly imprecise can sometimes be accommodated with the use of classifiers that effectively model the probability distribution of a map theme, such as maximum likelihood estimation. Of course, the weakness of maximum likelihood is in the assumption of Gaussian distributions, a condition often not exhibited in ecological features and fuzzy sets. Also, when representing problems of mixed pixels, fuzzy classification rules are structured to optimize the precision in classifier results relative to problems of gradual membership, both in ground conditions and in the continua that occur among map themes, and in the spatial resolution of remote sensing imagery and other predictor layers. Where membership is most gradual among themes, additional variables and more complex rule structure may be needed to optimize differentiation. Third, when imprecision is a significant issue and a probability distribution does not fit the natural pattern, the data represent a fuzzy set and the need for an alternative classifier. For such a mapping problem, fuzzy classifiers are typically represented by rule sets and often in the form of decision trees that subdivide feature space into progressively finer extents in a top-down approach (Safavian and Landgrebe 1991). A decision tree approach lessens the magnitude of image classification problems with the creation of more numerous but less complex rules with each additional layer in a decision hierarchy. As mentioned, decisions at each node of a rule set can be either crisp or fuzzy, automatically or manually generated (Gong et al. 1996), and be informed by continuous or categorical data (Nauck and Kruse 1999). Decisions, in turn, can reflect both expert and statistical relationships between observations and predictor data. In a study by Triepke et al. (2008), decision tree classifiers were used to predict the landscape distribution of Alliances and Associations of the US National Vegetation Classification (Jennings et al. 2009). The resulting decision tree represented a multi-classifier approach (Bonissone et al. 2010), combining manually generated fuzzy and crisp rules into one rule set, with nearest-neighbor classifiers applied at terminal nodes of the decision tree. In this study predictor layers were comprised of both continuous and categorical data to inform what Nauck and Kruse (1999) refer to as mixed fuzzy rules. While the development of most decision tree classifiers is automated (Nauck and Kruse 1997), the manual development of rules can be time consuming but the access to and understanding of decision rule structure offers extraordinary flexibility in addressing fuzzy sets and inputting expert knowledge of a landscape into a mapping problem.

Since the difference in fuzzy membership among a set of map categories is often subtle, as in the case of ecotones, small changes to fuzzy rules can produce significantly different classification results. Contrast among map unit concepts themselves may be weak, but greater problems may be in the lack of contrast in imagery and biophysical data or in the spatial resolution of the image data or other issues for which fuzzy systems can provide a superior solution (Cai et al. 2009). It is in these circumstances where fuzzy systems possess strengths over other classifier techniques since even slight differences expressed for a particular variable can be leveraged synergistically when variables are effectively combined to leverage the collective strength of multiple predictor data. Fuzzy classifier approaches can offer better representations of land cover features than crisp methods for the simple fact that much of the landscape is fuzzy and not described well by single map categories (Wang 1990). Even the most basic approaches for rendering geometrically fuzzy surfaces, such as interpolation and distance functions (Lowell 1994; Wang and Hall 1996), are likely to offer advantages in accuracy over crisp mapping at least for heterogeneous extents. For some types of map units such as glaciers, parking lots, lakes and ponds, riparian corridors, or scree slopes which have discrete edges, membership can change abruptly over short distances to limit the performance and suitability of a fuzzy solution (Zhang and Foody 1998). Additionally, discreteness is scale sensitive given the spatial resolution of the natural feature relative to the spatial resolution of the input imagery.

Back to automated rule development: fuzzy decision systems can and usually are generated from automated methods including fuzzy cluster analysis (Hoppner et al. 1999), neural networks (Nauck et al. 1997), and multi-classifier methods (Nauck and Kruse 1999; Bonissone et al. 2010), which have been shown to produce results that are better than individual classifier methods. In their study of fuzzy classifiers and the random forest algorithm, Bonissone et al. (2010) nicely summarize the evolution of ensemble techniques used to build decision tree rules and fuzzy systems by first describing bootstrap aggregating (Breiman 1996). Bootstrap aggregating, or bagging, is a machine learning algorithm designed to maximize the accuracy of decision tree classifiers. This approach results in an ensemble of classifiers that have been generated by resampling and replacement of individual training data in turn, where the final predictions are made by a vote of the most repeated classifications. Boosting, on the other hand, results in an ensemble of classifiers that have been added one at a time through iterative learning based on weaker classifications, with the eventual outcome of the strongest classifier (Schapire 1990). In the process, classifiers resulting in misclassification gain weight while strong classifiers lose weight, resulting in an ensemble of classifiers and the identification of their relative strength. Other key decision tree building classifiers have since been offered (e.g., Amit and Geman 1997; Ho 1998; Dietterich 2000) that lent to the development of random forest ensembles (Breiman 2001).

Random forest is another machine learning algorithm used for classification that "learns" from data (data mining) (Breiman 2001). Random forest operates through ensemble learning, based on patterns among training samples and predictor data and the construction of multiple decision trees. Classifications are ultimately assigned based on the most votes (mode) among outputs from the multitude of decision trees. As a classifier, random forest is a combination of bootstrap aggregating and the random selection of sample sets from the suite of training data through bootstrap selection and replacement (Amit and Geman 1997; Ho 1998). The only adjustable parameter of import is the out-of-bag error rate, which is used in determining the optimal range of predictor variables for inclusion in the model. Reducing the number of variables reduces both the strength of individual decision trees and the correlation (redundancy) between any two trees. Increasing the number of variables has the reverse effect. The out-of-bag error estimation is the proportion of decision tree scenarios that do not result in a correct classification according to the samples held back from selection for a given scenario—that is, not within the bag of selected samples. The out-of-bag error rate is an unbiased error output with random forest that can assist in determining the optimal range of predictor variables. Random forest can be effective and robust with the default settings, making it ideal for non-statisticians and image analysts given the limited parameters for tuning, including the number of variables tested at each split in the tree and the number of classification trees in the model. The decision trees generated by random forest are typically made up of crisp rules though fuzzy systems are a logical extension for some classification problems (Marsala 2009; Bonissone et al. 2010). Although fuzzy random forest has been used in other fields of science (Bonissone et al. 2008; Kulkarni and Sinha 2013; Lasota et al. 2013), the approach is not as yet a convention for land cover mapping. In the final section of this chapter, "A Look to the Future," the potential application and advantages of fuzzy random forest are explored.

Fuzzy Approaches for Identifying and Utilizing Uncertainty

Uncertainty determinations form a basic component of any scientific tool or product (Congalton 1991; Congalton and Green 1999; Brohman and Bryant 2005), most commonly expressed in statistical estimates of confidence. Different techniques have been applied to assess the uncertainty of map data, most often taking the form of thematic accuracy assessments summarized on confusion matrices. Objective determinations of map uncertainty are critical for informing end users, who may exacerbate error by the ways map data are misunderstood and applied (Bailey 1988; Cowling et al. 2005).

Uncertainty can be both thematic, dealing with extents that have shared characteristics of multiple map categories, or geometric, where the same spatial extents may be more or less homogenous but reflect zones of intergradation among disparate map units that make up a minor proportion of an analysis area. Put another way, the problem of shared characteristics to a mapping specialist is either one of mapped extents with attributes of more than one map theme or one of the transitional nature of ecological features across horizontal distances (Zhang and Stuart 2001). Added to the complexity in ecological patterns, and the ability to capture those patterns in a map model, is the question of model accuracy. For this question, the reader is referred to Chapters "Fuzzy Classification of Vegetation for Ecosystem Mapping" and "Portraying Wildfires in Forest Landscapes as Complex Objects" with the acknowledgement that the uncertainty in a map product in part reflects the capability and shortcomings of the underlying models used to generate map surfaces. Also, both thematic and geometric ambiguities can be augmented by, or even stem from, the relative coarseness in the spatial or thematic detail of predictor data—also a problem of model performance—not to mention possible registration errors in the various predictor layers. Chapter "Mapping Forest Landscapes: Overview and a Primer" introduces important concepts of uncertainty and poses the fundamental question: How confident can we be in the accuracy of map data? For now, the focus is on the specific problems of thematic and geometric uncertainty that can exist and be dealt with using fuzzy methods, either at the front end by improving map classification and the separation of map units or in the assessment of results and the characterization of map accuracy and fuzziness.

First, transitional areas of the landscape can share properties of multiple themes in a legend and difficult to map accurately to the most suitable theme(s), as in the case of mixed pixels (Li et al. 2014). Still, the problem may not necessarily be fixed by, say, acquisition of imagery with higher spatial resolution (e.g., Landsat TM versus RapidEye imagery, 30 m vs. 5 m resolution). In fact, the issue may be compounded if the higher resolution source is too sensitive to within-theme features of little interest that reflect greater detail than the map scheme—higher spatial resolution imagery is not always better. For example, small tree patches of a few square meters within a grassland matrix may be mappable with the high-resolution source, but nevertheless undesirable depending on the objectives and specifications of the project. In this case a satellite sensor of coarser spatial resolution may effectively blend the responses of tree and understory components to create a useful image source for mapping grassland or savannah systems without expressing within-theme features such as small shrub or tree patches (Fig. 6).

Nor would it be desirable or practical to map transitional areas of the landscape if they are only a minor element in the overall ecological pattern, unworthy of distinction in a map legend. Yet transitional areas may also occur over extensive expanses and protracted gradients, as with the zone between boreal forests and tundra on the northern Canadian Shield, where individual plant communities support

Fig. 6 Comparison of spatial resolution of satellite image sources, Landsat TM (30 m) versus RapidEye (5 m), from a grassland system with juniper encroachment east of Grants, New Mexico, USA (Earth 2014a)

both woodland and tundra components over vast areas (Barbour et al. 1998). In these environments trees and tundra co-occur with regularity and at common map scales with relative uniformity that warrant imagery and mapping systems that capture these features within the same categories. Categorical mapping assumes that, as with effective vegetation classification, the homogeneity within map units is maximized while simultaneously maximizing heterogeneity among map units. The maps then generated by these categories pose the same assumption in a spatial context. The conventional response to significantly large areas of intergradation and ambiguity is to create additional map units that capture those areas as themes unto themselves, the result being a map legend of lesser contrast but abiding mutual exclusivity. For example, in northwestern Montana of the USA there are significant swaths of forests heavily dominated by western larch (*Larix occidentalis*) and areas dominated by subalpine fir (*Abies lasiocarpa*). There are also extensive areas where these two constituents are intermingled as codominants of the same object, necessitating the representation of a mixed cover type in both the classification and mapping of vegetation (Leavell 2000; Triepke et al. 2008), albeit at the cost of additional complexity in vegetation classification and map development to address the practical needs of forest practitioners.

Fuzzy approaches can be helpful in forming vegetation classes and map units, in building maps, and then in assessing uncertainty of end products. As stated, while conventional mapping perspectives impose classical set theory on the assumption that map units are mutually exclusive rather than on continua within the landscape (Woodcock and Gopal 2000), fuzzy theory allows us to evaluate ambiguity in an ordered way and then to analyze uncertainty, both in terms of area estimates of mixed conditions and fuzziness and in terms of accuracy assessment itself. And fuzzy approaches can be applied both for thematic and spatial entities, the uncertainty of which is often intertwined (Aspinall and Pearson 1995).

Thematic Uncertainty

To explore thematic uncertainty, we will again use the example of codominant ponderosa pine and Douglas-fir and assume relative percentages of 60% and 40%, respectively, within a plant community. If an object representing the community is mapped as "Douglas-fir," a fuzzy analysis would show that the degree of misclassification is less than an object misclassified as "Douglas-fir" but made up entirely of ponderosa pine. Even a fuzzy assessment at a more superficial level, say broadleaf versus conifer, has utility over a crisp perspective that indicates total error without nuance, as in the case of two misclassified objects that are dominated by ponderosa pine in reality but classified as Douglas-fir and quaking aspen (*Populus tremuloides*). A fuzzy approach may allow some credit for the misclassified Douglas-fir conifer object over the object misclassified as a broadleaf "Aspen" type, versus a crisp approach that would show the two objects in complete and equal error. A key advantage of a fuzzy technique is in being able to assign the degree of deviation from truth, which may be defined simply as the theme of highest membership according to an observation (accuracy sample). This theme may, in fact, hold marginal leads over other themes of partial membership. It can be very much worth knowing, for example, the degree of deviation in a misclassified pixel relative to the theme of highest membership and whether the deviation is small in comparison to other themes that, under a fuzzy approach, would reflect a much higher amount of error (Wang 1990).

In another perspective of uncertainty, Woodcock and Gopal (2000) demonstrate a means of area estimation using a vegetation map of the Plumas National Forest of northern California, to contrast the amount of extents that, respectively, fall within classic and fuzzy sets. Woodcock and Gopal applied Card's method (Card 1982) and assigned degrees of map unit membership to each accuracy sample, and then integrated membership probabilities with area-weighting to determine the extent of the map that was relatively ambiguous for six map themes—water, barren-grass, meadow, brush, hardwood, and conifer. They found that aside from the water and conifer units, very few samples from other units reflected full membership to any particular theme, further implying that much of the map extent occurred as fuzzy sets. The general relationship is that as the threshold membership value for any particular map unit is reduced more map area is represented by that map unit (Fig. 7). The area estimate of brush in particular had a strong inverse relationship with membership. Using chaparral vegetation to illustrate the relationship, the authors describe a pattern where shrub species dominate the understories of plant communities with low tree cover that still meet the chief criterion of the conifer unit (>10% canopy cover of conifer trees), representing the circumstances of about a quarter of the entire analysis area. In these communities, shrubs often have an aerial extent on par or exceeding that of conifers so that they possess strong affinity to the shrub theme. While 61% of the map area was represented by conifer, only about half that area is estimated to have full membership to the theme. Figure 7 shows that

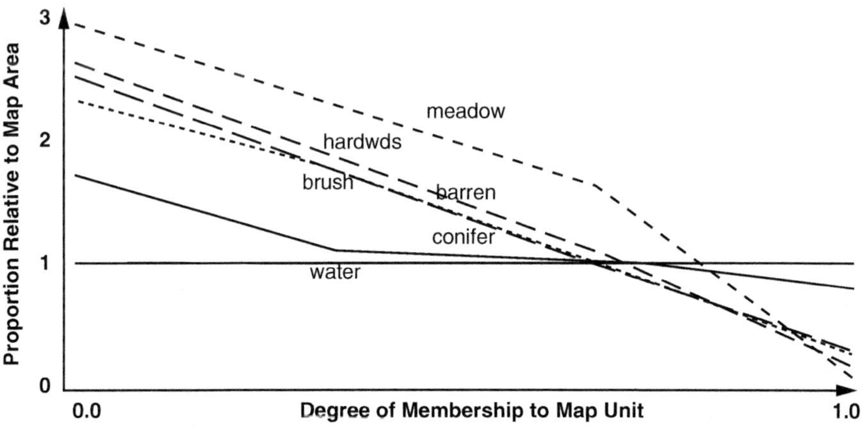

Fig. 7 Showing the relationship of fuzzy membership for six map units of the Plumas National Forest to the area estimated for each unit. Modified from Woodcock and Gopal (2000)

most of the six map units in the project represent an increasing amount of area with reduced level of membership. Approximately 49% of the analysis area is represented by degrees of membership less than full membership, so that about half the map surface occurs as fuzzy sets.

The Woodcock and Gopal (2000) study illustrates an area estimation approach based on fuzzy sets that quantifies the amount of area for each map theme by degrees of membership. This type of area estimation provides important details to end users and those developing map unit descriptions. Such an approach allows for a considerable range of analyses and queries within a GIS, and may be vital for accurate change detection and monitoring (Álvarez-Martínez et al. 2010).

Another study by Gopal and Woodcock (1994) provides an example of the use of fuzzy methods to assess thematic uncertainty for purposes of accuracy assessment. The study demonstrates a means of fuzzy accuracy assessment for crisp image classification outputs of map categories that reveals additional information about error structure, including the four dimensions of error given by the authors—the map categories in error, frequency (rate of error), magnitude (degree of confusion), and error source. As before, their approach provides a solution commensurate with inherently fuzzy ecological features and allows for variable membership by accuracy samples to each map category. They point to three main issues of conventional accuracy assessment:

- The premise of crisp assessments is that each object is unambiguously assigned to one map category.
- Accuracy results regarding the magnitude of error can only be inferred by the pattern of confusion between observations and predictions among map categories.
- The nature of accuracy results limits the ability of producers to interpret and respond to error, and limits the user's ability to effectively apply map data.

In response to these issues, Gopal and Woodcock (1994) decompose the basic question "how accurate is the map?" into two more exacting inquiries: "How frequently is the mapped category the best choice for the site?" "How frequently is the mapped category acceptable?" As mentioned, thematic accuracy assessment is most often facilitated with an error matrix or confusion matrix (Card 1982; Congalton and Green 1999), if map accuracy is evaluated at all. Accuracy results by this approach may only be derived for outputs of crisp classifiers and do not provide information on the true proximity of any map categories let alone the assigned category. The lack of such information impedes the examination of the characteristics and sources of map error. As important, the lack of interpretation also limits awareness of the error structure by end users, and precludes their ability to construct other outputs from the map data representing a posteriori mixed categories.

Accordingly, Gopal and Woodcock (1994) generated a scale, from 1 to 5, to rate the level of agreement in conditions at each accuracy sample site to each map unit concept, where 1 is "absolutely wrong" and 5 is "absolutely right." This expert assessment resulted in a set of membership values for each accuracy sample site that represented every map category, while the mapped value remained unknown.

Table 1 summarizes the accuracy assessment that they generated from a hypothetical image classification with four crisp map categories, A, B, C, and D. The image classification was intersected by 40 accuracy samples that had each been rated on the scale of agreement between 1 and 5 for every map category.

For accuracy assessment, there is more value in knowing both crisp and fuzzy agreement than in knowing either one alone. Case in point, at only 40% agreement the crisp assessment of category C shows this map unit to be a considerable issue in terms of accuracy of the image classification (Table 1). But by considering the amount of agreement occurring at an acceptable level, the fuzzy assessment indicates more promising results at 80% agreement. In fact, at the acceptable level of accuracy category C shows greater agreement than category B (60%) even though more samples in B are in agreement at the level of being "absolutely right."

To compute the magnitude of error, shown in the final column of Table 1, Gopal and Woodcock (1994) constructed a difference Table 2, by tabulating the magnitude of error within each map category. Error was calculated by comparing the fuzzy rating of each accuracy sample of each assessment site to the highest agreement level assigned to all other map categories, generating a simple and relative index of error severity. If the agreement level given to the map category was higher than the highest rating for all other labels, the resulting difference value was positive. A negative value resulted when the agreement level for the map category was lower than the highest level assigned for a differing map category. Difference values of −1 through 4 generally corresponded to the correct map labels. All difference values were averaged for each sample and map category, resulting in the values shown in the last column of Table 2.

Table 1 Accuracy assessment summary table adapted from Gopal and Woodcock (1994) showing agreement for both crisp (and absolutely right) and fuzzy (acceptable) assessments

Map category	Accuracy samples	Samples absolutely right (frequency)	Samples acceptable (frequency)
A	10	10 (100%)	10 (100%)
B	10	6 (60%)	6 (60%)
C	10	4 (40%)	8 (80%)
D	10	6 (60%)	8 (80%)
	40	26 (65%)	32 (80%)

Table 2 Difference table adapted from Gopal and Woodcock (1994) used to detect the magnitude and source of error

		Mismatches				Matches					
Map category	Accuracy samples	−4	−3	−2	−1	0	1	2	3	4	Mean
A	10	0	0	0	0	0	1	0	1	8	3.60
B	10	1	1	0	2	2	1	0	2	1	0.20
C	10	0	0	2	4	0	2	1	1	0	−0.10
D	10	0	0	1	3	4	0	0	2	0	0.10

The resulting differences between fuzzy ratings and the associated map categories are summarized in the columns and then averaged

In terms of magnitude, Tables 1 and 2 show that map categories B and D have similar rates of error (60%) but vary substantially in the magnitude of error. Category D has a greater number of 0 differences as a general indication that the magnitude of error for B exceeds that of D. The last column of Table 2 has the arithmetic mean of all difference values for each map category, so that the categories with the least magnitude of error would have the highest corresponding mean. For example, the mean once again shows greater magnitude of error in D than in B. The most accurate class, A, also has the least magnitude of error, with a mean difference of 3.60. The least accurate category, C, also shows the highest severity of error. Being able to convey the magnitude or seriousness of error is one of the main advantages of a fuzzy accuracy assessment.

Finally, the sources of error can also be explored using fuzzy accuracy assessment techniques given by Gopal and Woodcock (1994). The frequency of matches and mismatches given in a difference table can give clues about ecological complexity and error sources for map producers. In general, if there are greater numbers of mismatches in the single membership sites, then mapping issues may be concentrated in unambiguous locations (e.g., pure ponderosa pine stands of uniform structure). Conversely, if the mismatches are concentrated in the multiple membership sites, resources to improve map performance are better concentrated in areas of environmental heterogeneity. For instance in Table 2, notice that the frequency of multiple memberships is not the same for all categories, and that the proportion of matches for single membership sites, as in category A, is greater than in multiple membership sites (e.g., category B), contrary to a pattern expected by random effects. If the accuracy samples revealed single membership patterns across map categories, suggesting minimal ambiguity, there would be no need for a fuzzy accuracy assessment. Conversely, the number of accuracy samples representing multiple membership provides an indication of the extent of fuzzy sets on the landscape. The error structure expressed in a difference table can indicate the sources of map error among categories to map producers (Sarmento et al. 2010, 2013).

The studies cited in this section suggest several advantages of fuzzy techniques over more conventional crisp methods in assessing thematic uncertainty. Not only is it more problematic to explore and determine error patterns with crisp methods (Wang 1990), but fuzzy techniques also set the stage for a more functional integration of remote-sensed imagery with other ancillary census data. Techniques reflected in the Gopal and Woodcock (1994) study have since been applied over broad extents in the USA (Brewer et al. 2006). Though the Gopal and Woodcock approach is ideal for illustration, more sophisticated means of fuzzy assessment of both fuzzy and crisp outputs have since been created and applied (Zhang and Foody 1998).

Spatial Uncertainty

As with solutions for thematic uncertainty, fuzzy approaches can assist in the formation of map units, map development, and assessment of uncertainty of map products. Though fuzzy sets often represent an intertwining of thematic and spatial uncertainty (Aspinall and Pearson 1995), the two dimensions are first treated separately in this chapter for a clearer explanation of some of the concepts important within each dimension.

Conceptually, mapping that is produced by crisp classifiers could account for gradual boundaries and ambiguity among map themes if the grain of the objects, or pixels, is fine enough. That is, individual pixels can locally express the transitional nature of ecological features if class gradations are captured in the legend, and barring the commensurate spatial resolution of the imagery and effectiveness of the classifiers. In reality, mappers are often faced with relatively coarse imagery and census data, training data which are essentially crisp, and classifiers that may be limited in their ability to accurately portray fine gradation. The spatial uncertainty of mapping, in both developing and assessing the vagueness of boundaries, has been accommodated through other means, with some examples to follow.

Again from the perspective of thematic uncertainty, accuracy assessment is most often expressed in an error matrix without regard for spatial accuracy, where calculations for producer and user accuracies are generated on opposing axes based on a set of independent accuracy samples and the difference between predicted and observed outputs. By this conventional approach, overall accuracy, area-weighted user accuracy, and kappa statistic are often the measures of most interest (Rosenfield and Fitzpatrick-Lins 1986; Congalton 1991; Janssen and van der Wel 1994). The spatial uncertainty of boundaries between objects has been evaluated for positional error using different methods such as the epsilon band (Perkal 1956; Chrisman 1989), likely the most used error model for map delineations themselves (Leung and Yan 1998). The epsilon band allows for the characterization of spatial uncertainty and is most often directed at crisp map products. For purposes here, spatial uncertainty refers to the fuzziness and breadth of boundary conditions, and not to boundary location error for which the epsilon band is often applied (Shortridge and Shi 2012).

In its simplest form, the epsilon band produces a rectangular distribution of width 2ε imposed on a mapped line to indicate an area of uncertainty, manifested in vagueness of the map data and controlled by specifications of the program or project. The epsilon band was originally devised by Perkal as deterministic models represented in parallel-sided polygons such as running the length of map boundaries. The epsilon band has since been superseded by probabilistic data models (Leung and Yan 1998; Kronenfeld 2011), with the most recent adaptations of the model taking on irregular shape complexities to more realistically depict the uncertainty of horizontal transition zones between opposing map themes otherwise assumed as mutually exclusive (Fig. 8).

Whether deterministic or probabilistic, uncertainty banding offers a means of objectively characterizing ambiguity in boundaries among map categories. Advanced shape analysis tools, such as *ShrinkShape2* (Remmel 2015), further give users the means to quantify boundary complexity. This tool creates internal polygon buffers iteratively, generating summary metrics with each shrinking iteration for the characterization of spatial structure and complexity. Such methods can in turn be used to refine classifier outputs by the way map features are depicted in a product.

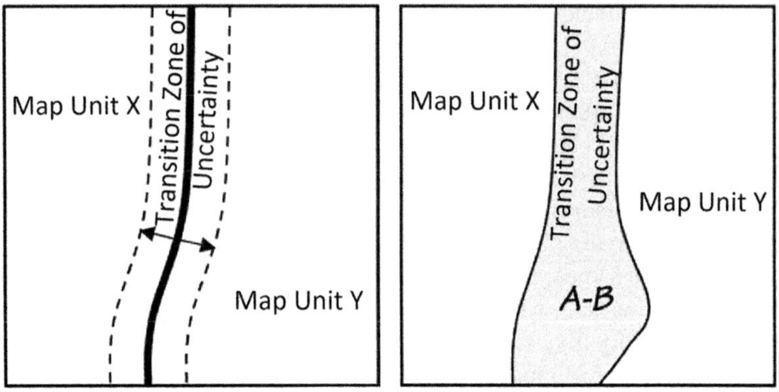

Fig. 8 Spatial uncertainty at the boundary of two opposing map themes, *X* and *Y*, represented by a deterministic epsilon band of standard width (*left*), and a probabilistic model of variable width (*right*). Modified from Kronenfeld (2011)

Crisp maps can be rendered fuzzy in part by redrawing delineations or using symbology to infer the fuzziness of boundaries according to distance-based functions (Lowell 1994).

Remmel (2009) introduced another means for assessing spatial uncertainty by applying coincidence matrices for separate map products of the same area. Coincidence matrices are often used for assessing map accuracy based on a set of observations, but can also be used in the comparison of different map products. In this way, Remmel applies the matrices to assess thematic uncertainty, and to assess spatial uncertainty by the geographic configuration of map features. Essentially the method uses the coincidence matrix as an expression of spatial complexity by applying two or more map products and representing the potential complexity by an accounting of the number of possible configurations in the matrix. A fixed amount of agreement can be reflected in multiple spatial configurations for the same set of map objects. The process results in a quantification of uncertainty that can be applied at local or full extents to determine both spatial and thematic uncertainty.

Simultaneous Consideration of Thematic and Spatial Uncertainty

As mentioned the nature of thematic and spatial uncertainties can be intertwined (Aspinall and Pearson 1995) though most published works have treated the two entities separately. What follows is a brief summary of a study by Zhang and Stuart (2001) that demonstrates a means of concurrent assessment for thematic and spatial uncertainty, and how that uncertainty is characterized while producing spatial outputs that optimally balance uncertainty with utility. In this study, the authors developed a geodatabase of suburban land cover classes. They created a technique for capturing the uncertainty of those map units using aerial photo interpretation and image classification procedures that produced a series of fuzzy surfaces for each map category, rather than establishing the fuzziness of individual sample points and then assessing the map product as in the case of Woodcock and Gopal (2000). These surfaces allow the user to quantitatively and graphically determine the underlying patterns of thematic and spatial uncertainty before assigning map categories to each object.

Zhang and Stuart (2001) began by summarizing the means of image classification that results in a set of fuzzy surfaces that each represents a map category. That is, each surface is developed by classifier methods that result in membership values for each category and image object. Fuzzy surfaces can be derived through any number of classifiers including manual rule building within eCognition (Definiens 2003), distance measures for spectral channels or other census data (Knick and Rotenberry 1998), from the Random Forest algorithm (Bonissone et al. 2008), or other classifier conventions. To be sure, both classic and fuzzy classifiers can result in fuzzy outputs, and the intention of this chapter is to highlight some advantages of both fuzzy classifiers and fuzzy map renderings. By the approach that Zhang and Stuart propose, it is necessary to use a classifier approach that produces membership values for every pixel or object, produced in surfaces of all potential land cover categories through the classification of spectral and other census data.

As a basis of the method proposed by Zhang and Stuart (2001), the most obvious land cover assignment within uniform extents of a landscape has more certainty than the most obvious assignment in transition zones, where vegetation conditions are heterogeneous and there is greater parity among potential categories. Their approach bears on the capacity to make fuzzy determinations in the development of training data that are very certain, a requirement that is operationally demanding but supports the creation of multiple fuzzy surfaces based on fixed points of knowledge. From there, a spatial technique of interpolation such as kriging (Krige 1951; Cressie 1990) is applied to generate membership values across unsampled zones of greater uncertainty, anchored by sampled areas of relative certainty. With this type of intermediate product, categorical maps can then be produced to the satisfaction of end users. For any given object, this process of post-classification typically represents a maxima among fuzzy membership values (Leekwijck and Kerre 1999; Islam and Metternicht 2005) from the underlying fuzzy surfaces to arrive at one land cover category—that is, a defuzzification resulting in one crisp value. Object attribution includes fuzzy membership values from all category surfaces, which can help to form various error models, not the least of which is the previously summarized epsilon band. Again, spatial uncertainty can be an expression of positional error as with orthorectification, but as mentioned the related focus of this chapter in on the uncertainty of classification.

In their approach to defuzzification, Zhang and Stuart (2001) demonstrate a thresholding technique that allows for the analysis of transitional zones among land cover types. Rather than simply classifying each object according to the maximum membership value among category surfaces, the application of thresholds provides a means of identifying areas of high thematic certainty to nominal map categories

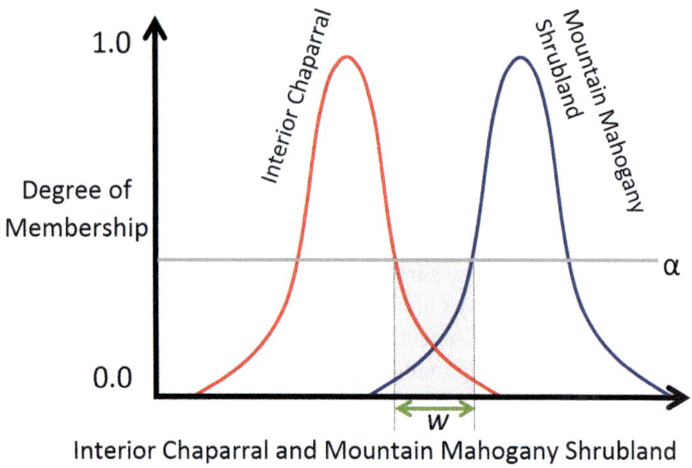

Fig. 9 Fuzzy operators showing the hypothetical relationship between two neighboring shrubland units. Here, a threshold value of α has been added to Fig. 4 to discern extents of high and low uncertainty, with a resulting unclassified region of width "w"

while simultaneously characterizing uncertainty spatially according to tolerance thresholds of the producer (Islam and Metternicht 2005). In converting raw classifier outputs to crisp map products, thresholding is achieved when all extents that meet the value of the threshold, α, are included within the given map categories. Given the capacity for quantitative analysis and solutions, this type of processing is compatible with the complex and nonuniform ecological patterns that make up fuzzy sets and that would otherwise require subjective responses or indifference. Extents that do not meet classification thresholds can simply be coded as highly uncertain. Figure 9 represents two contiguous shrubland units and the application of an uncertainty threshold. The region of uncertainty that exists between the two categories is shown by the width w in the figure and may be labeled as unclassified, or as a possible third mixed category not apparent in the initial map legend (Zhang and Foody 1998). To sum up, a thresholding technique can lead to zones of uncertainty that may either be represented thematically, by a legitimate mixed map unit (e.g., "Mixed Interior Chaparral-Mountain Mahogany Shrubland"), or spatially by an epsilon band of fixed uncertainty.

Zones of high uncertainty between map categories that are identified by thresholding can form the basis of epsilon models and offer a means to generate widths as a quantitative and geometric approach to characterizing error (Fig. 8) and refining crisp map units. By varying the uncertainty threshold, α, the epsilon band can be widened or narrowed to generate a range of epsilon models according to expectations of producers or clients. This approach may be particularly useful for assessing habitat quality including those species with affinities towards ecotones (Lloyd et al. 2012). In a sense the epsilon width can be varied continuously to reflect the relative certainty of all competing categories. In their study of suburban land cover mapping, Zhang and Stuart (2001) determined that epsilon band widths corresponded to standard deviation values and resulted in sizable spatial variation as uncertainty thresholds were altered. The degree to which epsilon bands varied locally according to threshold values suggests that deterministic epsilon models, of constant width, may be less suited to spatially depict land cover patterns. The reader is again referred to Fig. 8 for a visual comparison of deterministic and probabilistic epsilon bands. Using classifiers that result in multiple fuzzy surfaces allows mappers the ability to build probabilistic epsilon models, and to characterize uncertainty and leverage the knowledge in many constructive ways.

Multiple Outputs: Fuzzy Geodatabase

As examined in Chapter "Fuzzy Classification of Vegetation for Ecosystem Mapping," the variability in fuzzy membership for a given pixel or image object suggests that fuzzy approaches lend themselves to a range of mapping and analysis applications for natural features. Membership values can be used to inform the development of more precise map themes, or conversely to suggest more general themes of higher accuracy. As previously discussed the development of geodatasets that are comprised of fuzzy membership surfaces for each map unit offers the greatest flexibility in balancing map accuracy and precision for any given application, and for analyzing thematic and spatial uncertainty in the data.

Fuzzy surfaces can be constructed from satellite imagery, aerial photography, or other census data alone or in combination, using automated classifiers or manual interpretation to generate outputs that represent the ecological features of interest. The classification of map objects or pixels from the geodataset is usually a matter of assigning map units according to the surface of maximal membership value (Leekwijck and Kerre 1999; Islam and Metternicht 2005), so that each object is attributed by both the most likely category resulting in a conventional crisp map rendering. The thresholding technique described previously allows for outputs to be controlled by uncertainty criteria, with the possibility of disqualified extents that are significantly uncertain, or that comprise candidates for additional "mixed" feature classes.

Such a geodataset also permits a data-driven approach to describing uncertainty and estimating error. The overall effect is to allow substantial flexibility in generating multiple outputs from one spatial dataset, laying the groundwork to empower end users to generate map themes on their own terms in a GIS that are specific to a given purpose. The user is able to co-analyze uncertainty for an optimization of particular outputs along with a comprehensive characterization of uncertainty. In this way, the end user can develop rules and thresholds to produce tailored outputs for particular spatial applications and in response to their own uncertainty criteria. In his project on urban mapping, Hansen (2003) provides a helpful example of a fuzzy geodatabase using the case of urban land-use mapping in Denmark (Fig. 10). In this study, Hansen notes that fuzzy modeling offers a more useful product than a crisp map since it can simultaneously show the primary, secondary, etc. land use

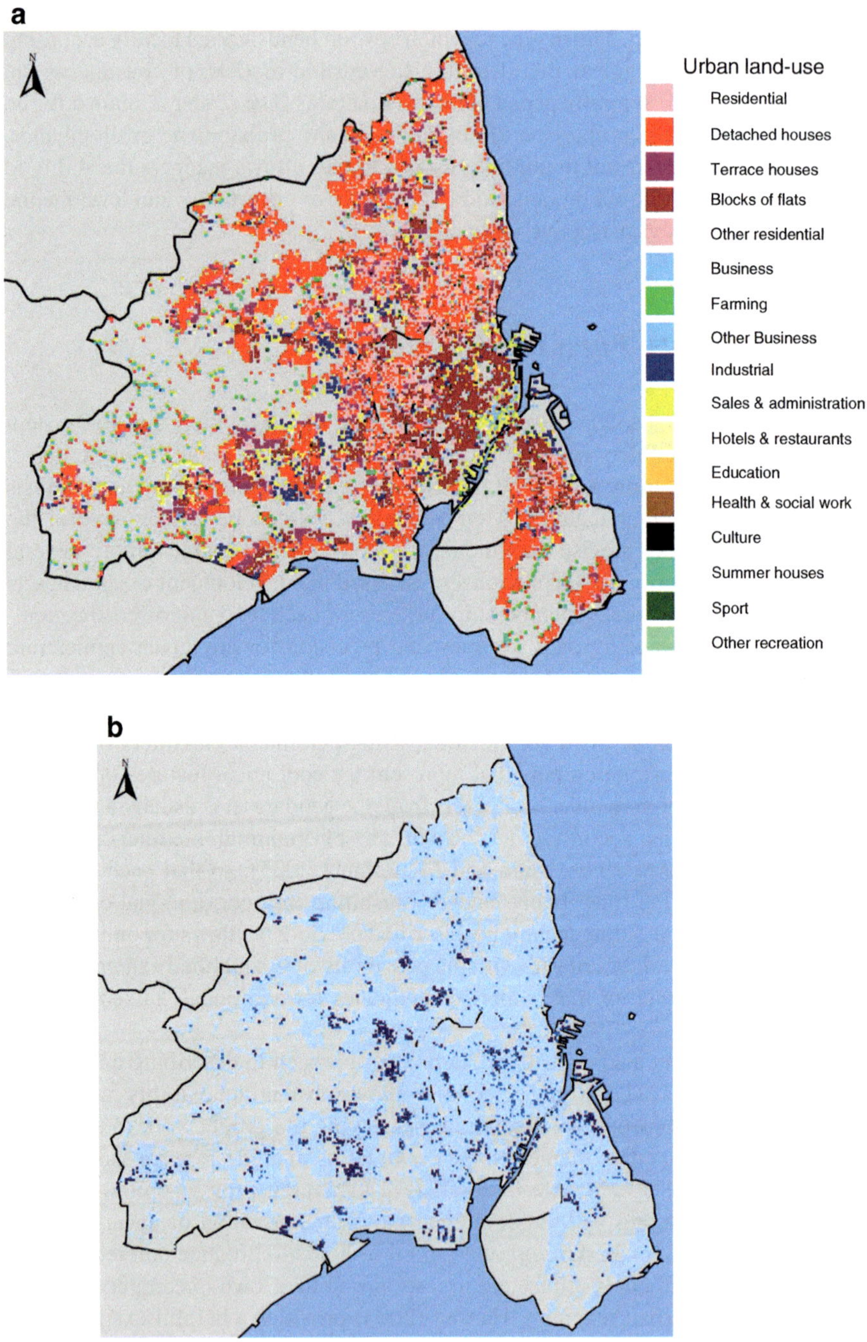

Fig. 10 Urban land use for Copenhagen, Denmark, including crisp (**a**) and fuzzy (**b**) maps. The fuzzy map is of one class, "Industrial," showing increasing levels of membership with darker shades of blue. Modified from Hansen (2003)

categories. At the top of Fig. 10, the map gives the dominant land use category for each 100 m pixel across greater Copenhagen. This crisp map was developed through defuzzification of other map unit values, assigning each pixel a map unit according to the map unit of highest membership for that pixel. As with the second map at the bottom of Fig. 10, individual fuzzy membership surfaces for each map category can be displayed separately. The second map has one theme, "Industrial," and shows that surfaces of continuous values for one theme reveal much more information about that theme than can be offered in the corresponding crisp output. In a variation on this approach, geodatasets comprised of the most basic themes, such as "Impervious Surface" or "Vegetation Life Form," can be used as primitives from which to combine and produce any number of map typologies as part of a "legend-less" mapping system (ADPC 2016), giving end users substantially more ability to determine and customize viewing and analysis.

It also bears mentioning that many legacy raster map, generated through conventional image classification, can be augmented by combining the data with image segments using zonal processing techniques. In this way, multiple outputs can be generated from many existing ecological map datasets, not only by subjecting raster inputs to different rules and thus generating different outputs, but also by linking rule sets across more than one dataset or feature class to create polythematic outputs such as state-class units representing combinations of dominance, canopy cover, and size class (Westoby et al. 1989; Steele 2000). Though legacy raster mapping usually lacks fuzzy membership attribution, the pixels that fall within a given image object can be weighed collectively using crisp or fuzzy rules to classify the objects. The technique assumes that the imposing objects have ecological features that are relatively uniform, spatially and thematically, so that pixels are grouped in a meaningful way (USDA Forest Service 2012). Depending on the rule set, this post-processing method can be used to improve data accuracy beyond the underlying raster data by using majority, average, or other statistics that effectively combine pixel values within an object to reduce noise and improve accuracy with the exchange of thematic or spatial precision of the raster data. Alternatively, the approach can be used to derive additional map themes: the following section details a hypothetical example where fuzzy rules are used to develop an additional map theme from legacy raster data.

Vertical Structure Mapping

Until now, the chapter has focused on fuzzy applications within two horizontal dimensions, thematic and spatial, where membership is determined by the degree of alignment of an object to multiple map themes of an area, or geometrically according to the horizontal irregularities or gradation of boundary zones between map themes (Remmel and Perera 2009). This section focuses on another dimension based on vertical vegetation structure, where membership is determined for canopy layering, tree height stratification, or other local habitat attributes of canopy architecture. In this dimension, the membership of an object bears on its affinity to one or more vertical features, such as canopy layering or "storiedness," as in the example to come. The affordability of producing digital surface models (DSMs) from stereo imagery or from LiDAR data to detect vertical features, and the demand for information on vegetation structure, will only improve the technology and availability of these data sources with time. Hirschmuller (2005) and others have developed semi-global matching, or "phodar," and other technologies to efficiently build digital surface models (DSMs) from high-resolution stereo imagery (Gehrke et al. 2010; Clark et al. 2016), thereby allowing for image classification of vegetation composition and dense terrain extraction for vegetation structure from the same data. Chapter "Mapping the Abstractions of Forest Landscape Patterns" provides further details of LiDAR for forest mapping.

What follows is a synopsis on how outputs for vertical diversity were rendered from raster map data of tree size class, leveraging size-height relationships in combination with the heterogeneity among contiguous pixels representing the same forested stand (Helms 1998) (Fig. 11). In this example, the pixels retained their original classification for tree diameter size class from a previous mapping effort (USDA Forest Service 2014) and were expressed in different vector outputs for vertical diversity, according to the relationships of neighboring pixels and the local inferences of tree diameter on vertical diversity. As is often the case, size class data, from forest inventory or map models, are more affordable and available than tree age data that are derived from tree coring and intensive field sampling (e.g., Triepke et al. 2012). Tree size is often used as a surrogate for tree age for locally characterizing cohort patterns

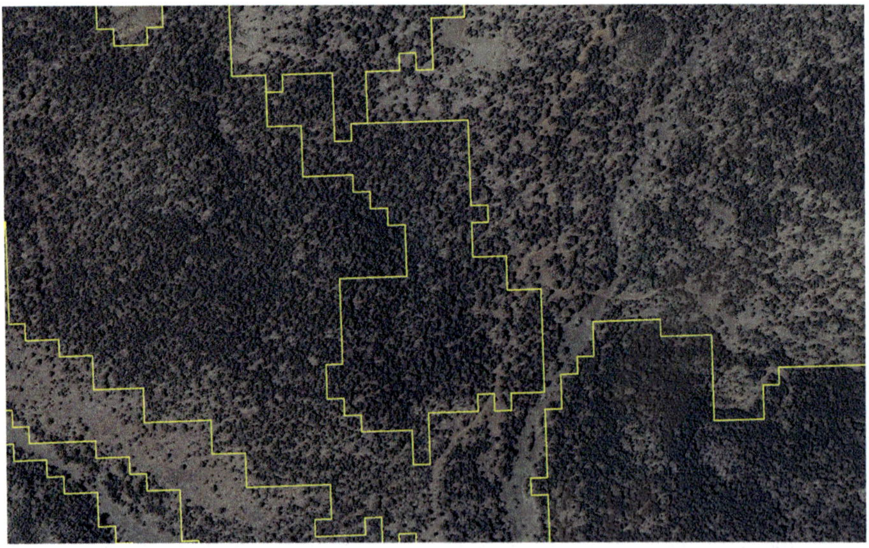

Fig. 11 Polygon configuration generated from segmentation of Landsat ETM+ to represent existing vegetation and forested areas of similar tree composition and structure. Excerpt of 2011 aerial photography taken of the National Agriculture Imagery Program (NAIP 2014), for the northern San Mateo Mountains of New Mexico, USA

(Curtis 1967; Huang et al. 1992; Schmidt et al. 2011), as with silvicultural applications that require the assessment and management of tree age diversity (Triepke et al. 2011). Where the relationships between size and age are marginal at broad scales, the relationship is strengthened at more local scales (Ferguson and Carlson 2010).

Fuzzy math lends itself to analyzing the vertical complexity or "storiedness" of forest communities where vertical structure can be difficult to characterize objectively. As a first step in mapping vertical diversity of forested systems in Arizona and New Mexico, a segmentation layer was generated from Landsat 7 ETM+ imagery to depict existing (actual) vegetation at the spatial scale of plant communities or stands (USDA Forest Service 2012). In this way, communities of similar tree dominance, size, canopy cover, and vertical diversity were delimited to the form of a polygon configuration of plant communities with similar vegetation pattern (Fig. 11).

With a polygon configuration of plant communities across Arizona and New Mexico in place, zonal processing techniques were used to vectorize raster mapping of existing vegetation and diameter class from the Integrated Landscape Assessment Project (ILAP) (USDA Forest Service 2014). Polygons averaged from 10 to 20 ha. In the zonal processing of raster data, various rule sets were applied to collectively assess the pixel values within each polygon using ILAP themes of tree size, canopy cover, and dominance to produce a range of feature class outputs for vegetation composition and structure. Rules were also generated for storiedness mapping, utilizing the inference of tree diameter on tree height, and then assessing the pattern of height conditions among contiguous pixels of the same image segment (Fig. 12).

The rule set used in this case relates the variability and abundance of tree size classes, number of canopy layers, and fuzzy membership to the storiedness theme (USDA Forest Service 2012). Similar to the canopy cover class scenario illustrated in Fig. 5, fuzzy membership was expressed in categories, either 0.2 (one story), 0.4 (two story), or 0.6 (three-plus story). In this case, forested image segments with a membership of 1 would represent layering maxima for a storiedness theme. The perspective of fuzzy logic assumes that a given object can have membership to more than one class of storiedness, which is fitting given the considerable structural variability of natural communities and the complexity in conceptualizing, interpreting, and conveying structure conditions meaningfully to biologists and land managers. Accordingly, storiedness rules have been written to accommodate different end-user needs and to produce different spatial and tabular outputs from the same input data, as in the case of tree canopy layering (Vandendriesche 2011). Canopy layering is but one possible variable of the third dimension of vertical structure (Fig. 13). Other structural variables, such as above-ground biomass, canopy base height, canopy texture, and stream embeddedness, likewise lend themselves to fuzzy classification algorithms for the interpretation of map data and vertical features. Also, vertical information can be stored in image stacks where each layer, either a crisp or fuzzy surface, represents an upward sequence of height classes that collectively reflect the vertical profile at any given point.

Fuzzy membership values and rules can simplify the integration of vertical and horizontal features and represent map objects with multiple attributes simultaneously. Membership thresholds among fuzzy surfaces can be used in image classifi-

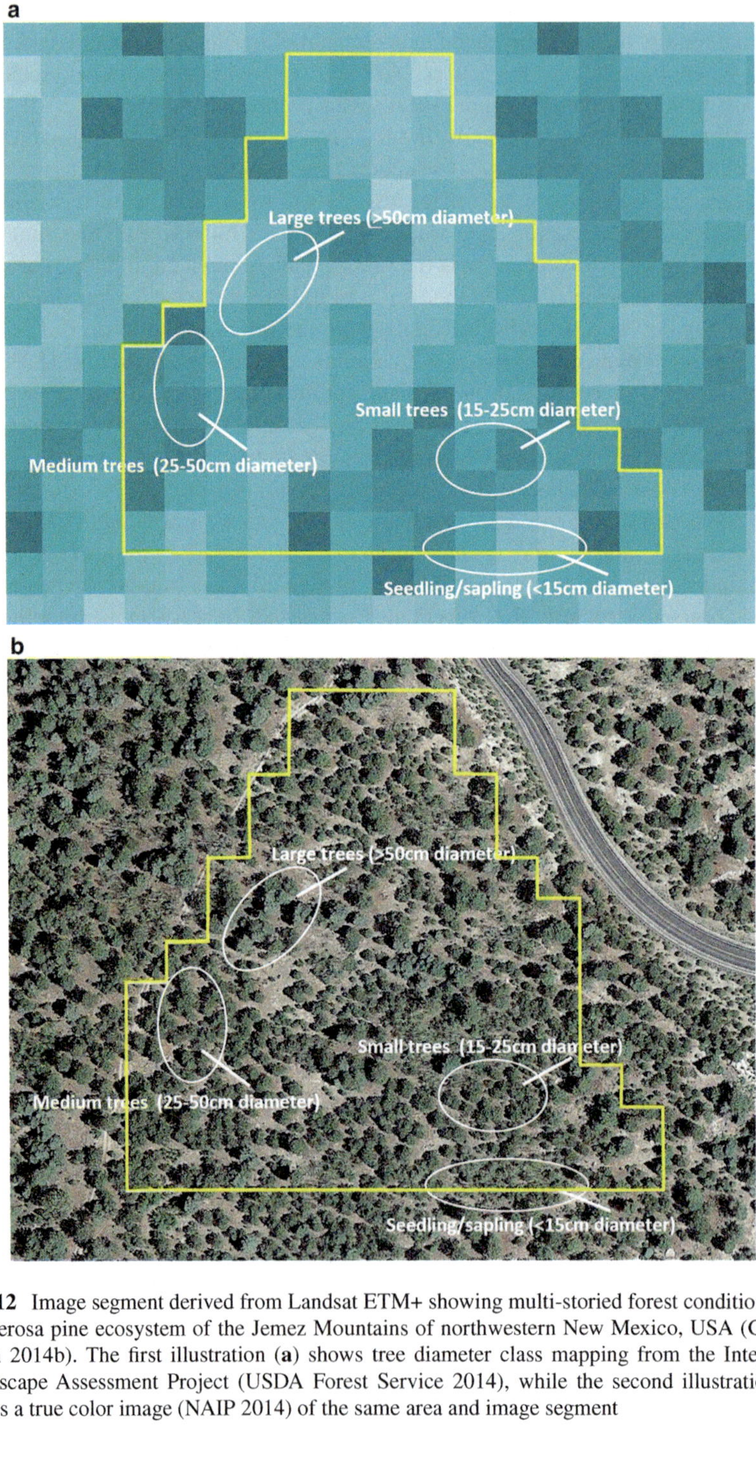

Fig. 12 Image segment derived from Landsat ETM+ showing multi-storied forest conditions in a ponderosa pine ecosystem of the Jemez Mountains of northwestern New Mexico, USA (Google Earth 2014b). The first illustration (**a**) shows tree diameter class mapping from the Integrated Landscape Assessment Project (USDA Forest Service 2014), while the second illustration (**b**) shows a true color image (NAIP 2014) of the same area and image segment

Fig. 13 Pine flatwoods community in northern Florida, USA, showing longleaf pine (*Pinus palustris* Mill.) of similar height in a single-story forest structure (photo by Jack Triepke)

cation (see section "Fuzzy Representation with Continuous and Categorical Data") or to delineate areas of interest, like old growth forest, with rule sets that combine key surfaces. Having fuzzy membership values further allows users an easy means to apply compensatory factors, where appropriate, among the habitat variables at play (e.g., storiedness for tree cover). As with the scenario involving storiedness, fuzzy techniques give us many opportunities to integrate information sources and to make map data go further.

A Look to the Future

This closing section briefly explores some of the immediate opportunities for fuzzy applications. By the innate potential of fuzzy systems, not to mention the accessibility and rate of related technology evolution, many advances remain underexploited and some latent applications for fuzzy systems are considered here.

Current mapping applications that output fuzzy membership values, some covered in this chapter, allow for end-user computation according to specific needs. The generation of fuzzy surfaces, as with the study by Zhang and Stuart (2001), offers the most obvious example of spatial outputs that can be readily interpreted and post-processed into a tailored deliverable. In this case every object has a fuzzy membership value for every established map category. Although outputs are still constrained by the makeup of the original legend and the training data, the resulting fuzzy surface outputs offer what is otherwise a geodataset that is neutral of classification schemes and a priori stratifications. End users would, for example, be able to map a ponderosa pine cover type according to a given project design and specific class concepts of tree species proportions, and to assess thematic and spatial uncertainty according to the same fuzzy membership values. Having access to these values allows clients the power to post-process spatial information in GIS in response to specific uncertainty specifications. In the case of raster data, having an associated segmentation layer that depicts patterns of structure and composition at somewhat coarser scales (Figs. 11 and 12) would allows users yet another level of capacity in forming products by the rules used to classify parent segments by the makeup of resident pixels. Such an approach has also been used to generate a vertical dimension to mapping according to the relationship in structural attributes among pixels (USDA Forest Service 2012) (Fig. 12). Including fuzzy membership values in classifier outputs allows the generation of a range of map products from the same dataset within a common GIS platform according to the mapping and uncertainty analysis needs of natural resource managers and researchers.

Web-based interfaces are the obvious next step for user-derived map products, once spatial data are attributed with fuzzy membership values. While GIS environments offer an accessible means for organizations and specialists with software and training to manage and analyze spatial information, Web-based applications are the logical means of circulating spatial information and cultivating crowd-sourcing and the development of post-products, analyses, and tools. OpenStreetMap (OSM 2014), Google Earth Plug-in (Earth 2014c), and other map viewers and data management applications employed by lay people offer potential outlets for sharing, viewing, and processing geodata that has been attributed with fuzzy, continuous, or categorical values (ADPC 2016). In this environment, end users can extract fuzzy data from Web resources, not only in terms of a specific extent, but also according to the precise characteristics of ecological features. While one user, for instance, interested is wildlife habitat is able to output a map of forest plant communities of 20–45% tree cover with at least three canopy layers, the next user can generate an output for communities with 10–30% tree cover and four canopy layers from the same dataset without constraints of predetermined map categories (e.g., Zabihi et al. 2017). And by combining multiple datasets likewise attributed by fuzzy membership, the potential for developing wide-varying map themes becomes even less limited. Also, based on a range of potential outputs on available map themes, users will be able to game membership scenarios interactively on each of the themes to generate very precise outputs for a particular purpose. Finally, users will be able to build products that express the desired relationship of accuracy and precision, make adjustments to membership thresholds to satisfy uncertainty requirements, and characterize uncertainty similarly as spatial analysts in a GIS lab.

The adoption and evolution of advanced classifier methods also hold promise for map development and spatial analysis. In addition to the classifier technology surveyed in section "Mapping with Fuzzy Classifiers," other image classification methods that build on machine learning include neuro-fuzzy classifiers (Sun and Jang 1993; Nauck et al. 1997; Nauck and Kruse 1997). Neural networks, inspired by the nervous systems of animals, offer supervised learning ability to generate classification algorithms based on potentially large numbers of inputs (Aitkenhead and Dyer 2007). Neuro-fuzzy classifiers, that combine neural networks with fuzzy systems, have been applied in various fields for over a decade and have been used to map ecological features. Neuro-fuzzy systems combine the learning power of neural networks with the knowledge represented in fuzzy inferences, integrating key advantages of neural networks and fuzzy systems (Hosseini and Zekri 2012). Despite their obvious strengths, neuro-fuzzy classifiers have yet to realize their potential for purposes of mapping in natural resources.

Some discussion is warranted regarding the use of fuzzy random forest as a classifier approach for making maps. Although fuzzy random forest classifiers (Bonissone et al. 2008) have been used in other scientific fields (Bonissone et al. 2010; Kulkarni and Sinha 2013; Lasota et al. 2013) it is as yet conventional for land cover mapping. The classification trees in conventional random forest classifiers are unpruned trees in that each terminal node is represented by one observation, leading to a crisp ruleset within each tree—i.e., one answer only. Yet, the amalgamation of outputs for multiple crisp trees results in information that is inherently fuzzy because of the disagreement among votes (Grossmann et al. 2010). In their 2010 study, Bonissone and others (Bonissone et al. 2010) combined random forest classifiers made up of fuzzy decision trees to build classification outputs for various types of data. Their work included image segmentation, but not specifically ecological feature extraction. In short, the approach combines the flexibility found with fuzzy systems with the efficiency and interpretability of decision tree classifiers, with the robustness provided in a multiple-classifier approach, and the ability of randomness to build tree diversity and the most plausible range of outputs. The advantages of decision tree methods, fuzzy classifiers, and random forest were summarized in section "Mapping with Fuzzy Classifiers." Prior to the application of random forest (Breiman 2001), fuzzy systems had been combined with decision trees for classifier applications (Lee et al. 1999; Mendonça et al. 2007). Given the innate elasticity of fuzzy logic, the fuzzy component has the key advantage of bringing stability to noise, gaps, and incongruences in input data (Bonissone et al. 2008). It was with this premise that Bonissone and others generated a fuzzy random forest as a base classifier (Bonissone et al. 2010). Several classifier methods were compared, all based on the "majority vote" for random forest ensembles, and including an algorithm using fuzzy membership values to weight decisions among classification trees. In the latter case, the underlying premise was that since classifiers of the ensemble are not of identical accuracy, the more capable classifiers would be weighted to reflect a greater competency. They found that the classification approach using weighted functions provided better accuracy in comparison to the non-weighted method typical of random forest ensembles. Overall the study showed that the fuzzy random forest systems produced accuracy on par with the best classifiers when applied to the range of conventional datasets in the test. But unlike the non-fuzzy approaches, the fuzzy random forest classifiers had consistent accuracy results when faced with datasets of noisy and missing values. Fuzzy random forest classifiers will likely continue to improve technologically and see growing application for mapping purposes in natural resources.

Fuzzy classifiers, online tools, and other technologies are advancing existing applications and creating new possibilities. Online tools that are accessible to the masses are the future in the way that open-source technology, citizen science, and big data are used to support shared environmental goals and concerns. Climate change, loss of habitat and biotic diversity, and other global issues are propelling both research and development and the widespread application of remote sensing technologies that were until understood and used by a relative few until recently. While fuzzy methods offer solutions to the ambiguities of natural landscapes, they are likewise suited to future problems, temporal analysis, and "what-if" scenarios involving drought, fire, sea-level rise, temperature and precipitation regimes, and other factors that, when considered as fuzzy parameters, can lead to output ranges that may be correlated with measures of prediction, uncertainty, and opportunity. Recent Web-based tools such as Collect Earth (FAO 2016) give users with minimal experience simple means to quickly generate reference data for land cover mapping as well as for monitoring and capturing conditions at multiple points in time with readily available archived imagery. Collect Earth is a free open-source solution that integrates with Google Earth, Google Earth Engine, and other Web-based tools to gather, analyze, and display geographic information and ecological features. These tools can be easily structured to capture current land cover and land-use attributes as well as change mechanisms depicted or inferred by multiple archived satellite scenes. Such technology enables fuzzy membership to be represented simultaneously across map themes, spatial scales, and temporal scales, giving fuzzy methods a growing role in addressing environmental challenges into the future.

Summary

Widespread concerns about biodiversity and ecosystem integrity worldwide have spurred natural resource monitoring (Miura et al. 2008) and development of analysis tools to assess current conditions and risk (Suter 2006). Emphasis on ecological restoration, climate change, and other important aspects of conservation has driven the gathering of observational data, analysis and interpretation, and development of technology such as advanced land cover mapping applications (Lillesand and Kiefer 2000; Kulkarni and Sinha 2013; Li et al. 2014). Map data generated to depict various ecological features are being used increasingly for purposes of research, analyzing ecosystems, land management, and planning (Goetz and Maus 2006; Friggens et al. 2013; USDA Forest Service 2014; Triepke 2016). Ecological mapping can be developed through image classification or from mining of existing map sources as with the approach described in section "Multiple Outputs: Fuzzy Geodatabase." Fuzzy approaches have given map producers advanced solutions for not only generating map outputs but also characterizing them through the analysis of uncertainty (Lowell 1994; Sarmento et al. 2010; Kronenfeld 2011).

Conventional mapping and uncertainty analysis has relied on crisp approaches, based on mutually exclusive hard ecological categories among the objects or pixels of a given extent. As described, these objects may reflect multiple categories (Wood and Foody 1993) as in the case of tree size (Fig. 2), or may reflect unrecognized categories of a more general nature as in mixed types (Fig. 9). Fuzziness can be expressed in mixed pixels, where the spatial resolution of the base models units is coarse relative to the resolution of ecological features being mapped (Foody 1997; Zhang and Foody 1998; Campbell 2002). Such objects may be an expression of a fuzzy set or simply be a mixture of map themes with hard boundaries at a local scale. In the latter situation, the pixel size is greater than the spatial extent of map themes so that multiple themes are expressed within the space of the same pixel. In either case, a crisp solution may be to create a mixed type in the map legend, provided that the type occurs often enough to warrant distinction. The problem may be remedied more efficiently and precisely with a fuzzy approach where partial membership is assigned to each theme present in the pixel.

In contrast to crisp approaches, and the constraints they impose on responding to ambiguity, fuzzy system classifications respond through assignment of varying levels of fuzzy membership (Zadeh 1965) for each object and apparent map category. That is, fuzzy systems represent gradual change from membership to non-membership among objects among multiple categories and dimensions. From the standpoint of logic, fuzziness is expressed in one of the two ways: first, ambiguity and the difficulty in classification may stem from the vagueness and authenticity among available categories (Rocchini and Ricotta 2007). Second, ambiguity may stem from the lack of distinctiveness in the object itself; or it may stem from both. The replacement of classical set theory with fuzzy set theory marks an advance in our ability to deal with ambiguity and to depict or analyze ecological features which are often inherently fuzzy (Rickel et al. 1998). In addition, fuzzy systems offer more wide-ranging and flexible solutions for representing geographic information.

As detailed in Chapter "Fuzzy Classification of Vegetation for Ecosystem Mapping," the flexibility of fuzzy approaches demands clear spatial and thematic specifications in the development of map products. Their development and application will only become more flexible in time, as specialists take advantage of the increasing availability of ecological mapping and the power of fuzzy operators in building and applying map data. Clear specifications and definitions are essential for consistency across themes in mapped data.

References

ADPC (2016) Regional land cover typology workshop report summary. Asian Disaster Preparedness Agency Resource report available www.adpc.net/igo/contents/Publications. Accessed May 2016

Aitkenhead M, Dyer R (2007) Improving land-cover classification using recognition threshold neural networks. Photogramm Eng Remote Sens 73:413–421

Álvarez-Martínez JM, Stoorvogel JJ, Suárez-Seoane S, Calabuig EL (2010) Uncertainty analysis as a tool for refining land dynamics modelling on changing landscapes: a case study in a Spanish Natural Park. Landsc Ecol 25:1385–1404. doi:10.1007/s10980-010-9492-z

Amit Y, Geman D (1997) Shape quantization and recognition with randomized trees. Neural Comput 9:1545–1588

Aspinall RJ, Pearson DM (1995) Describing and managing uncertainty of categorical maps in GIS. Innov GIS 2:71–83

Bailey RG (1988) Problems with using overlay mapping for planning and their implications for geographic information systems. Environ Manag 12:11–17

Barbour MG, Burk JH, Pitts WD, Gilliam FS, Schwartz MW (1998) Terrestrial plant ecology, 3rd edn. Benjamin Cummings, Menlo Park

Bonissone PP, Cadenas JM, Garrido MC, Dıaz-Valladares RA (2008) A fuzzy random forest: fundamentals for design and construction. In: Proceedings of the 12th international conference on information processing and management of uncertainty in knowledge-based systems (IPMU'08), Malaga, June 2008. pp 1231–1238

Bonissone P, Cadenas JM, Carmen Garrido M, Andrés Díaz-Valladares R (2010) A fuzzy random forest. Int J Approx Reason 51:729–747. doi:10.1016/j.ijar.2010.02.003

Breiman L (1996) Bagging predictors. Mach Learn 24:123–140

Breiman L (2001) Random forests. Mach Learn 45:5–32

Brewer CK (2007) Remote sensing applications to support sustainable natural resource management. Dissertation, University of Montana

Brewer C, Bush R, Berglund D et al (2006) Integrating vegetation classification, mapping, and strategic inventory for forest management. In: Aguirre-Bravo C, Pellicane PJ, Burns DP, Draggan S (eds) Monitoring science and technology symposium: unifying knowledge for sustainability in the Western Hemisphere, Denver, September 2004. USDA Forest Service proceedings RMRS-P-42CD. Rocky Mountain Research Station, Fort Collins

Brohman R, Bryant LD (eds) (2005) Existing vegetation classification and mapping technical guide: version 1.0. USDA Forest Service, Washington DC

Brown DE, Reichenbacher F, Franson SE (1998) A classification of north American biotic communities. University of Utah Press, Salt Lake City

Cai Z, Li Z, Kang Z, Liu Y (eds) (2009) Computational intelligence and intelligent systems: 4th international symposium on intelligence computation and applications (ISICA 2009), Huangshi, October 2009. Springer, New York

Campbell JB (2002) Introduction to remote sensing, 3rd edn. Guildford Press, New York

Card DH (1982) Using known map category marginal frequencies to improve estimates of thematic map accuracy. Photogramm Eng Remote Sens 48:431–439

Chrisman NR (1989) Modeling error in overlaid categorical maps. In: Goodchild MF, Gopal S (eds) Accuracy spatial databases. Taylor & Francis, London

Clark A, Stam C, Goetz W, Maus P, Megown KA, Triepke FJ, Matthews B, Muldavin E (2016) Mapping riparian vegetation on the Gila National Forest using photogrammetric techniques. USDA Forest Service technical report RSAC-10121-RPT1 available www.fs.fed.us/eng/rsac/documents/pdfs. Remote Sensing Applications Center, Salt Lake City

Comer PJ, Faber-Langendoen D, Evans R et al (2003) Ecological systems of the United States: a working classification of US terrestrial systems. NatureServe. Available www.natureserve.org/publications. Accessed Sept 2011

Congalton RG (1991) A review of assessing the accuracy of classifications of remotely sensed data. Remote Sens Environ 37:35–46
Congalton RG, Green K (1999) Assessing the accuracy of remotely sensed data: principles and practices. Lewis Publishers, Boca Raton
Costa H, Foody GM, Boyd DS (2017) Using mixed objects in the training of object-based image classifications. Remote Sens Environ 190:188–197
Cowling RM, Kerley GIH, van Cauter A (2005) The consequence of inaccuracies in remote-sensed vegetation boundaries for modelled mammal population estimates. South Afr J Wildl Res 35:1–7
Cox E (1992) The great myths of fuzzy logic. AI Expert 7:40–45
Cressie N (1990) The origins of kriging. Math Geol 22:239–252. doi:10.1007/BF00889887
Curtis RO (1967) Height-diameter and height-diameter-age equations for second-growth Douglas-fir. For Sci 13:365–375
Daubenmire R (1966) Vegetation: identification of typal communities. Science 151:291–298
Daubenmire R (1978) Plant geography with special reference to North America. Academic Press, New York
Definiens (2003) eCognition object-oriented image analysis – user guide, 3rd edn. Definiens Imaging, Munich
Dietterich TG (2000) An experimental comparison of three methods for constructing ensembles of decision trees: bagging, boosting, and randomization. Mach Learn 40:139–157
Google Earth (2014a) Grassland system east of Grants, New Mexico, USA. 34°59'18″ N and 107°08'47″ W. Accessed Aug 2014
Google Earth (2014b) Ponderosa pine forest system northwest of Jemez Springs, New Mexico, USA. 35°52'29″ N and 106°40'02″ W. Accessed Aug 2014
Google Earth (2014c) Google Earth Plug-In. In: Google Earth products. Available www.google.com/earth/explore/products. Accessed Nov 2014
FAO (2016) Food and Agriculture Organization of the United Nations, Open Foris tools for environmental monitoring. Available www.openforis.org. Accessed June 2016
Ferguson DE, Carlson CE (2010) Height-age relationships for regeneration-size trees in the northern Rocky Mountains, USA. Res Pap RMRS-RP-82WWW. USDA Forest Service Rocky Mountain Research Station, Fort Collins
Foody GM (1997) Fully fuzzy supervised classification of land cover from remotely sensed imagery with an artificial neural network. Neural Comput Appl 5:238–247. doi:10.1007/BF01424229
Friggens M, Bagne K, Finch D, Falk D, Triepke FJ, Lynch A (2013) Review and recommendations for climate change vulnerability assessment approaches with examples from the Southwest. USDA Forest Service Gen. Tech. Rep. RMRS-GTR-309. Rocky Mountain Research Station, Fort Collins
Gehrke S, Morin K, Downey M, Boehrer N et al (2010) Semi-global matching: an alternative to LiDAR for DSM generation. Int. Arch Phot & Rem Sens, Calgary 38:Part B1
Goetz W, Maus P (2006) LANDFIRE existing vegetation products – How useful for midlevel forest needs? USDA Forest Service, Remote Sensing Tips 0074-TIP1, available www.fs.fed.us/eng/rsac/documents/pdfs. Washington Office, Remote Sensing Applications Center, Salt Lake City
Goguen JA (1969) The logic of inexact concepts. Synthese 19:325–373. doi:10.1007/BF00485654
Gong P, Pu R, Chen J (1996) Elevation and forest-cover data using neural networks. Photogramm Eng Remote Sens 62:1249–1260
Gopal S, Woodcock C (1994) Theory and methods for accuracy assessment of thematic maps using fuzzy sets. Photogramm Eng Remote Sensing (United States) 60:181–188
Grossmann E, Ohmann J, Kagan J, May H, Gregory M (2010) Mapping ecological systems with a random forest model: tradeoffs between errors and bias. Gap Anal Bulletin 17:16–22
Hansen HS (2003) A fuzzy logic approach to urban land-use mapping. In: In: proceedings of the 9th Scandinavian research conference (ScanGIS) on geographical information science, June 2003. Espoo, Finland, pp 41–56
Helms JA (1998) The dictionary of forestry. Society of American Foresters, Bethesda

Hirschmuller H (2005) Accurate and efficient stereo processing by semi-global matching and mutual information. In: Proceedings of the 2004 IEEE Computer Society conference on computer vision and pattern recognition, San Diego, California, June 2005
Ho TK (1998) The random subspace method for constructing decision forests. IEEE Trans Pattern Anal Mach Intell 20:832–844
Hoppner F, Klawonn F, Kruse R, Runkler T (1999) Fuzzy cluster analysis: methods for classification, data analysis, and image recognition. John Wiley & Sons, New York
Hosseini MS, Zekri M (2012) Review of medical image classification using the adaptive neuro-fuzzy inference system. J Med Signals Sens 2:49–60
Huang S, Titus SJ, Wiens DP (1992) Comparison of nonlinear height-diameter functions for major Alberta tree species. Can J For Res 22:1297–1304
Ishibuchi H, Nakashima T (2001) Effect of rule weights in fuzzy rule-based classification systems. IEEE Trans Fuzzy Syst 9:506–515. doi:10.1109/91.940964
Islam Z, Metternicht G (2005) The performance of fuzzy operators on fuzzy classification of urban land covers. Photogramm Eng Remote Sens 71:59–68
Janssen LL, van der Wel FJ (1994) Accuracy assessment of satellite derived land-cover data: a review. Photogramm Eng Remote Sens 60:419–426
Jennings M, Loucks O, Peet R et al (2003) Guidelines for describing associations and alliances of the U.S. National Vegetation Classification, version 3.0. Ecological Society of America. Available http://www.vegbank.org/vegdocs/panel. Accessed Feb 2013
Jennings MD, Faber-Langendoen D, Loucks OL et al (2009) Standards for associations and alliances of the US National Vegetation Classification. Ecol Monogr 79:173–199
Klaua D (1966) Über einen zweiten Ansatz zur mehrwertigen Mengenlehre. Monatsber Dtsch Akad Wiss Berl 8:161–177
Knick ST, Rotenberry JT (1998) Limitations to mapping habitat use areas in changing landscapes using the Mahalanobis distance statistic. J Agric Biol Environ Stat 3:311–322. doi:10.2307/1400585
Kosco B (1995) Combining fuzzy systems. In: Fuzzy systems. Proceedings of the 1995 international joint conference of the fourth IEEE international conference on fuzzy systems and the second international fuzzy engineering symposium, Yokohama, Mar 1995, vol 4. IEEE, pp 1855–1863. doi: 10.1109/FUZZY.1995.409933
Krige D (1951) A statistical approach to some mine valuation and allied problems on the Witwatersrand. Dissertation, University of the Witwatersrand
Kronenfeld BJ (2011) Beyond the epsilon band: polygonal modeling of gradation/uncertainty in area-class maps. Int J Geogr Inf Sci 25:1749–1771. doi:10.1080/13658816.2010.518317
Kulkarni VY, Sinha PK (2013) Random forest classifiers: a survey and future research directions. Int J Adv Comput 36:1144–1153
Lasota T, Telec Z, Trawiński B, Trawiński G (2013) Investigation of random subspace and random forest regression models using data with injected noise. In: Graña M, Toro C, Howlett RJ, Jain LC (eds) Knowledge engineering, machine learning and lattice computing with applications. Springer, Berlin
Leavell DM (2000) Vegetation and process of the Kootenai National Forest. Dissertation, University of Montana
Lee KM, Lee KM, Lee JH, Lee-Kwang H (1999) A fuzzy decision tree induction method for fuzzy data. In: Fuzzy systems conference proceedings, Seoul, Aug 1999, vol 1. IEEE, pp 16–21. doi:10.1109/FUZZY.1999.793199
Leekwijck WV, Kerre EE (1999) Defuzzification: criteria and classification. Fuzzy Sets Syst 108:159–178. doi:10.1016/S0165-0114(97)00337-0
Leung Y, Yan J (1998) A locational error model for spatial features. Int J Geogr Inf Sci 12:607–620
Li M, Zang S, Zhang B et al (2014) A review of remote sensing image classification techniques: the role of spatio-contextual information. Eur J Remote Sens 47:389–411
Lillesand TM, Kiefer RW (2000) Remote sensing and image interpretation, 4th edn. John Wiley & Sons, New York

Lloyd H, Sevillano RíOs S, Marsden SJ, ValdéS-VeláSquez A (2012) Bird community composition across an Andean tree-line ecotone. Austral Ecol 37:470–478. doi:10.1111/j.1442-9993.2011.02308.x
Lowell K (1994) An uncertainty-based spatial representation for natural resources phenomena. In: Waugh TC, Healey RG (eds) Advances in GIS research: proceedings of the sixth international symposium on spatial data handling, Edinburgh, 1994. International geographical union commission on GIS and the Association for Geographic Information, vol 2. Taylor & Francis, London, pp 933–944
Lu D, Weng Q (2007) A survey of image classification methods and techniques for improving classification performance. Int J Remote Sens 28:823–870. doi:10.1080/01431160600746456
Mansoori EG, Zolghadri MJ, Katebi SD (2007) A weighting function for improving fuzzy classification systems performance. Fuzzy Sets Syst 158:583–591. doi:10.1016/j.fss.2006.10.004
Marsala C (2009) Data mining with ensembles of fuzzy decision trees. In: IEEE symposium on computational intelligence and data mining, Nashville, Mar–Apr 2009. IEEE symposium, pp 348–354. doi: 10.1109/CIDM.2009.4938670
Mendonça LF, Vieira SM, Sousa JMC (2007) Decision tree search methods in fuzzy modeling and classification. Int J Approx Reason 44:106–123
Miura T, Yoshioka H, Fujiwara K, Yamamoto H (2008) Inter-comparison of ASTER and MODIS surface reflectance and vegetation index products for synergistic applications to natural resource monitoring. Sensors 8:2480–2499
Morin NR (ed) (1993) Flora of North America north of Mexico, volume 2: pteridophytes and gymnosperms. Oxford University Press, New York
Myint SW, Gober P, Brazel A et al (2011) Per-pixel vs. object-based classification of urban land cover extraction using high spatial resolution imagery. Remote Sens Environ 115:1145–1161. doi:10.1016/j.rse.2010.12.017
NAIP (2014) National Agricultural Imagery Program, Salt Lake City. http://www.lta.cr.usgs.gov/node/300. Accessed Sept 2014
Nauck D, Kruse R (1997) What are neuro-fuzzy classifiers? In: Proceedings of the seventh international Fuzzy Systems Association world congress, Prague, 1997, vol. 97. IFSA, pp 228–233
Nauck D, Kruse R (1998) How the learning of rule weights affects the interpretability of fuzzy systems. In: Fuzzy systems proceedings, 1998. IEEE world congress on computational intelligence, Anchorage, May 1998, vol 2. IEEE, pp 1235–1240
Nauck D, Kruse R (1999) Fuzzy classification rules using categorical and metric variables. In: Proceedings of the 6th international workshop on fuzzy-neuro systems, Leipzig, Mar 1999
Nauck D, Klawonn F, Kruse R (1997) Foundations of neuro-fuzzy systems. John Wiley & Sons, New York
de Oliveira JV (1999) Semantic constraints for membership function optimization. IEEE Trans Syst Man Cybern Part Syst Hum 29:128–138. doi:10.1109/3468.736369
OSM (2014) OpenStreetMap. Available http://www.openstreetmap.org/copyright. Access Nov 2014
Pajares G, Guijarr M, Herrera PJ, Ribeiro A (2009) Combining classifiers through fuzzy cognitive maps in natural images. Comput Vis IET 3:112–123. doi:10.1049/iet-cvi.2008.0023
Perkal J (1956) On epsilon length. Bull Acad Pol Sci 4:399–403
Remmel TK (2009) Investigating global and local categorical map configuration comparisons based on coincidence matrices. Geogr Anal 41:144–157
Remmel TK (2015) ShrinkShape2: a FOSS toolbox for computing rotation-invariant shape spectra for characterizing and comparing polygons. Canadian Geogr 59:532–547
Remmel TK, Perera AH (2009) Mapping natural phenomena: boreal forest fires with non-discrete boundaries. Cartogr Int J Geogr Inf Geovisualization 44:274–288. doi:10.3138/carto.44.4.274
Rickel BW, Anderson B, Pope R (1998) Using fuzzy systems, object-oriented programming, and GIS to evaluate wildlife habitat. AI Appl 12:31–40
Rocchini D, Ricotta C (2007) Are landscapes as crisp as we may think? Ecol Model 204:535–539. doi:10.1016/j.ecolmodel.2006.12.028
Rosenfield GH, Fitzpatrick-Lins K (1986) A coefficient of agreement as a measure of thematic classification accuracy. Photogramm Eng Remote Sens 52:223–227

Running SW, Loveland TR, Pierce LL (1994) A vegetation classification logic based on remote sensing for use in global biogeochemical models. Ambio 23:77–77

Safavian SR, Landgrebe D (1991) A survey of decision tree classifier methodology. IEEE Trans Syst Man Cybern 21:660–674

Sameen MI, Pradhan B (2017) A two-stage optimization strategy for fuzzy object-based analysis using airborne LiDAR and high-resolution orthophotos for urban road extraction. J Sensors. doi:10.1155/2017/6431519

Sarmento P, Carrão H, Caetano M et al (2010) Incorporating uncertainty in the accuracy assessment of land cover maps using fuzzy numbers and fuzzy arithmetic. In: Tate NJ, Fisher PF (eds) Proceedings of the 9th international symposium on spatial accuracy assessment in natural resources and environmental sciences, Leicester, July 2010

Sarmento P, Fonte CC, Caetano M, Stehman SV (2013) Incorporating the uncertainty of linguistic-scale reference data to assess accuracy of land-cover maps using fuzzy intervals. Int J Remote Sens 34:4008–4024

Schapire RE (1990) The strength of weak learnability. Mach Learn 5:197–227

Schmidt M, Kiviste A, von Gadow K (2011) A spatially explicit height-diameter model for scots pine in Estonia. Eur J For Res 130:303–315

Shortridge A, Shi M (2012) Changing the TIGER's stripes: detecting road network change under positional uncertainty. In: Vieira C, Bogorny V, Aquino R (eds) Proceedings of the tenth international symposium on spatial accuracy assessment in natural resources and environmental sciences, Florianópolis, July 2012

Steele BM (2000) Combining multiple classifiers: an application using spatial and remotely sensed information for land cover type mapping. Remote Sens Environ 74:545–556

Sun CT, Jang JS (1993) A neuro-fuzzy classifier and its applications. In: Proceedings of the second IEEE international conference on fuzzy systems, San Francisco, Mar 1993, vol 1. IEEE, pp 94–98. doi: 10.1109/FUZZY.1993.327457

Suter GW (2006) Ecological risk assessment, 2nd edn. CRC Press, Boca Raton

Triepke, FJ (2016) Assessing the climate change vulnerability of ecosystem types of the southwestern U.S. Dissertation, University of New Mexico

Triepke FJ, Brewer CK, Leavell DM, Novak SJ (2008) Mapping forest alliances and associations using fuzzy systems and nearest neighbor classifiers. Remote Sens Environ 112:1037–1050

Triepke FJ, Higgins FB, Weisz RN et al (2011) Diameter caps and forest restoration—evaluation of a 16-inch cut limit on achieving desired conditions. USDA Forest Service Forestry Report FR-R3-16-3 available www.fs.usda.gov/Internet/fse_documents. Southwestern Region, Albuquerque

Triepke FJ, DeMeo TE, Al Otoum M, Al-Azzam L (2012) Composition and structure of Aleppo pine (Pinus halepensis) communities in the Dibeen Forest reserve, Jordan. Nat Areas J 32:356–366

USDA Forest Service (2012) Existing vegetation map of the Southwestern United States. In: Southwestern Region GIS Datasets. Available www.fs.usda.gov/detail/r3/landmanagement/gis. Accessed Feb 2014

USDA Forest Service (2014) Integrating, social, economic, and ecological values across large landscapes. Gen Tech Rep PNW-GTR-896. Pacific Northwest Research Station, Portland

USGS (2014) US Geological Survey western U.S. 250m eMODIS remote sensing phenology data, Sioux Falls. http://www.phenology.cr.usgs.gov/ndvi_foundation.php. Accessed Jan 2014

Vandendriesche DA (2011) A compendium of NFS regional vegetation classification algorithms. USDA Forest Service, Southwestern Region. Available http://www.fs.fed.us/fmsc/ftp/fvs/docs/gtr. Accessed Jan 2014

Wang F (1990) Fuzzy supervised classification of remote sensing images. IEEE Trans Geosci Remote Sens 28:194–201

Wang F, Hall GB (1996) Fuzzy representation of geographical boundaries in GIS. Int J Geogr Inf Syst 10:573–590

Westoby M, Walker B, Noy-Meir I (1989) Opportunistic management for rangelands not at equilibrium. J Range Manag 42:266–274

Wood TF, Foody GM (1993) Using cover-type likelihoods and typicalities in a geographic information system data structure to map gradually changing environments. In: Haines-Young R, Green DR, Cousins SH (eds) Landscape ecology and GIS. Taylor & Francis, London

Woodcock CE, Gopal S (2000) Fuzzy set theory and thematic maps: accuracy assessment and area estimation. Int J Geogr Inf Sci 14:153–172

Zabihi K, Paige GB, Hild AL, Miller SN, Wuenschel A, Holloran MJ (2017) A fuzzy logic approach to analyse the suitability of nesting habitat for greater sage-grouse in western Wyoming. J Spatial Sci. doi:10.1080/14498596.2017.1292965

Zadeh LA (1965) Fuzzy sets. Inf Control 8:338–353

Zhang J, Foody GM (1998) A fuzzy classification of sub-urban land cover from remotely sensed imagery. Int J Remote Sens 19:2721–2738

Zhang J, Stuart N (2001) Fuzzy methods for categorical mapping with image-based land cover data. Int J Geogr Inf Sci 15:175–195. doi:10.1080/13658810010005543

Portraying Wildfires in Forest Landscapes as Discrete Complex Objects

Tarmo K. Remmel and Ajith H. Perera

Abstract Boreal wildfires are characterized by internal heterogeneity that arises from variations in fuel availability, fuel moisture content, and weather conditions during a fire event. This heterogeneity extends from an uneven burn intensity that affects the degree of forest disturbance to inconsistency in boundary abruptness at the fire perimeter, in spot fires associated with the main fire, and in areas internal to the fire where residual vegetation and unburnable land cover types are encountered. We begin with a brief discussion of wildfire anatomy and how fires burn to create new and complex landscape patterns. We then describe some of the common approaches that are used to map wildfires, paying particular attention to the importance of scale in the mapping process. We address the complexities and heterogeneities of wildfire boundaries and internal structures by consistently linking their characterization and interpretation to spatial scale and statistical characteristics of mapping. Having outlined the variability in the formation and mapping of wildfire complexity, we propose a standardized terminology for describing these phenomena and provide some thoughts on the future of efforts to map dynamic and complex landscape entities such as wildfires.

Abbreviations

0D, 1D, 2D, 3D	Zero-, one-, two-, and three-dimensional
FD	Fractal dimension
GPS	Global positioning system
LiDAR	Light detection and ranging

T.K. Remmel (✉)
Department of Geography, York University,
4700 Keele Street, Toronto, ON, Canada, M3J 1P3
e-mail: remmelt@yorku.ca

A.H. Perera
Ontario Forest Research Institute, Ministry of Natural Resources and Forestry,
1235 Queen Street East, Sault Ste. Marie, ON, Canada, P6A 2E5
e-mail: ajith.perera@ontario.ca

T.K. Remmel, A.H. Perera (eds.), *Mapping Forest Landscape Patterns*,
DOI 10.1007/978-1-4939-7331-6_3

MMU	Minimum mapping unit
NDVI	Normalized-difference vegetation index

Introduction

Mapping natural entities is a different task than cartographic efforts that focus on human-constructed objects or features, which have clearer and more predictable boundaries. For instance, researchers must revise how they approach the tasks of entity definition (i.e., defining a "thing" that they want to map), boundary delineation, and thematic classification (i.e., grouping parts of a map into different categories, which are called "themes") when they address natural entities that are inherently fuzzy, complex, and scale dependent. In this chapter, we address the topic of mapping discrete but complex objects, using wildfire "footprints" (the area directly affected by the fire) in boreal forests to demonstrate the problems and their potential solutions.

Wildfire is a natural disturbance agent that is prevalent in most forest landscapes, but especially in the boreal biome. Simultaneously, there are many anthropogenic influences on boreal fire regimes: people ignite many fires in boreal landscapes, and make concerted efforts to manage the economic and other impacts of fires by means of aggressive fire suppression. Regardless of the ignition source or the suppression effort applied, numerous environmental factors also influence fire behavior, leading to highly complex disturbance footprints within forested landscapes. These footprints have a highly heterogeneous internal structure due to spatial and temporal variations in burn intensity and in the subsequent response of vegetation.

The footprint of a typical boreal forest fire, even when it results from an intense wildfire, will be spatially heterogeneous with respect to burn severity within its interior, and that interior will contain some amount of unburned vegetation with a varying species composition and a range of age- and height-class distributions (Perera and Buse 2014). Spatial heterogeneity of terrain and of the water table in boreal forest landscapes, even when subtle, adds to the complexity of the structural mosaic that is seen before, during, and after wildfires. Furthermore, the boundary of the burned area is convoluted and nonlinear, as it is formed by fire behavior that evolves in response to varying fuel, weather, and terrain conditions during the fire (Remmel and Perera 2009) in both the temporal dimension and the horizontal and vertical spatial dimensions. In addition, fire spread is not always contiguous. Wildfires may generate "spotting," in which burning material is carried by the wind and jumps in advance of the fire to cause satellite burn patches that may merge with the main fire or remain disconnected, forming a complex mosaic of multiple burn patches that collectively define a single fire (Remmel and Perera 2009). For these reasons, boreal wildfire footprints represent great examples of discrete and complex entities that pose significant mapping challenges.

Our use of wildfires as an example is not merely an academic pursuit. The scientific literature is rich with examples of wildfire mapping because understanding of boreal fire footprints, and especially their patterns of spatial extent, is of increasing ecological, economic, and social importance (e.g., see the review by Morgan et al. 2001). Many forest ecological processes are influenced by wildfires, including carbon and hydrologic cycling, creation and elimination of habitat, improvement or deterioration of biodiversity, and regeneration of forests. As well, the economic effects of wildfires include concerns such as changes in the timber supply, damage to property and communities, and evaluation of fire management and suppression investments. Spatial information on fires is collected and reported both at a regional level (e.g., in Ontario, https://www.ontario.ca/page/forest-fires) and at a national level (e.g., in Canada, http://cwfis.cfs.nrcan.gc.ca/home). These data are also used in continental- and global-level assessments, as in the case of EarthData (https://earthdata.nasa.gov/earth-observation-data/near-real-time/firms/active-fire-data) and Global Forest Watch Fires (http://fires.globalforestwatch.org/home/), which are used to develop important forest management policies.

Our discussion of the approaches to mapping complex entities will follow a progression from points, lines, and areas to thematic classification (see Chapter "Mapping Forest Landscapes: Overview and a Primer"). Complex entities in a landscape, such as forest fires, can be mapped in ways that range from a collection of simple, discrete points or objects to continuous and complex objects with fuzzy boundaries (see Chapter "Fuzzy Classification of Vegetation for Ecosystem Mapping") and heterogeneous internal structures. The approach typically depends on the goals of the mapping and the scale of the representation. We begin by describing the characteristics of a wildfire (its "anatomy") and briefly describing the factors that determine their development and how this leads to the development of the complex and heterogeneous phenomena that we wish to map. We then focus extensively on the scale dependence of these entities and how changes to the scale of measurement and mapping will alter the final characterization of these entities on the map. We reflect on the assessment of mapping accuracy in remote areas and discuss emerging methods for quantifying and characterizing complexity.

Wildfire Initiation and Anatomy

To understand the challenges involved in the mapping of wildfires in forest landscapes, it is necessary to understand how they form and spread. Even though this description will be, of necessity, somewhat rudimentary, we will provide an overview of the key mechanisms that cause the initiation, growth, and termination of wildfires. This will provide valuable context for perceiving the spatial complexity of the resulting objects and the mapping challenges.

Initiation and Growth

Most wildfires start as a point-source ignition, typically caused either by a lightning strike or by human activities such as failure to extinguish a campfire or throwing a cigarette butt into a pile of dry wood at the side of a road. Wildfires initiated by anthropogenic sources are spatially biased, since they most commonly occur near human settlements, transportation corridors, and recreation areas. In contrast, natural ignitions are mostly random in forest landscapes, although there are spatial biases for lightning strikes (e.g., higher areas tend to be hit more often). In addition, trees in certain forest types and at certain topographic positions are more likely to burn than trees in other forest types and locations.

Once ignited, forest fuels will continue to burn until the fuel is consumed or weather conditions change enough to stop the fire. If the fire can spread into new fuel, the wildfire will grow outwards (radially) from the ignition point; each part of the fire that is expanding outwards is referred to as a "front." Sometimes, this spread will cease due to lack of fuel or extinction by fire suppression activities; if this occurs sufficiently rapidly, the fire's spatial signature will have a very small extent, easily represented at scales of interest as a point on a map. If not, the wildfire will spread by thermal radiation (transmission of heat through the air), conduction (transmission of heat through physical contact), convection (transmission of heat through moving air), and mass transport (movement of firebrands, which are burning pieces of wood, by the wind). These processes are influenced by a three-way spatial interaction among the forest fuel, weather, and topography. The spatial and temporal variation and the characteristics of these factors will determine the direction, distance, and rate of spread of the fire, as well as the thermal energy it releases.

In the simplest hypothetical case, in which the fuel is homogeneous, the terrain is flat, and there is no wind, the fire spread will be isotropic (i.e., equal in all directions) and will result in a circular wildfire, with the ignition point perfectly at its center (Fig. 1a). If wind is introduced, the fire spread will remain radial, but the fire will spread faster along the direction of the wind (the "head fire"), slower perpendicular to the wind (the "flank fires"), and slowest in the direction opposite to the wind (the "backfire"). The result is an approximately elliptical wildfire with the ignition point no longer at its center (Fig. 1b), and the length-to-breadth ratio of the ellipse will increase with increasing wind speed. As the wind speed increases, flaming pieces of fuel can become airborne; as we noted earlier, they can be transported and deposited ahead of the fire front. This spotting mechanism can ignite additional fires downwind, and these may eventually merge with the main body of the wildfire (Fig. 1c), though this is not inevitable.

In reality, fire spread is hardly ever this simple or even mechanistically deterministic, nor is the resulting shape a regular geometric object (e.g., an ellipse). Several factors are responsible for this:

- First, the forest fuel characteristics (e.g., flammability, mass, bulk density, moisture content) vary widely, forming a mosaic of highly heterogeneous fuel types. Juxtaposition of different fuel types in a forest landscape causes a spreading

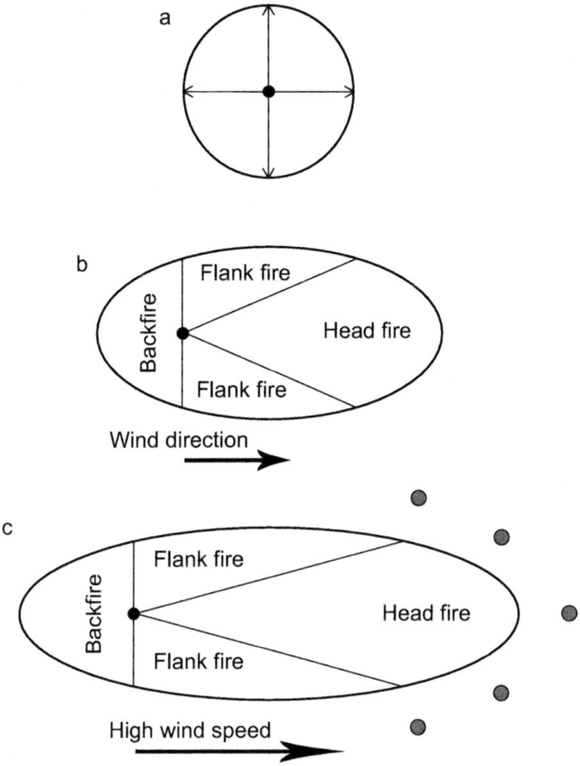

Fig. 1 Illustration of how fires grow in a simplified scenario that assumes homogeneous fuel and flat terrain: (**a**) In the absence of wind, fires spread radially from the ignition point, resulting in a circular shape. (**b**) In the presence of a wind, the fire spreads at different rates on different sides, resulting in an approximately elliptical shape. (**c**) In the presence of high wind speeds, the shape becomes more elongated due to faster travel of the head fire, and spotting occurs downwind of the fire (*gray* spots). Dark spots within the circle or ellipse indicate the original ignition point. Modified from van Wagner (1969)

wildfire to vary its flame length, rate of spread, direction of travel, and thermal energy as it moves between fuel types.

- Second, terrain is typically not homogeneous, and that also affects the spread of fire. For example, changes in slope will affect spread rates, since fires spread faster uphill and slower downhill, and concave terrain such as the bogs that form in depressions tend to have a higher water table that slows the fire's spread because water absorbs heat and lowers the fire's intensity. Slope and the slope aspect relative to solar radiation and prevailing winds have effects (e.g., south- and west-facing slopes in the northern hemisphere tend to be drier than north- and east-facing slopes). The presence of bodies of water (e.g., lakes) and exposed bedrock creates natural barriers to the fire's spread, but rapidly moving fires can jump vast distances, possibly even across a lake, to continue burning on the other side, and sufficiently hot fires can burn even wetlands and islands within lakes.

- Third, the wind and local weather change spatially and temporally during a wildfire. Both wind direction and speed vary spatially and temporally during the day, and are in turn affected by the microscale weather created by intense fires; for example, horizontal and vertical vortices (whirlwinds) are often generated by wildfires. Such changes directly affect the fire's rate of spread, its direction, and the thermal energy it generates.
- Fourth, these factors interact, leading to cumulative or even multiplicative influences on the expansion and intensity of a wildfire. For example, the effect of changing wind speed on fire spread may differ among fuel types, and the impact of the interaction will depend on the slope of the terrain and the depth of the water table.

Thus, the real-world physical processes that govern a fire's spread are highly complex, as are the spatial patterns of the damage caused by the fire's thermal energy, leading to a correspondingly complex burned area. The formation of a burned area is even more complex because it represents an emergent property of the spatial interaction between the heat generated by the fire and the ability of the forest fuel to tolerate the heat and avoid or delay combustion. These interactions are both deterministic and stochastic (i.e., subject to random or probabilistic factors).

Fires do not progress continuously. A spreading wildfire will often stop locally, and will eventually be extinguished altogether even in the absence of human intervention. The stoppage occurs when a fire front encounters fuel with low flammability (e.g., high moisture content), a place with no fuel (e.g., bedrock), or an unburnable barrier (e.g., a lake). In this case, the fire may stagnate and continue to burn slowly (i.e., smolder) until it consumes all its fuel or weather conditions change, in which case the fire is extinguished locally or diverted to spread in a different direction. Complete extinction of a wildfire occurs when progression stops along all fire fronts and all smoldering ceases. In the case of managed fires, local or complete extinction may occur due to active fire suppression (e.g., dropping water from water-bomber aircraft), creation of fire barriers, and fuel reduction by means of controlled burns. The position where a fire front finally stops progressing forms a discrete or fuzzy local boundary between the burned area and the adjacent unburned area; that boundary outlines the area affected by the wildfire. The outermost extent of these boundaries represents the outer edge of the wildfire (i.e., the perimeter of the fire's footprint).

In summary, a wildfire "event" (a temporally discrete and spatially stochastic process during which a fire grows from a point to an area) produces a spatially heterogeneous footprint (a surface that is interspersed with burned, partially burned, and unburned areas) that is demarcated by a perceived perimeter (the abovementioned outer edge, across which a discrete or fuzzy transition occurs from a burned area to an unburned area). Figure 2 presents two examples of simplified boreal wildfire footprints (i.e., with no depiction of internal heterogeneity, with a smoothed outer perimeter, and without considering satellite fires ignited by spotting). It nonetheless illustrates the complexities of the fire's initiation and subsequent spread. First (Fig. 2a) is a 37,000-ha wildfire that resulted in an elongated shape due to high and sustained wind speeds for 2 days along its 40-km primary axis, and the many spot fires on the right flank that eventually merged with the main body. Second

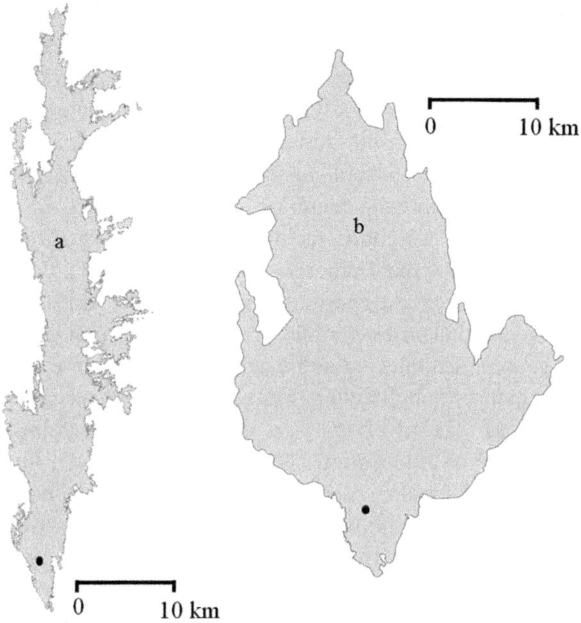

Fig. 2 Examples of the footprints of boreal wildfires, showing only the ignition points and boundaries: (**a**) A 37,000-ha fire that evolved under conditions with high wind speeds from a single direction over 2 days. (**b**) A 56,000-ha fire that evolved under conditions with variable wind speeds and directions over 23 days. Dark spots at the bottom of each footprint indicate the original ignition point

(Fig. 2b) is a 56,000-ha wildfire that burned for 23 days under conditions with variable wind directions and speeds, but also with merger of spot fires on both flanks.

Again, we want to stress that the description in this section is highly generalized in terms of how wildfires initiate, spread, and extinguish. If you want to learn more about the many exceptions to our generalizations, and intricate details of the sophisticated and evolving field of fire behavior research, there are many excellent descriptions of wildfire behavior (e.g., van Wagner 1969; Rothermel 1972; Chandler et al. 1983; Johnson 1992; Whelan 1995; Johnson and Miyanishi 2001; Sullivan 2009; Albini et al. 2012; Finney et al. 2013). Nonetheless, our simplified representation is sufficient to illustrate that the key processes involved in a wildfire and their consequences are highly complex, and produce complex and spatially heterogeneous objects that become the landscapes that we will attempt to describe using maps.

Descriptors of Footprints

In this section, we describe some of the geometric and other descriptors that are used to characterize wildfire footprints.

Point-Based Mapping

Points are often mapped when the goal is to identify the locations and positions of active or smoldering fires ("hot spots") rather than the complete perimeter of a fire. One particularly useful type of point to be mapped is a "centroid." Unlike the geometric center, which can be difficult to identify for large objects with complex shapes, the centroid represents the center of mass; that is, it represents a weighted average. Because different weighting techniques exist, the location of the centroid will depend on the technique that is chosen (Farmer et al. 2011). Point indicators are typically selected to represent the centroid of some feature. Although the concept of a centroid is simple, various means for identifying its position (i.e., different weighting methods) have been established and used, making even the seemingly simple decision to map a centroid challenging. For regular shapes (e.g., circles, squares), the geometric centroid is relatively easy to identify, and that point provides a reasonable approximation for the location of that feature. However, for more complex shapes (e.g., a U-shaped feature, such as a fire that burns around two sides of an unburnable rock outcrop), the geometric centroid may exist outside of the shape itself. This leads to debate over whether the centroid should be constrained to exist within the boundaries of the feature or whether a more abstract definition should be permitted, in which the true geometric center reflects the properties and maximum extents of the shape in all directions or only in a certain key direction.

The approach can be extended to mapping extensive and complex fire damage using a single point to which attributes (e.g., time, burn intensity) can be attached. In this context, it is less important to accurately map the overall footprint than it is to map key points within the footprint such as the location of a burning activity (e.g., the initiation of a new fire). This method is commonly implemented to identify new fires or fires that have not yet been fully extinguished, or for isolated locations where detailed mapping may be difficult. It is also used when the outlines of a fire cannot be seen at the resolution of the map or other display being used to show the fire's location. In the United States, the National Interagency Fire Center (https://www.nifc.gov/index.html) records the locations of all fires that have burned in recent decades along with their size and cause (Morgan et al. 2001). Since complete mapping of the footprint is not required, this point-based approach can be relatively rapid and can be used as a monitoring tool that will (when appropriate) trigger further mapping and updating of active fire records, as well as direct fire suppression efforts.

The earliest organized activity related to fire detection and mapping relied on human sightings of smoke or flames from observation points, which were typically tall towers that provided unobstructed views over large areas such as watersheds. In North America, most of these towers were erected between 1920 and 1950; eventually, thousands of these towers were in operation in the United States and Canada, with 325 in Ontario, Canada, alone (http://ontarioftl.bravehost.com/index.html).

Sometimes, active fires and their dynamic elements are of more interest than the footprints the disturbance leaves behind. In these cases, the ability to map high-temperature anomalies (hot spots with a temperature significantly greater than the average) is important and relies heavily on remote sensing technology capable of observing radiation in the infrared through thermal infrared wavelength range (0.7–13.0 μm) of the electromagnetic spectrum. However, thermal remote sensing, particularly using data with coarse spatial resolution (e.g., NOAA-AVHRR, MODIS), is often used to guide subsequent analysis with higher spatial resolution (Siegert and Hoffmann 2000).

Thermal imaging is often conducted at times when the contrast with the ambient air or ground temperature is maximized (e.g., at night) to improve the detection of subtle thermal differences (Spichtinger et al. 2004). Generally, mapping of hot spots yields a map of points where heat was greater than normal, but the concept has been extended to predict the areal extent of a fire (Hessl et al. 2007) based on interpolation between points or extrapolation beyond the points to generate boundaries that enclose the mapped hot spots (i.e., fire perimeters). That is, mapping of thermal hotspots can extend the analysis beyond instantaneous conditions. Pozo et al. (1997) assessed the thermal regimes in successive images and used the changes to map the growth of fires and ultimately produce maps of a burned area. Locating hot spots generally relies on thermal measurements, with the positions identified for optically bright cells or areas in satellite images or air photos. The thermal images often produce better results than optical approaches, since smoke and haze generated by a fire can obscure the position of hot spots in optical approaches.

Although point mapping can be used to identify the locations of instantaneous or short-term phenomena at appropriate scales, the points are also commonly used to identify larger areal features at scales that cover larger areas. In this case, the points become indicators of larger phenomena.

Area-Based Mapping

Area-based mapping can be achieved in many ways, particularly when classifying gridded data such as satellite images or scanned aerial photographs. Although simple image classification algorithms that rely on discrimination of differences in per-pixel spectral or tonal density are a straightforward option, more sophisticated options exist. Examples include methods that rely on "region growing" (Kettig and Landgrebe 1976), in which an area is defined moving outwards from a central point based on calculation of the degree of homogeneity; when the degree of homogeneity of the enclosed area decreases beyond a certain threshold value, the area stops growing (Hirata and Takahashi 2011). Such an approach has become part of the evolving group of methods for inventories of individual tree crowns (Breidenbach et al. 2010). Depending on the spatial resolution and the scale of the data acquisition, an individual tree crown can be considered as a homogeneous unit that forms the basis of a segmentation operation (in which the goal is to divide an image into individual components, such as trees) and classification (to define the characteristics of those components). Jakubowski et al. (2013) compared vector- and raster-based segmentation methods to delineate the crowns of individual trees based on point cloud data generated by a light detection and ranging (LiDAR) scanner and canopy height models. (Here, for simplicity, we define raster data as gridded collections of point data, and vector data as collections of lines. Both types of data can be used to define an area, as in a cluster of points or the region inside a boundary line, respectively.) Several examples in the literature identify combinations of methods (i.e., hybrid methods) for mapping complex dynamic features. In this chapter, we focus on examples that relate to wildfire mapping. For example, Fraser et al. (2000) used a combination of changes in the normalized-difference vegetation index (NDVI), statistical thresholds, clustering, and filtering to determine the area of a boreal fire and map its perimeter. Holden et al. (2005) used a series of spectral indices to map fires and then compare them with the rudimentary boundaries presented in fire atlases.

Remote sensing technologies that represent a wide spectrum of sensors, wavelength sensitivities, orbital characteristics, and resolutions have been used to map both active fires and footprints of extinguished fires in forest landscapes (Remmel and Perera 2001; Merino-de-Miguel et al. 2010; Oliva et al. 2011; Schroeder et al. 2011; Ruiz et al. 2012b). Yet despite decades of research and development in this domain, no clear approach has emerged as inherently superior. The enduring use of a variety of approaches to mapping likely has as much to do with the many different reasons for performing the mapping as it does with the rich availability of (or access to) data and the complexity of the landscapes to be mapped. Nonetheless, there have been numerous attempts to automate and standardize the methodology, coupled with arguments for using specific spectral indices such as NDVI (Kasischke and French 1995) or the differenced normalized burn ratio (Murphy et al. 2008), and studies that compare satellite-mapped fires with hand-digitized images or other reference data have continued to emerge (Remmel and Perera 2002; Holden et al. 2005).

Fires are dynamic (they change over time), but eventually they extinguish. Simultaneously, landscapes are dynamic because of biomass accumulation (e.g., forest growth), but fire disturbance regimes make them even more dynamic. Forest landscapes are mapped over decades to monitor disturbances and trends and to allow improved consistency in their mapping. Although the remote-mapping tradition stems from older photogrammetric methods devised for use in forest inventory (Leckie and Gillis 1995) or ecosystem mapping (Baker et al. 1995), much of the satellite-based effort has focused on remote boreal forests where field access is difficult and wildfire disturbances are both numerous and extensive (Kasischke et al. 1993). Much of the effort has relied on data with coarse spatial resolution but coverage of wide areas to minimize the need to create image mosaics from a series of smaller images (French et al. 1995), with a specific emphasis on detecting changes to vegetation indices such as NDVI (Remmel and Perera 2001). Other researchers have used multiple seasons of satellite data to detect NDVI changes for identifying wildfire disturbances (Kasischke and French 1995). Efforts using imaging radar (Bourgeau-Chavez et al. 1997, 2002) are also finding a niche use due to the ability of this technology to detect differences in surface roughness (which affects the image texture), soil moisture (due to radar being affected by soil dielectric properties), and atmospheric effects (e.g., clouds, aerosols, and fine particulate material).

Interfaces Between Burned and Unburned Areas

As we noted earlier, mapping often begins with point data; these are considered to be zero-dimensional (0D) features because they do not extend in any dimension from a given point and therefore have no length, area, or volume. But it is also possible to directly measure and map features that extend in one dimension (1D features), thereby creating lines that have length but no area or volume. If those lines are extended far enough to enclose an area, they produce a two-dimensional (2D) feature that has a length in each dimension and an area, but no volume. Three-dimensional (3D) features have length in each dimension, but they also have a surface area (both for the whole surface and for individual facets of that surface), a projected area (based on the cross section of the object), and a volume, and each facet of the feature represents a 2D surface with its own properties. Lines (1D features) often represent interfaces between adjacent areas; thus, they may be either sharp demarcations (e.g., the stumps that represent the last row of harvested trees in a plantation) or 2D transitional *areas* that have a width, and can represent land cover gradients and transitions. Identifying such interfaces requires examination of the available data to detect a substantial contrast between neighboring cells in a grid (or cells separated by longer distances) with respect to some measured parameter value (or combination of values). The contrast is used to decide whether two areas with significantly different characteristics exist; in that case, a boundary should exist at the specified location or range of locations between them.

The spatial resolution, data format, classification scheme, and method for assessing whether a boundary exists all affect whether a boundary will be identified. This approach is fundamentally and theoretically different from an area-centric approach, in which homogeneous regions are usually defined by growing outward from a central point until the underlying characteristics change so much that they can no longer be considered homogeneous; at that point, new area stops being added and a boundary is considered to exist at that location. In mapping of boundaries, the contrasts are identified first and may eventually enclose homogeneous regions or simply identify linear features within the landscape matrix that identify regions with structurally different characteristics.

The remote-sensing and image-processing literature refers to searches for linear elements as "edge detection" (Heath et al. 1998). Edge detection algorithms can take several forms, including the use of thresholds applied to images that record the high-frequency components of the image to identify areas with substantial rates of change between adjacent pixels; the use of Laplacian and Sobel filtering (Gonzalez and Woods 2008), which seeks to emphasize contrasts between neighbors; and the use of wombling, a technique for identifying points or edges where the rate of change in characteristics is particularly high. Within the ecological literature, categorical and lattice wombling (Fortin 1994; Fortin 1999; Philibert et al. 2008; Oden et al. 2010) have gained substantial traction due to their ability to assess the impact of spatial scale and intra-patch heterogeneity. The lattice wombling approach is particularly favored due to its simplicity. Wombling involves fitting a plane through a 2×2 matrix of cells that exist within a larger grid to identify potential boundaries (Fig. 3) and then apply-

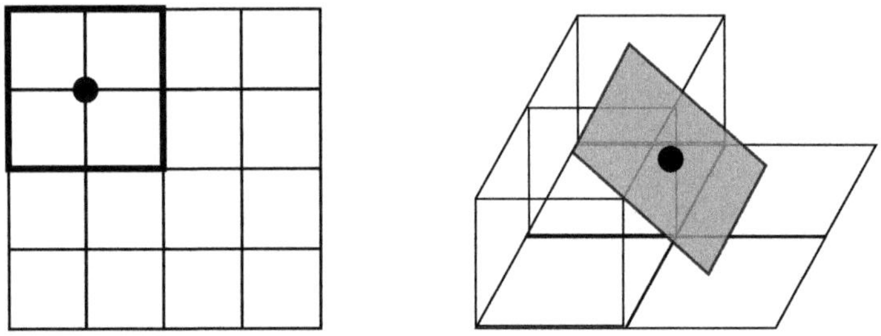

Fig. 3 Lattice wombling is used to identify rates of change across an image. Here, the technique uses (*left*) a 2 × 2 grid of cells (*black square*) within a larger grid, and (*right*) calculates the slope of a plane (the *grey* rectangle) that connects the centers of the two halves of the 2 × 2 grid. The *black* square is then moved one position to the right and the slope calculation is repeated. The approach can be used across scales (by changing the number of cells in the grid) to detect potential interfaces (boundaries). Reconstructed from Fortin (1999)

ing statistical analyses to assess the significance of the slope (rate of change between adjacent cells) and identify the presence of a true local contrast or edge (Hall and Maruca 2001). By altering the spatial resolution (e.g., aggregating cells), the computed slopes can provide a cross-scale analysis. Boots (2001) developed the theoretical background for using local statistics to map the dominant boundaries between mapped polygons and their surroundings across multiple scales.

Increasingly modern modes of fire perimeter mapping use the global positioning system (GPS), which relies on a constellation of satellites whose real-time positions are precisely known and that broadcast unique signals that can be interpreted by receivers at or near the Earth's surface. The lag between the time of each emitted satellite signal and the reference time recorded by the receiver is used to compute the distance from each satellite to the receiver. The distance between the satellite and the receiver represents the radius of a sphere centered on each satellite; the intersections of the spheres from multiple satellites with the sphere that represents the Earth's surface ("trilateration") identify a unique point on the Earth's surface that corresponds to the position of the receiver. This position is represented by a coordinate pair (or a triplet if elevation is also calculated) that can be subsequently mapped. By collecting positional data continuously as the receiver moves along the perimeter of a surface feature such as a fire footprint (i.e., the interface between burned and unburned trees), a researcher establishes a collection of positions that can be connected by line segments to form a line with multiple vertices; when that line is closed by returning to the starting point of the survey, it forms a polygon that encloses the area of interest and defines the perimeter of the fire footprint. Note that although this approach is possible, it may not be feasible; in practice, difficulty of access to remote sites or sites with difficult terrain conditions often makes it more of a theoretical possibility than a routinely implemented methodology (Gillis and Leckie 1996).

Even in a situation where an extinguished fire's perimeter could be walked with a GPS receiver or sketched onto a map of the region, accuracy could still be problematic. Walking the perimeter can be very difficult if the terrain is challenging, as in the case when there are many fallen trees after a fire, since the fallen trees often create a tangle of logs and branches that intertwine at multiple heights and make

Fig. 4 Fallen trees and debris along the interface between a boreal wildfire's footprint and the surrounding forest in northwestern Ontario show the difficulty encountered when trying to traverse the fire's perimeter on foot. In addition, the presence of unburned trees standing among burned trees shows that the boundary is sometimes "fuzzy" and difficult to precisely define. Photo by Tarmo Remmel

walking nearly impossible (Fig. 4). Simply traversing the perimeter can be difficult enough; attempting to discriminate the boundary's location and simultaneously following it is a much more difficult task. When walking a fire perimeter is difficult due to obstacles, the mapped boundary will be horizontally displaced from its true position. In addition, if the surveyor walks at different speeds along different parts of the boundary, the horizontal spacing between points (i.e., between the vertices that define the boundary's position) will vary since the temporal measurement interval remains consistent, leading to inconsistent spatial variation in the map's precision. Furthermore, when the surveyor stops (or begins to move very slowly) while continuing to collect positional measurements, the inherent horizontal positional error within GPS positioning can exceed the vertex spacing and produce vertices that loop boundary lines back on themselves, thereby causing a variety of topological errors that need to be eliminated prior to use of the data.

When a boundary is defined as a single "object" defined by a contiguous set of lines that connect vertices, this represents an object-based view of reality. Unfortunately, this perspective forces us to represent the mapped entities as sets of discrete points, lines, and areas to form nonoverlapping spatial units that are separated by crisp boundaries. This can be misleading because fires (and geographical phenomena in general) show inherent spatial variation, and boundaries are not always clearly demarcated, as shown in Fig. 4. Boundaries of natural phenomena are often better described as being "fuzzy," a descriptive term that was chosen to distinguish these boundaries from the "crisp" geometric edges of human-made objects (see Chapter "Fuzzy Classification of Vegetation for Ecosystem Mapping"). In this sense, *fuzzy* means diffuse, uncertain, or extending over an appreciable distance, with categorization of whether any given point belongs to one side or the other of the boundary defined using a spatially varying membership function (Zhang and Stuart 2001). It has been demonstrated that forest patch interiors and interior edges ("perforations") differ in many respects, including species composition, stem diameter, and shrub height (de Casanave et al. 1995), and that the implications of these differences for ecological fragmentation (in terms of the abundance of transitions called "ecotones" between ecosystems) can be substantial; thus, interpreting these conditions from mapped boundaries can be biased if the interfaces are always considered to be abrupt and absolute.

In geographical information system software, it is much more convenient to represent boundaries as absolute lines. Given this tradition, ecotones, fuzzy boundaries, and inherent complexity of patch edges have often been oversimplified. Although tools and approaches for handling fuzzy boundaries exist (Wang and Hall 1996), their application requires more data than many practitioners are willing to obtain and more effort than they wish to expend. Even relatively simplistic diagnostic methods, such as the implementation of "epsilon bands" that characterize the fuzziness of line segments (Dunn et al. 1990), are not commonly implemented. Methods have emerged that indirectly consider positional errors by considering statistics on the overlap between mapped entities (Remmel and Perera 2002), or by assessing mapping errors by classifying mapped elements as either errors of commission or errors of omission (Oliva et al. 2011).

In an attempt to characterize the abruptness of a wildfire interface, Remmel and Perera (2009) counted the neighboring cells that belonged to the fire footprint at each location using a spatial kernel (i.e., a fixed-size area that was iteratively moved through a grid that represents the landscape). This approach led to computation of a membership probability at each pixel; this represented the probability that a given pixel belonged to the wildfire footprint class or the unburned class. Transects measured across the interface between the unburned matrix in which the fire occurred and the burned wildfire footprint allowed the membership strength to be plotted as a function of distance along the transect. In this analysis, steep slopes along subsets of these plots represented more abrupt interfaces, and were generally found where fires had burned right up to the edge of nonburnable land cover types (e.g., bedrock), whereas gradual slopes represented locations where the fire slowly burned out to form a more gradual transition zone. The interface's width and abruptness were clearly influenced by the fire's behavior and can thus be mapped as properties of the interface itself.

Simplification

The mapping of complex entities, such as wildfires, generally requires some level of generalization and simplification; if nothing else, the map must be smaller than the real-world area it represents, and that requires the elimination of some amount of detail. Wildfire events can be represented as simplified scale-independent objects for which a perimeter encompasses a singular entity (the footprint), implying that all of the area within that perimeter is burned and that the internal state is homogeneous. However, it has been well documented (Perera and Buse 2014) that these assumptions are not true and that wildfires contain a rich mixture of live and dead materials that have been heterogeneously affected by the fire. Similarly, the detail in which wildfire interfaces are recorded can vary substantially among practitioners and among the methods used to collect the data used for mapping.

For larger and less accessible fires, aerial mapping from fixed-wing aircraft has been implemented to generate maps that depict the location of the fires and their boundaries. Aerial mapping is typically conduced in one of the two ways: (1) an observer draws the boundaries onto a topographic map or photograph of the area (nowadays, perhaps displayed on a tablet computer equipped with a stylus), or (2) the observation platform (usually an aircraft) flies above the approximate perimeter while collecting GPS point data. The former method requires the observer to quickly assess the correspondence between the observed landscape and the medium onto which spatial data is being recorded; the latter requires fast reflexes of the pilot to ensure that the aircraft closely follows the perimeter. Obviously, larger and more distinct fires with clear and abrupt boundaries will be easier to map using both methods. However, aerial views are not necessarily effective for mapping local irregularities and highly detailed shapes of the burn interface. Helicopters offer some improvements over fixed-wing aircraft, since they allow slower travel along the boundary and therefore permit greater detail. However, helicopter time is substantially more expensive, and financial considerations are often more important than increased precision if precision is not absolutely necessary.

Regardless of the mapping platform, larger features will be mapped more accurately, since flying along (or even identifying) boundaries will be easier when the boundaries pass through vast tracts of land rather than through subtle and irregular local patterns. Moreover, general sketches of fire boundaries created from fast-moving aircraft tend to record more simplistic boundaries and areas than do methods that follow boundaries with a GPS unit that automatically collects approximately one geographic location each second. The latter approach may falsely increase the perceived level of detail due to the number of vertex positions that are collected and used to define the interface line; in a digital representation, these differences are difficult to discern without additional contextual information.

Detailed lines created in this way can be simplified by using well-known algorithms such as the Douglas–Peucker line-simplification algorithm or its derivatives (Wu and Marquez 2003). This results in boundary lines with fewer vertices and segments, but also distorts the representation because the elimination of detail is based on mathematical simplification rather than on actual measurements, and the loss of detail may be problematic for some applications (Smith 2010). Here again, the consideration of an appropriate and consistent scale for mapping complex landscape features becomes inescapably tied to the data collection methods.

Scale Dependence

Whether mapping is conducted by sketching visual interpretations of the landscape while flying over an area affected by a wildfire or by collecting airborne or even field-level GPS positional information related to the boundary location, numerous factors will influence the scale of the data acquisition or the final representational scale of the maps that are produced. Variation in scale will lead to differences in interpretation, in utility, and in what can be depicted by maps. An understanding of scale and the scale dependence of mapping based on decisions made by the cartographer is fundamental to having a solid foundation in forest landscape mapping.

Scale itself can be considered from multiple perspectives, including the cartographic scale of sketched or vector data described in the previous section, the spatial resolution (grain size) of remote-sensing data, the minimum mapping unit (MMU, which represents the smallest resolvable unit on the ground) in aerial photographs, or the spacing between adjacent GPS positional markers. In Canada, National Topographic Series base maps (https://www.nrcan.gc.ca/earth-sciences/geography/topographic-information/maps/9767) are produced at 1:250,000 and 1:50,000 (cartographic) scales. Although local maps are sometimes produced at 1:20,000 or even 1:10,000 scales, even large map sheets express relatively small spatial extents and thus are not used for mapping wildfires with vast extents (which would often extend beyond a single map sheet). The cartographic scale of base map chosen by an aerial observer above a wildfire footprint will substantially affect the resulting mapping scale. Similarly, the abruptness of changes in the boundary and the observer's ability to match the real landscape they are observing to the topographic features of the map will influence how accurately the footprint can be sketched and the level of detail recorded.

In automated GPS data collection, the acquisition of positions (points) is done at a fixed time interval. Thus, if the GPS platform or surveyor changes velocity or altitude, then the spacing of points will change, thereby affecting the mapping scale, and the line's complexity will change in response. Such variations in velocity will also affect the MMU, and will therefore affect the area and perimeter measurements being made. Fires mapped with different scales cannot be directly compared, other than in general terms, and will require some form of normalization to make the two images easier to compare. For example, Remmel and Perera (2001) coarsened hand-digitized wildfire perimeters by rasterizing them at a spatial resolution that would override any local effects that resulted primarily from the image scale; this facilitated comparisons with the same fire footprint mapped using low spatial resolution AVHRR satellite images.

As the scales of measurement and mapping change, the effects on the MMU, spacing between points, and spatial resolution will also change. Assuming, for the sake of simplicity, that data is represented in raster format (i.e., as gridded points), then a coarser spatial resolution will result in a smoother surface, whereas a finer spatial resolution will result in a surface that becomes increasingly ragged and heterogeneous as the resolution increases. There will always be a trade-off between spatial resolution and the level of observable detail; the increased image heterogeneity associated with higher spatial resolution can cause processing considerations that would otherwise be avoided. For example, greater storage capacity and faster computers are required to manipulate the larger amounts of data generated at high resolutions. A simple doubling of the spatial resolution will quadruple the number of cells and hence the storage and processing requirements. Of greater concern are the influences on parameters such as area, perimeter, and land-use types that result from changes in the MMU or the spatial resolution.

Using data with coarser spatial resolution than the phenomenon of interest (e.g., a poorly selected MMU) can conceal variations such as multiple land use or cover types residing within the area represented by a single pixel (i.e., a mixed pixel; Cracknell 1998). In such cases, it's necessary to "unmix" the mixed pixels in end-member analyses (Yu et al. 2017), which are intended to reveal where one land use or cover type transitions into another, to make sense of otherwise potentially confusing information of multiple land cover types inhabiting the area of a single pixel. Although such techniques are powerful ways to improve the apparent resolution of the data, they are unable to reveal features that cannot be inferred without additional data; for example, in images with a pixel size of 250 m (e.g., MODIS satellite data), a narrow stream that runs through vegetated pixels will be invisible, and its presence cannot be inferred unless the vegetation type in those pixels is strongly associated with riparian habitats. It is also important to consider the choice of what will constitute an edge and how that definition will affect the assessment of a landscape. For example, Zipperer (1993) notes the influence of spatial processes on the ratio of interior (core) forest areas to edges (ecotones); detection of these relationships will be affected by the choice of scale. Mapping approaches work for a given scale, a given goal, and a given method, but the resulting products and their interpretations are not necessarily globally accurate.

The English mathematician Lewis Fry Richardson demonstrated that the measured length of natural features (e.g., coastlines) increases without limit as the resolution of the measurement increases; this is called the Richardson effect (Richardson 1961). When the patterns can be defined mathematically by an equation that reveals self-similarity at all scales, the pattern is described as *fractal*. One way to describe this effect is by means of fractal analysis (McAlpine and Wotton 1993; Sun and Southworth 2013). In particular, a parameter called the "fractal dimension" (FD) can be mathematically computed for any pattern as a ratio that quantifies the change in complexity of the pattern in response to a change in the scale at which it was measured (Mandelbrot 1967). From a cartographic perspective, this leads to the seemingly paradoxical condition in which the measured length is directly influenced by the MMU or the spatial resolution (Fig. 5). Thus, calculations (e.g., perimeter-to-area ratios), comparisons between or among data collected or presented at differing scales, and conclusions based on these calculations or comparisons may at best be inaccurate (because they change as the scale changes) and at worst completely meaningless.

FD values can be computed when self-similarity is evident in a boundary, and they fall somewhere between 1D and 2D representations. Thus, as a line becomes more complex and tends to contain self-similar pattern elements at multiple scales,

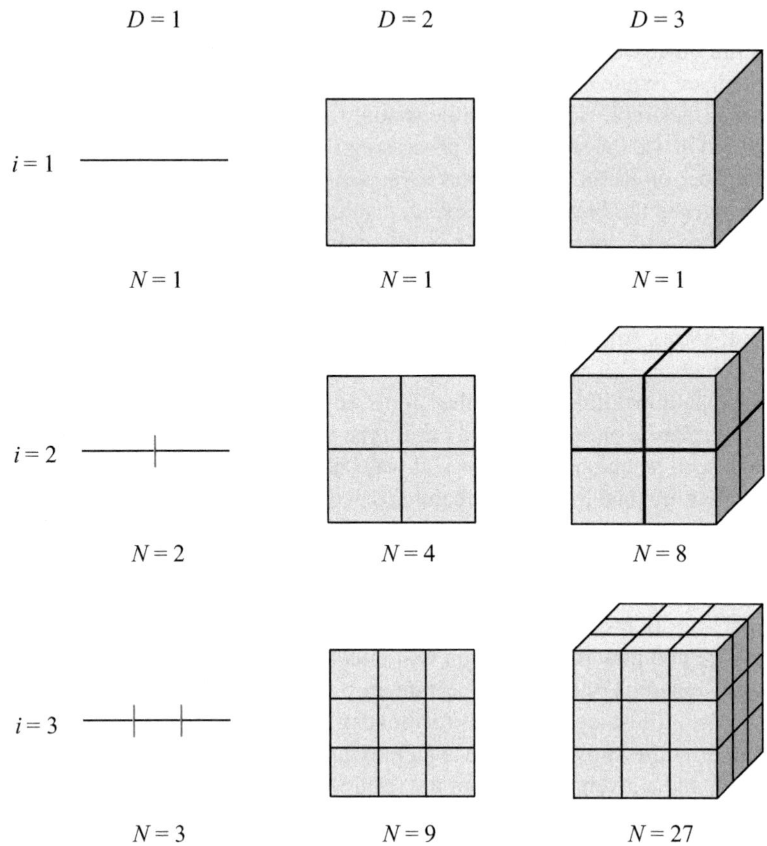

Fig. 5 An illustration of geometric dimensions. *D* is the geometric dimension, *i* is the number of repetitions of the basic unit (here, a line segment for $D = 1$, a square for $D = 2$, and a cube for $D = 3$), and *N* is the number of units used to create the shape

the fractal dimension increases toward that of a planar shape. Simpler features approach and approximate a straight line. This fractal notion can be extended from horizontal to vertical dimensions and used to characterize the complexity of landscape features. Figure 5 demonstrates the partitioning of space (and volume) into smaller but identical segments, highlighting a fractal property but also the effects of the MMU or the scale of representation.

Imagine a coastline, the edge of a lake, or the boundary of a wildfire footprint; in each case, there are elements of the pattern (the undulations, incisions, or protrusions) that repeat at multiple spatial scales. Thus, wildfire disturbances in boreal landscapes tend to produce complex fractal patterns. Replicating, measuring, and quantifying such landscape entities create many complexities.

The boundary between burned and unburned locations increases in complexity as we examine it more closely: it changes from being a simple line along which all boundary characteristics remain consistent as the boundary length increases but changes to a more complex pattern (perhaps a fractal) as the level of mapping becomes more detailed. As a boundary winds across a landscape to separate disturbed from undisturbed locations, the stand and terrain conditions (e.g., burn intensity or topographic variability) that define the boundary are likely to vary. The concept of the MMU is relevant, since the spatial resolution (for raster data) or the vertex spacing (for vector data) will alter the level of spatial detail available to represent the boundary. The selection of an appropriate MMU will depend on the ability to measure the boundary at a specific level of detail, the reasons for mapping the boundary, and the data-handling capabilities of the surveyor. Often, when the complexity of the line is very high but the quality of the data is suspect, the line can be simplified by implementing an appropriate algorithm (Cromley 1991; Shi and Cheung 2006; Park and Yu 2011). Such algorithms simplify lines by eliminating inessential or misleading vertices, often by enforcing a minimum spacing between vertices along the boundary.

The abruptness of a boundary, which can be measured at any given point along the boundary, is both a defining property and an emergent property of the boundary. On the one hand, a high abruptness indicates that a change exists at that location and that a boundary should exist; on the other hand, the scale of the mapping will determine the abruptness that is measured at any given location.

The effects of scale on mapping and representation have been extensively studied (Turner 2005), but no universal or consistent understanding of its effect on patterns or processes has been determined (Rouget et al. 2006; Peters et al. 2008). It is clear however that as the spatial scale changes, this produces noticeable effects on the area, perimeter, and thematic components (e.g., land-use classes) of maps. Remmel (2015) describes the *ShrinkShape2* method, which incrementally shrinks a planar shape that was created to enclose a region of interest in a map, and then records the area, perimeter, and number of contiguous parts created from the original planar shape as a result of this change. Once the shape has been eliminated (i.e., reduced to the point of extinction) by successive iterations, the software can display the changes in the area, perimeter, and counts of the shape's features to provide detailed structural information about the shape. The goal of this analysis is to characterize polygon shape complexity, and scales at which this complexity change.

Simple shapes tend to decrease in area and perimeter in a relatively smooth and monotonic fashion (i.e., there are no abrupt changes); irregular shapes will be characterized by more complex functions that describe the decrease (e.g., slopes can change rapidly between iterations of the shrinking). Planar shapes that have lobe-like characteristics (i.e., that have protrusions from the main body of the shape) will reveal pinch-points, which are areas (usually the "stem" of a lobe) that narrow until they disappear, causing the original shape to split into two or more parts. This can be seen in a plot of the frequency of contiguous parts during the shrinking process; an increase in the number of separate parts provides important clues to the shape's overall complexity, particularly with respect to the existence of structural lobes. These characteristics are all scale dependent; the slopes (rates of change) and frequencies of characteristics such as the number of parts will change as the amount of shrinkage is adjusted. Therefore, the spatial scale at which certain characteristics manifest themselves will be revealed by differences in the plots of slopes and frequencies over a range of shrinking distance settings. Topographic settings strongly influence the shapes of 2D spatial processes such as wildfire that act on landscapes, and thus the linkages among area, perimeter, and their associations with scaling will be affected by topography; however, although this hypothesis is logical, it has not yet been demonstrated conclusively in the field.

Wildfires as Discrete and Complex Objects

As we described earlier in section "Initiation and Growth", wildfire formation is a highly stochastic and complex process that results in spatially heterogeneous and complex objects. However, it is common in mapping approaches to perceive and represent wildfires as vector polygons associated with various attributes (e.g., fire intensity). The first step in such an analysis is to demarcate the fire's perimeter and then denote the internal composition of those polygons as "burned." In this section, we explain why that approach oversimplifies the complexity of fire footprints.

The Outer Edge of a Wildfire is Scale Dependent

The wildfire interface (i.e., the boundary between burned and unburned areas) forms when spread of the fire stops due to local extinction. It is logical to perceive the outermost edge of a wildfire as its "perimeter," but some characteristics of wildfire behavior complicate this choice. First, most wildfires, and especially the largest and most intense fires, create satellite fires a few meters to several kilometers from the main body due to spotting. Unlike the most distant spot fires, which form spatially distinct entities, many spot fires arise close to the main body of the wildfire, making it challenging to determine the outermost edge of the main fire perimeter: Should these spot fires be delineated as part of the main fire, or as separate fires? The answer depends, in part, on the scale at which we ask the question (Fig. 6). Figure 7 summarizes the visual data in Fig. 6 to show the relationships between the spatial resolution and the corresponding number and area of the spot fires and the total area of the fire footprint. In such instances, the location and properties of the fire perimeter, as well as the area of the wildfire polygon demarcated using that perimeter, will vary considerably in response to changes in the spatial resolution of the mapping exercise.

A second problem is that a spreading wildfire will not necessarily be extinguished abruptly (i.e., with a sharp spatial boundary) except at locations where fuel is absent or when burning conditions suddenly change (e.g., during a heavy rainfall). Even then, spotting could occur when the fire jumps over fire breaks such as bodies of water, exposed bedrock, and built-up areas. Instead, most spreading wild-

Fig. 6 The same fire footprint depicted at spatial resolutions ranging from 0.4 to 1024 m. In each case, estimates of the area, perimeter, number of spot fires, and the area of the spot fires are affected. Figure 7 depicts these changes graphically

fires will extinguish gradually over a certain distance, resulting in a gradient from completely burned, through partially burned, to unburned (Fig. 8).

When a fire's interface is fuzzy, the resulting fire perimeter is not a discrete line; instead, it becomes a fuzzy line (i.e., a boundary whose width is not crisp or does not have a consistent width) whose location will depend on the spatial resolution of the mapping (Kim and Cho 1994; Pourghasemi et al. 2016). The fire's perimeter will therefore be scale dependent.

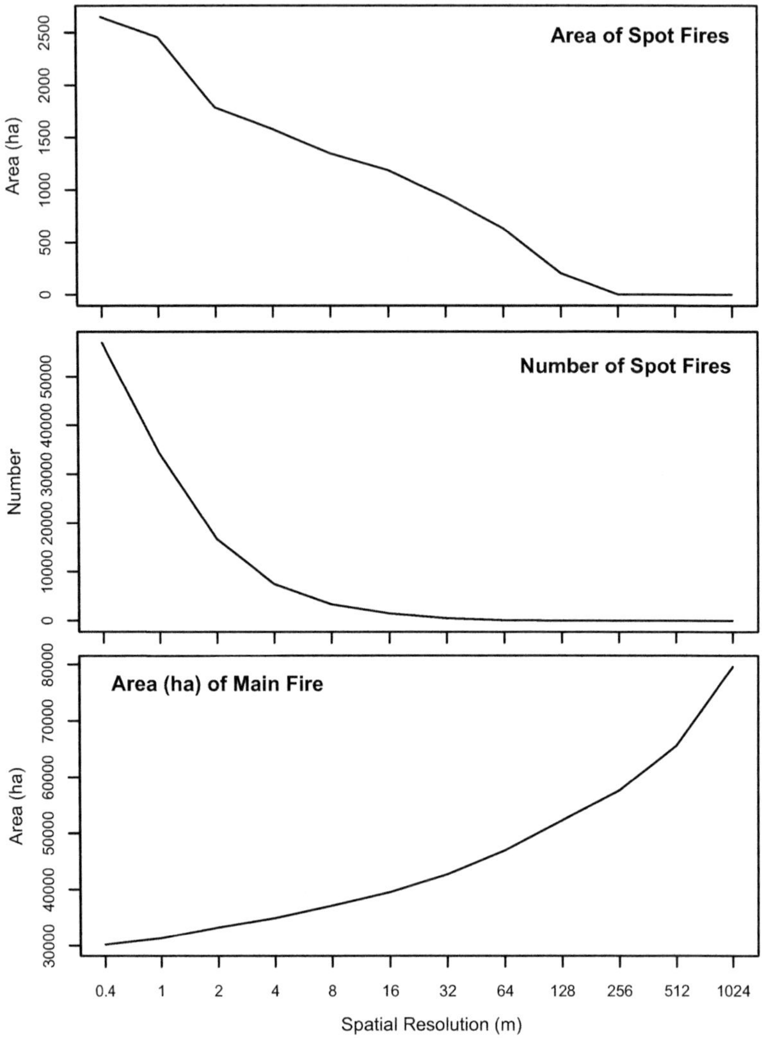

Fig. 7 Changes in the size and number of spot fires and in the total area of the main fire footprint (based on the images shown in Fig. 6) as a function of the spatial resolution of the analysis, which ranged from 0.4 to 1024 m

Width of the Ecotone

Ecotones represent regions of transition between ecosystems or between communities. In the context of wildfire, the ecotone width is defined perpendicular to the fire's boundary. It represents the distance over which the boundary influences the landscape characteristics and therefore has a meaningful real-world existence. Over

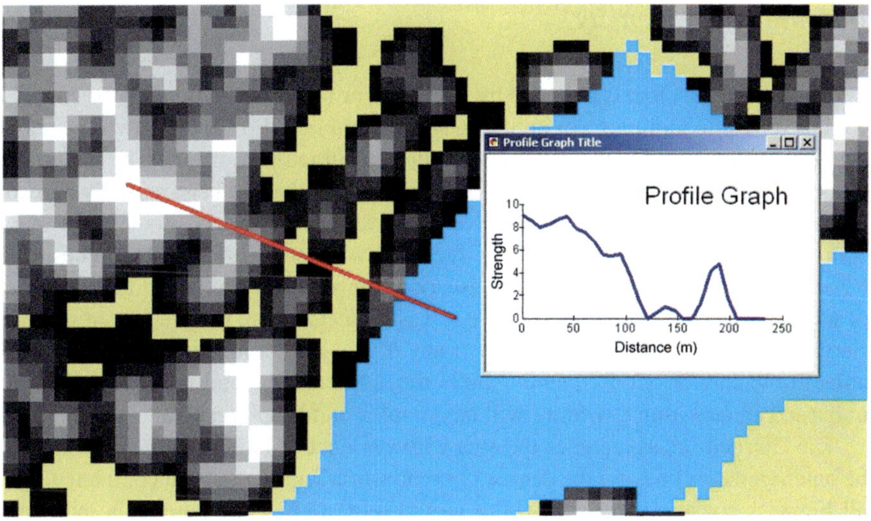

Fig. 8 Wildfire membership strength (*white* = high probability of having been burned, *black* = low probability) within the landscape mosaic (*tan*), with a lake (*blue*) acting as a firebreak. The transect through the burned area (*red* line) represents the probability of vegetation having been burned (i.e., the strength of membership in the burned class), which decreases moving from inside the wildfire (*top left*) toward the lake. The inset graph shows the changes in membership strength along this transect. This function is irregular and captures the complexity of the interface between the burned and unburned areas

this distance, land cover types and other conditions change sufficiently (from a practical or theoretical perspective) to warrant the use of different labels on different sides of the boundary. Ecotones can be treated as if they are boundary segments with a defined (nonzero) width or as distinct land cover units themselves (if the width is large). If we consider the degree to which an area belongs to a land cover class as a mathematical function, that function measures the value of some attribute (e.g., the proportion of burned stems) relative to the distance from the boundary. This membership function can be abrupt, in which case it exhibits a drastic change (a crisp transition) over a short distance, or it can occur gradually over a long distance (a fuzzy transition). However, there can be many different functional forms between these extremes. In terms of the slope (rate of change) of the two membership functions, a crisp transition may be represented by a vertical line, with an undefined slope, whereas a horizontal line with a slope of zero would represent no change; a fuzzy transition would have a slope somewhere between these extremes. Here, a range of possibilities exist, from shallow to steep simple linear slopes to more complex (e.g., logarithmic or logistic) functions. Fuzzy transitions also blur the distinction between a weak boundary and no boundary. Ecologically, the presence and characteristics of ecotones can have an important meaning; mapping them therefore requires an appropriate spatial resolution coupled with a technique that can detect subtle land cover state changes across the landscape at that scale.

Internal Heterogeneity

The mapped object that is bounded by the fire's perimeter can only rarely be considered as a simple and uniform entity. It will be composed of at least two thematic classes: unburned land outside the footprint and burned land inside. The unburned class is not homogeneous: in addition to including fuel that has been partially burned to some extent, it will be composed of areas with no fuel (e.g., bodies of water, bedrock) and areas with flammable fuel but with no evidence of combustion. Potentially flammable fuel may have escaped combustion within a wildfire because of its high moisture content or because it was protected from the fire's front by barriers, shifts in the wind, precipitation, or any of many other factors that lead to variable fire behavior. Consequently, both the composition and the properties of unburned areas within a wildfire will be complex and heterogeneous.

The "burned" class includes the area with evidence of combustion, and will also be heterogeneous because the degree of combustion will range from complete (e.g., all biomass consumed by the fire) to partial (e.g., only a fraction of the biomass consumed by the fire). This heterogeneity results initially from the type, arrangement, and dryness of the fuel and the environmental conditions (especially wind) and their local interactions between fire intensity and the susceptibility of a given fuel to fire. As fires become increasingly intense, the fire itself becomes a greater factor in its own propagation. Furthermore, the variability and heterogeneity over time and space of these factors lead to stochastic effects that make fire behavior difficult to predict at all scales. Whether the degree of combustion is expressed as burn severity, fire damage, or vegetation mortality, it will vary spatially (Keeley 2009; Araya et al. 2016a).

When juxtaposed, the areas with different degrees of combustion form a complex and heterogeneous spatial mosaic within the wildfire's footprint, embedded within the spatial mosaic formed by the diversity of unburned areas (Fig. 9). As is the case with fire perimeters, the internal heterogeneity of a wildfire will also be scale dependent, and spatial properties will change when the spatial resolution of the representation changes.

Area Complexity

The area recorded for any mapped areal feature will be directly related to the size of the MMU. For raster representations, the MMU is often equivalent to the spatial resolution (pixel size), since it is the smallest resolvable unit that can be mapped. For vector representations, this concept is somewhat more complex, as it relates to a number of controlling factors, including the cartographic scale of the mapping and representation, the minimum spacing between vertices along linear features, and the attribution resolution that will be used for labeling mapped features. Both for raster and vector data, increasing the MMU (making the resolution coarser) will cause a loss of detail, resulting in a more generalized representation of the characteristics of

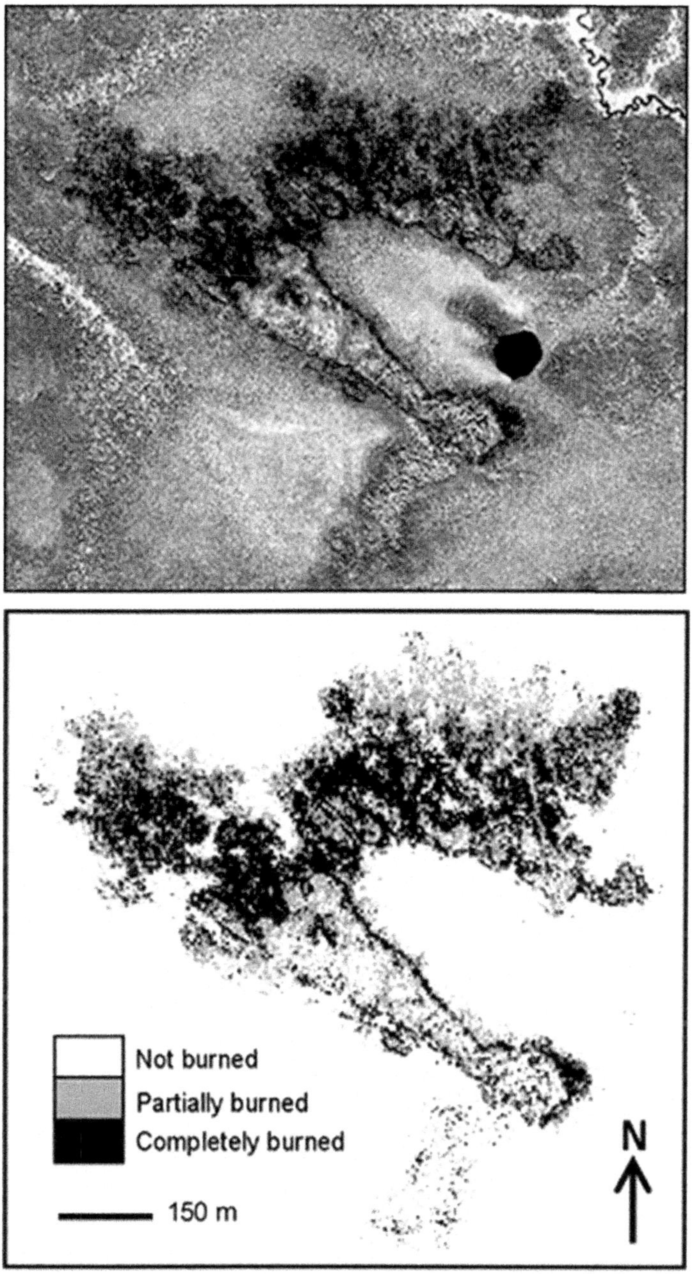

Fig. 9 Images showing the internal heterogeneity of a northwestern Ontario, Canada, wildfire and the internal complexity based only on the burn severity. The top panel shows the near-infrared band from the Ikonos satellite, with 3.2 m spatial resolution; the lower panel classifies the burn severity into three categories based on supervised image classification

most areas. Thus, selecting an appropriate MMU is critically important; to reduce information losses due to resampling, the MMU is generally selected to be close to the original data resolution (Knight and Lunetta 2003). The choice of an unsuitable MMU leads to errors in classifying the land cover composition, as well as to over- or underestimation of the diversity of land use and cover types; the problem is particularly serious for rare classes (Saura 2002). The choice of MMU also affects the metrics that are used to quantify landscape patterns (Remmel and Csillag 2003; Baldwin et al. 2004).

The complexity of a fire footprint is most clearly exemplified by the presence of fully contained residual patches of unburned vegetation. These patches can be visualized as holes or gaps within the matrix of burned areas inside the footprint, and typically comprise vegetated land, wetlands, open water, or exposed soil and rock. Although the identification and description of these residual patches can be scale and representation dependent (Perera et al. 2004; Ostapowicz et al. 2008), their boundaries are commonly referred to as "perforations" in the literature pertaining to spatial patterns (Vogt et al. 2007) to distinguish them from traditional edges such as those that form the perimeter of the overall footprint. As in the case of the individual spatial units that form the footprint, the individual residual patches can be further described based on their area, boundary, pattern, and geometry.

Thematic Contrasts

Upon defining a land cover classification scheme with a certain number of classes, a fire footprint and its surrounding area can be described using the frequency distribution of these classes. It is also possible to assess whether any land cover types are typically adjacent to other land cover types (i.e., whether there is a thematic contrast); taken together, the frequency distribution and adjacency relationships characterize both the composition and the configuration of the land cover types at or near a fire site. It is possible to assess the relative frequencies of specific contrasts between classes within the fire's footprint, such as defining the interfaces between burned and unburned pixels. These contrasts can be extended to describe the fire's interface based on the relationships among land cover classes, topographic settings, site characteristics, or proximity to unburnable features, as well as connections among common land cover types.

Transects can be established that run through fires and cross their perimeter (as in Fig. 8), or a series of concentric rings can be positioned running outward from the center of the footprint and past its edge. By examining how the characteristics of the land change along the transects or with increasing distance from the fire's center, we can examine changes in the thematic composition and configuration of the footprint and surrounding area. Thus, it becomes possible to assess whether the land cover composition or specific configurations of classes vary with respect to location. An unanswered question related to the composition and configuration of the resulting thematic map is whether the residual (unburned) patches within a fire's footprint resemble the "normal" characteristics outside the footprint; that is, it's unclear how

Table 1 Conditional probability distributions for the configuration of a composition class given the presence (1) or absence (0) of the same class in a given compass direction, summarized across all combinations

North	South	West	East	*P* (presence of class 0)	*P* (presence of class 1)
0	0	0	0	0.0282	0.0120
0	0	0	1	0.0564	0.0334
0	0	1	0	0.0098	0.0440
0	0	1	1	0.0016	0.0063
0	1	0	0	0.0581	0.0093
0	1	0	1	0.0591	0.0019
0	1	1	0	0.0101	0.0196
0	1	1	1	0.0005	0.0469
1	0	0	0	0.0540	0.0581
1	0	0	1	0.0417	0.0047
1	0	1	0	0.0096	0.0431
1	0	1	1	0.0387	0.0461
1	1	0	0	0.0209	0.0590
1	1	0	1	0.0597	0.0467
1	1	1	0	0.0005	0.0380
1	1	1	1	0.0308	0.0513

much the residual patches resemble the prefire conditions. Moreover, it's unclear how the composition and configuration of the landscape change with increasing distance from the fire's footprint. Analyzing these patterns can reveal the preference of fire for certain landscape structures, or the existence of certain structures that act as natural fire breaks.

Current (unpublished) research by the authors is developing a pattern learning algorithm capable of measuring and comparing binary landscape maps (i.e., maps with two land cover classes) and presenting the resulting compositional and configuration states as conditional probability distributions. Thus, the presence or absence of a specified condition (e.g., a land cover class) could be predicted by the spatial distribution of land cover types in the surrounding cells of the grid to the north, south, east, and west (Table 1). Conditional probabilities of presence and absence across all combinations form the expectation for composition and configuration on a landscape. Currently, this is only done in the four cardinal directions to reduce the demand on computational resources, but the approach could be extended to eight directions to include the northeast, southeast, southwest, and northwest directions. This approach is applicable at any scale, transferable to any landscape, and able to record numerous configurational blocks that combine to form higher order landscape patterns. The downfall of this approach is twofold: first, the large number of possible configurations means a high number of parameters and hence inherent complexity in summarizing the pattern, and second, it increases the likelihood of gaps in the distribution if sample sizes are small (e.g., not all pattern combinations exist in a landscape, meaning that cover types with near-zero probabilities lead to limitations in using the data). This approach uses a focal window of four orthogonal neighbors (north, south, east, and west) centered on each cell within a data layer. The values in each compass direction represent frequencies that can be converted into conditional probabilities (P) for the presence or absence of the same class given that the neighboring configuration is known.

The spatial distribution of land cover types will vary across spatial scales because some types will be smaller than the pixels and therefore not included in the analysis or not visible. The distribution will also be influenced by numerous factors. These include biological competition for nutrients, moisture, and sunlight; the formation of symbiotic relationships; the availability of seed and other propagules; and topography. As a result of these influences, certain species will colonize specific sites more readily than others. Depending on the states of these variables and their interactions, species distributions will evolve into patterns that range from clustered (with a high and positive spatial autocorrelation) to randomly dispersed or evenly distributed (with a high and negative spatial autocorrelation), and the degree of clustering will depend strongly on the scale of observation. Elkie and Rempel (2001) used spatial lacunarity methods (mathematical tools to describe the texture of fractal features or the distribution of "holes" in a map) to identify boreal landscape textures along a continuum of patch dispersion in Ontario, a method that uses windows of increasing size, with clustering reassessed at each new size.

Thematic complexity results from a number of land cover types being mapped in some configuration within a landscape. Because that configuration can be measured and mapped at a range of scales, this can lead to as many representations as there are measurements and scales, potentially creating a different representation at each scale. If the measurement framework does not align with the resolution of the actual field data, then mixtures of classes may occur within individual cells in gridded or raster data formats. To identify these mixtures, various forms of mixture analysis and various object-based classifications can be used (Fernandez-Manso et al. 2009; see also Chapter "Fuzzy Classification of Vegetation for Ecosystem Mapping" of this book). As the number of thematic classes increases, the possible number of spatial configurations will also increase, and landscapes will appear more complex than the same landscape with fewer classes. Such differences in thematic complexity make comparisons of patterns between disparate representations more complex due to the potential introduction of a bias that results from the difference in resolutions. To illustrate this point, consider a scenario in which pre-wildfire data on the land cover types and their distribution is available at a spatial resolution of 30 m for 21 cover types. If the post-disturbance landscape is mapped at a spatial resolution of 4 m and with 14 cover types, the two configurations cannot be directly compared. To allow such a comparison, both maps will need to be brought to a common spatial resolution and will need to use the same thematic classes.

Probability of Fire and Other Phenomena

Statistical mapping, which focuses on the probabilities of certain events or properties, has received much less attention than traditional land cover mapping due to its abstract nature (see Chapters "Regression Tree Modeling of Spatial Pattern and

Process Interactions" and "Mapping the Abstractions of Forest Landscape Patterns"). The thematic representations in land cover maps are highly intuitive, since the labels on the map match the physical conditions on the ground. In contrast, statistical maps range from descriptive summaries to probabilities, or to frequency and other distributions measured at or near the sites where labels appear in the map; as a result, they can be far more abstract than thematic representations. Perhaps the easiest way to visualize this is to consider a statistical map layer that summarizes the average annual temperature in each cell of a regular grid. If you were to visit that cell, it's unlikely that the temperature displayed for that cell would match the temperature during your visit; however, if you visited sufficiently often, you would eventually arrive on a day with the right temperature. Here, the mean temperature is a statistical summary that describes an expectation over time and not the actual value at a specific point in time.

Sometimes fire maps are produced by spatially explicit models (Parisien et al. 2011), and the mapped values represent the probabilities of a particular outcome from a process (e.g., the risk of a fire) rather than precise states at a specific time (Araya et al. 2016b). Dickson et al. (2006) used this approach to map fire probabilities in northern Arizona, USA. Liu et al. (2013) used this approach to assess wildfire susceptibility and risk under different fuel management scenarios within a basin in northeastern China. In these examples, the focus was on a map that depicted statistical probabilities to characterize the potential of some event occurring, rather than recording the absolute conditions at any given location at the time of mapping. Such representations can identify a landscape's susceptibility to disturbance or the likelihood of disturbance occurring given a suite of underlying factors (e.g., for fires, the Canadian Fire Weather Index system; http://cwfis.cfs.nrcan.gc.ca/background/summary/fwi). As such, they provide useful data for land management or the development of simulation scenarios.

Standardized Depiction of Wildfires as Discrete Complex Objects

As a collection of interacting dynamic processes, wildfire is a driver of landscape change that results in new spatial patterns. To begin studying, understanding, and mapping these processes, their complexity, and their results, we will propose standardized terminology for wildfires and aspects of the resulting modified landscape. Figure 10 illustrates the effects of a wildfire at a given point in space and time, and in the rest of this section, we refer to this illustration to support our proposal of standard terms for the various aspects of the affected area.

A fire "event" includes all burned areas (across a range of disturbance intensities) caused by a specific disturbance that originates from an external ignition source (e.g., lightning), including all unburned areas completely internal to the footprint. Internal ignitions, originating from the burning fire itself, could create new fires termed "spot fires." These may merge with the main body of the fire or remain disjoint. In some

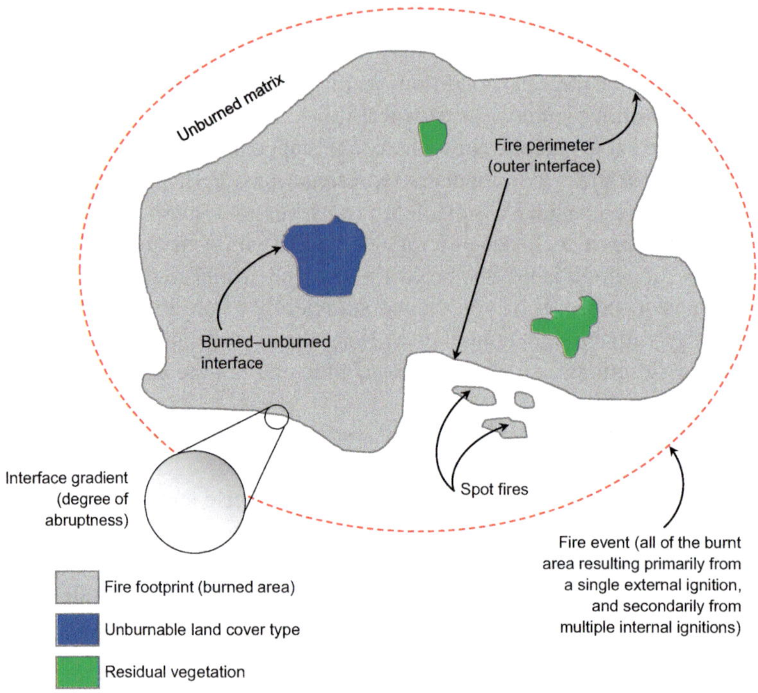

Fig. 10 Illustration of a typical wildfire event, which serves as a reference for defining the key descriptive terms

instances, multiple fire events may merge, thereby creating a single fire footprint and eliminating distinct inter-event boundaries. A fire event develops from the set of physical and direct short-term ecological interactions that affect the landscape during a period of burning. The map of a wildfire disturbance includes the area burned and all enclosed residual (unburned or partially burned) patches of vegetation along with all nearby spot fires that are attributed to the same burning event. In this context, the unburned background area in which the fire occurred represents the "matrix" (a term from the landscape ecology literature that describes the dominant land cover type in an area). The matrix is separated from the burned area by the fire perimeter—the interface between the burned and unburned land cover classes. The interface gradient represents how abruptly conditions change between the two sides of the interface; as we noted earlier, the gradient can occur over a short distance (crisp) or a longer distance (fuzzy). Perimeters can be either external, in which case they represent the outermost edge of the main fire polygon or polygons (possibly including one or more spot fires), or they can be internal, in which case they surround unburned areas within the burned area (including residual vegetation and unburnable land cover types).

The footprint can be visualized as the geometric union of the bounding polygons of all identified burned areas at the specified mapping scale (including all interior unburned patches and residuals) and all spot fire areas. Thus, the footprint is an instantaneous concept and is not necessarily spatially contiguous, since some areas may escape burning (e.g., if they are protected by an unburnable obstacle such as a lake) and since wildfire may also spread by spotting. Thus, footprints represent one or more spatial units that preserve the two-dimensional record of a single fire through its impacts on the landscape. This footprint (or its components) can be further described and quantified based on various spatial and geometric properties, such as its area, boundary length, pattern, or shape.

The footprint provides a means by which we can represent the extent of a complex process that occurred in both horizontal and vertical dimensions within a defined time period as a collection of one or more shapes in a map. The specific details of how these shapes are recorded will depend on whether data is represented in raster or vector format, and on the spatial scale used to construct the representation. Each decision regarding the representation will influence the final form of the footprint.

The choice of a classification scheme for the landscape attributes will influence the spatial patterns within the landscape, and hence the complexity of the representation. For example, a smaller number of land cover types to be mapped will lead to a simpler representation than if many land cover types were mapped for the same landscape. Using more classes permits division of the landscape into an increased number of smaller but more homogeneous areas, but related classes differ by a smaller amount than if there were fewer classes. Thus, the increased classification complexity is inevitably linked to greater potential complexity of the depiction of the landscape's composition and configuration and can help determine the scale required (e.g., the MMU) to adequately represent the landscape's complexity. Researchers commonly develop hierarchically nested classification schemes, such as the one shown in Table 2; these hierarchical levels can be collapsed or expanded to support assessments at different thematic scales (Baldwin et al. 2004; Remmel

Table 2 An example of a hierarchically nested land classification scheme with coarser (more inclusive) thematic classes at the left and increasingly finer (less inclusive) classes at the right

Level 1	Level 2	Level 3	Level 4
Land	Bedrock and non-vegetated		
	Burned	Completely burned	
		Partially burned	
		Old burn	
Vegetated	Forest	Coniferous	Dense conifer
			Sparse conifer
		Deciduous	
	Non-forest	Shrub	Tall shrub
			Low shrub
		Wetland	Open wetland
			Treed wetland
			Marsh
Water			
Other	Cloud and shadow		

et al. 2005; Nadeau and Englefield 2006; Remmel and Csillag 2006) or to draw equivalencies between different classification regimes (Remmel et al. 2005).

Fine divisions of thematic classes (e.g., vegetation density classes that result from canopy layering) can be difficult to discern from aerial photos or remote-sensing imagery, and at certain spatial scales may be impossible to discern; thus, some levels of thematic classification may be inappropriate if they are not supported by the image resolution. As the thematic scale or complexity changes, the classification's ability to reveal or mask trends and states also changes. In fire disturbance scenarios, the potential abruptness of the footprint's perimeter or the width of ecotones is particularly affected. Recent improvements to remote sensing technology have seen spatial resolutions increasing from kilometers to tens of meters, and now even tens of centimeters. Although these images provide beautiful contextual layers for use in interpretive mapping and offer the potential for detailed mapping of spatial and thematic classes, the increased heterogeneity they reveal can complicate the analysis more than the results justify. Higher spatial resolutions result in vastly more pixels and variability to contend with, image analysis algorithms need to be adjusted to handle this increased complexity, and the level of detail is often more than is required for basic forest landscape mapping tasks. Increased spatial detail also increases the need to quantify fragmentation or the porosity of landscape patches, which results from the ability to detect finer gaps in landscape classes that would appear homogeneous at coarser mapping resolutions. Furthermore, the assessment of accuracy at a very high spatial resolution becomes increasingly difficult compared to more general estimates of the composition and distribution of land cover classes.

The Future of Mapping Wildfires

Mapping and the related cartographic techniques have undergone great changes during the past two centuries. In the early phases, cartographic development was largely driven by human understanding of the Earth's surface and its shape. This eventually gave way to improved technology for measuring distances and angles, driven largely by the needs of maritime navigation and the ability to measure latitudes and longitudes at sea. However, truly radical changes began with technological advances in photography, starting in the 1800s, and twentieth-century digital tools for data acquisition and map representation have offered many innovative means of data acquisition, storage, encoding, and presentation. Although technology continues to drive development, the new methods and tools are also pushing cartographers to develop new ways of creating and presenting maps, including 3D, dynamic, and interactive maps; much of the need for these new modes of map construction and delivery has been created by new research questions and the needs for improved capture and use of spatial data. These forces will lead to exciting new tools in the years ahead.

Much of what is happening in the realm of forest mapping encompasses new sensors and technologies for landscape imaging and data capture, the assessment of sources of uncertainty and of the accuracy of derived products, and the integration of data from multiple sources to produce interconnected and dynamic maps of complex ecosystems and their functioning. Other researchers are working to make data and maps available online through portals that facilitate collaboration in the creation of huge geographical databases and improve our ability to interact with maps.

The importance of selecting an appropriate scale or MMU and its impact on the construction of maps of complex phenomena that cross multiple scales remains an area that requires continued attention. The constant struggle between increasingly available high-resolution data and complexity of handling and interpreting that information is likely to spur continued studies on how to identify the optimal mapping scale.

Accuracy Assessment in Remote Regions

Accuracy, uncertainty, and error assessment are central problems that must be solved to establish the credibility of scientific results and claims. These problems have been mitigated using well-developed techniques such as error matrices (Congalton 1991), yet they have not yet been solved; doing so is one of the more difficult and often-neglected aspects of geospatial work, particularly for remote locations. When access to sites is difficult, as is often the case with wildfires that burn in the far northern boreal forests, field validation efforts are both costly and difficult due to restrictions on reaching the site and exploring it once one arrives at the site. In such situations, alternative means for quantifying accuracy and uncertainty are desirable. Much attention has been turned to using data with a higher spatial resolution or a greater level of detail as a source of validation information. Unfortunately, this approach can be somewhat circular in terms of the logic, since the more detailed data would also require validation, negating its benefits.

Work by Mitchell et al. (2008) investigated the use of typically ignored statistical data from traditional satellite image classifications, such as the probability of a pixel belonging to land cover classes other than the most probable one, to assess the uncertainty of assigning class labels to pixels, and found that these data can be incorporated in an uncertainty assessment when true validation data is not available. Although additional statistical simulation methods exist (Csillag et al. 2006) to estimate confidence intervals around summary metrics used in classification, tools that focus on pattern variability (Remmel 2009) provide additional methods that can be implemented when it is necessary to compare data for the identical location from two periods. However, a comprehensive accuracy assessment method that would be applicable across spatial and temporal scales does not exist, and it would be valuable to develop such a method or perhaps a small suite of methods that are optimized for specific situations.

The quantification of spatial patterns to allow statistical comparisons among multiple landscapes has been studied from many angles, by many researchers, over several decades (e.g., O'Neill et al. 1988; Foody 2007; Ruiz et al. 2012a, 2012b), but no all-encompassing method has yet emerged. The ecological literature describes hundreds of metrics (McGarigal and Marks 1995) that assess specific aspects of spatial patterns, generally at patch, cover class, and landscape scales, and subsets of these metrics are often used as a suite to describe specific traits of certain landscapes (Riitters et al. 1995). Although these metrics can be excellent diagnostic indicators, a lack of knowledge of their underlying statistical properties makes it difficult to assess their value. Some more recent work, and software to facilitate the analysis, has begun to mitigate these limitations (Remmel and Csillag 2003; Remmel and Fortin 2013). There have also been attempts to assess the composition and configuration of landscape elements using common measurement units—"bits"—from information theory (Remmel and Csillag 2006). These units have been used to assess uncertainty in patterns using combinatorics computed from coincidence matrices to quantify pattern uncertainty (Remmel 2009), or to provide uncertainty maps along with a mapped variable (Wang et al. 2009), yet widespread adoption of these methods has not been achieved due to the complexity and effort involved.

Efforts to map spatial statistics and summarize pattern metrics (see Chapters "Regression Tree Modeling of Spatial Pattern and Process Interactions" and "Mapping the Abstractions of Forest Landscape Patterns" of this book) are providing visualization tools for complex data. New measurement and representation frameworks based on morphological pattern elements (Vogt et al. 2007) are spurring new conceptualizations in the domain of pattern analysis. We believe that the future of spatial pattern analysis and assessment lies in the fusion and seamless scalability of methods that measure and statistically compare spatial patterns across scales and produce spatial summaries of the comparisons. The ability to adjust the boundary abruptness, fuzziness of attributes, and permitted positional accuracy across a range of scales, combined with an improved understanding of the correlations with explanatory factors, will permit more powerful hypothesis testing than is currently possible with the current unstructured approach, which is based on an ad hoc and nonintegrated combination of methods for each research project.

Traditional forest inventory methods have assumed the mapping of measured variables onto a flat plane. Whether the mapped features represent forest stands, disturbed areas, or residual vegetation, there is an increasing trend to include a vertical component to the assessment and to explore new forms of data presentation, such as virtual reality or three-dimensional visualization. Forest landscapes are inherently complex in the horizontal plane, but the vertical dimension is likely to prove equally complex given how it evolves over time even in the absence of disturbance. However, coping with this complexity can reveal interesting aspects of biomass allocation, the distributions of taxonomic or functional groups, or structural aspects of forest stands (Ko et al. 2013).

Although 3D data can be obtained using photogrammetric methods such as traditional examination of stereophotos, more recent methods such as structure from motion analyses (Mancini et al. 2013) represent rejuvenated photogrammetric methods that allow extraction of LiDAR-style point cloud data (the *x*, *y*, and *z* coordinates of typically reflected laser pulses) from digital stereo-images. These point clouds can be further processed in a number of ways to extract vertical profiles of point density that let researchers define crown and stem positions and even identify a tree's species. This approach can also unlock information regarding tree heights, biomass, branching structures, and other biophysical metrics that improve the quality of forest inventory data (Lim et al. 2003). More recent developments in this technology take advantage of the full-waveform data obtained from interactions of the laser pulse with the forest canopy (see Chapter "Airborne LiDAR Applications in Forest Landscapes" of this book for details). Although this approach is potentially more informative, the dramatically increased data volumes lead to a requirement for more powerful computers with high storage capacity, and LiDAR surveys are constrained (for the moment) by reduced (horizontal) spatial resolution. Manipulation, handling, and visualization of these newer data types also require specialized software and data structures that let researchers effectively read, process, and display the vast quantities of data.

Conversion of point clouds into a series of rectangular solids or prisms called "voxels' (i.e., "volumetric pixels"; Wu et al. 2013) has let researchers simplify forest structures into standardized shapes (e.g., cones for spruce canopies, cylinders for tree stems, spheres on a stick to represent deciduous canopies and stems). This approach allows the resulting 3D models to describe forest aspects such as the vertical canopy distribution or even the branching structures of plants (Prusinkiewicz and Lindenmayer 1990). Similarly, the popularity of images with very high spatial resolution from unmanned aerial vehicles (also called "drones") is reinvigorating the use of digital photogrammetric methods; as a result, 3D topographic reconstructions, continuous orthomosaics, and hybrid products that fuse image data with LiDAR data are becoming more common (Alonso-Benito et al. 2016). These developments are producing new forms of mapping that resemble virtual reality more than traditional 2D cartography.

Landscape Persistence

Dynamic landscapes are often assessed to detect changes through time and across space. Whether these changes are brought about by wildfires, harvesting, or any other disturbance or regeneration process, it is the change that is typically detected, measured, quantified, and mapped. In contrast, the areas that are (as yet) unaffected by change exhibit "persistence," a state that is easily ignored due to the seeming lack of activity. However, a growing body of literature focuses on the characteristics of spatial persistence (Pontius et al. 2004). In particular, researchers have studied the stability of landscapes to describe the spatial legacies that have been imprinted on fire-dominated landscapes (James et al. 2007), many of which have persisted for hundreds of years and may serve as ideal locations for conservation efforts (Rayfield et al. 2007). Given that change is a minor factor in landscapes dominated by persistence (Flamenco-Sandoval et al. 2007), these inverted notions of landscape behavior and resilience are likely to attract increasing interest within the research community. The concepts of ecosystem resilience, recovery from disturbance, rates of change, and fluctuations within natural ranges of variability will all be combined to characterize landscapes better than if we were to only examine extreme changes.

Hierarchical Data Formats for Capturing Scale Effects

Forests form horizontally, vertically, and temporally complex and dynamic landscapes that can be conceptualized at multiple scales for competing, cooperating, and interacting reasons. As a result, forest landscapes have been and will continue to be studied across multiple scales. We foresee a substantive change in how research is conducted because of the need to account for the influence of scale. Rather than focusing on only one or a small subset of scales, analyses should increasingly move toward handling the full spectrum of scales (see Quattrochi and Goodchild 1997), as has been done in studies based on wavelets (Csillag and Kabos 2002) or quadtrees (Csillag and Kabos 1996). Some novel approaches to pattern analysis across spatial and thematic scales, as well as external attribute scales (e.g., grouping maps by an external factor such as regulatory jurisdiction) exist (Remmel and Csillag 2006), but are not yet ubiquitously implemented. As another example, Remmel (2015) developed the tools to run the *ShrinkShape* algorithm for decomposing areas, perimeters, and complexity of shapes across multiple representational scales.

The fusion of data from multiple sources, at multiple scales, will permit integrative data analyses, which will dominate the coming decades. The mapping of such complex and multidimensional results will keep cartographers occupied finding new means by which data and results can be processed, presented, consumed, and interacted with by the growing variety of audience who want to consume this information in real time, often on mobile application platforms such as smartphone-based GPS navigation systems, and who increasingly want to access the information remotely via Internet-based queries.

Landscape complexity forces us to deal with not only different scales of measurement, but also terminology adjustments to reflect changes to the context, scope, focus, and measurable details. At one scale, wildfire disturbances are merely points scattered within an extensive landscape. However, as one zooms in on that landscape, those points begin to evolve into patches that have an area and a perimeter, and begin to exhibit emergent properties that the individual points could not display. The point patterns therefore morph into more sophisticated spatial patterns that can be described using terminology developed to describe the patchy mosaics that form the landscape structure. However, the focus can also grow closer, allowing individual patches to be characterized as complex objects that comprise multiple contiguous and unconnected parts. Each part would then possess its own spatial, thematic, and temporal properties that permit more detailed characterization of patterns within the footprint of disturbances such as wildfire. This idea can be further extended to individual stands, trees, or even finer scales. This complexity raises an important philosophical and practical question: What is the optimal scale at which to study forest landscapes? Although it is not wrongs to suggest that the optimal scale depends on the purpose of the analysis, the potential of multi-scalar solutions suggests that the correct answer may become "all scales at the same time."

References

Albini FA, Alexander ME, Cruz ME (2012) A mathematical model for predicting the maximum potential spotting distance from a crown fire. Int J Wildl Fire 21(5):609–627

Alonso-Benito A, Arroyro LA, Arbelo M, Hernández-Leal P (2016) Fusion of WorldView-2 and LiDAR data to map fuel types in the Canary Islands. Rem Sens 8:669

Araya YH, Remmel TK, Perera AH (2016a) What governs the presence of residual vegetation in boreal wildfires? J Geogr Syst 18(2):159–181

Araya YH, Remmel TK, Perera AH (2016b) Spatially explicit prediction of residual vegetation patch occurrence within boreal wildfires. Int J Geoinform 12(3):1–15

Baker WL, Honaker JJ, Weisberg PJ (1995) Using aerial-photography and GIS to map the forest–tundra ecotone in Rocky Mountain National Park, Colorado, for global change research. Photogramm Eng Rem Sens 61:313–320

Baldwin DJ, Weaver K, Schnekenburger F, Perera AH (2004) Sensitivity of landscape pattern indices to input data characteristics on real landscapes: implications for their use in natural disturbance emulation. Landsc Ecol 19:255–271

Boots B (2001) Using local statistics for boundary characterization. GeoJournal 53:339–345

Bourgeau-Chavez LL, Harrell PA, Kasischke ES, French NHF (1997) The detection and mapping of Alaskan wildfires using a spaceborne imaging radar system. Int J Rem Sens 18(2): 355–373

Bourgeau-Chavez LL, Kasischke ES, Brunzell S, Mudd JP, Tukman M (2002) Mapping fire scars in global boreal forests using imaging radar data. Int J Rem Sens 23(20):4211–4234

Breidenbach J, Naesset E, Lien V, Gobakken T, Solberg S (2010) Prediction of species specific forest inventory attributes using a nonparametric semi-individual tree crown approach based on fused airborne laser scanning and multispectral data. Rem Sens Environ 114(4):911–924

de Casanave JL, Pelotto JP, Protomastro J (1995) Edge-interior differences in vegetation structure and composition in a Chaco semi-arid forest, Argentina. For Ecol Manage 72:61–69

Chandler C, Cheney P, Thomas P, Trabaud L, Williams D (1983) Fire in forestry: volume II. Forest fire management. Wiley-Interscience, New York, NY

Congalton RG (1991) A review of assessing the accuracy of classifications of remotely sensed data. Rem Sens Environ 37:35–46

Cracknell AP (1998) Synergy in remote sensing—what's in a pixel? Int J Rem Sens 19(11): 2025–2047

Cromley RG (1991) Hierarchical methods of line simplification. Cartogr Geogr Inf Syst 18(2): 125–131

Csillag F, Kabos S (1996) Hierarchical decomposition of variance with applications in environmental mapping based on satellite images. Mathem Geol 28(4):385–405

Csillag F, Kabos S (2002) Wavelets, boundaries, and the spatial analysis of landscape pattern. Ecoscience 9(2):177–190

Csillag F, Kabos S, Remmel TK (2006) Confidence in coincidence. Int J Rem Sens 27(6):1269–1276

Dickson BG, Prather JW, Xu YG, Hampton HM, Aumack EN, Sisk TD (2006) Mapping the probability of large fire occurrence in northern Arizona, USA. Landsc Ecol 21:747–761

Dunn R, Harrison AR, White JC (1990) Positional accuracy and measurement error in digital databases of land use: an empirical study. Int J Geogr Inf Syst 4(4):385–398

Elkie PC, Rempel RS (2001) Detecting scales of pattern in boreal forest landscapes. For Ecol Manage 147:253–261
Farmer E, Jones SD, Deakin RE (2011) Derivation of a resilient polygon centroid for natural resource management applications. Int J Geo Info Sci 25(9):1545–1558
Fernandez-Manso O, Quintano C, Fernandez-Manso A (2009) Combining spectral mixture analysis and object-based classification for fire severity mapping. Invest Agraria-Sist Recursos Forestales 18:296–313
Finney MA, Cohen JD, McAllister SS, Jolly WM (2013) On the need for a theory of wildland fire spread. Int J Wildl Fire 22(1):25–36
Flamenco-Sandoval A, Ramos MM, Masera OR (2007) Assessing implications of land-use and land-cover change dynamics for conservation of a highly diverse tropical rain forest. Biol Conserv 138:131–145
Foody GM (2007) Map comparison in GIS. Progr Phys Geogr 31:439–445
Fortin MJ (1994) Edge detection algorithms for two dimensional ecological data. Ecology 75(4):956–965
Fortin MJ (1999) Effects of quadrat size and data measurement on the detection of boundaries. J Veg Sci 10:43–50
Fraser RH, Li Z, Cihlar J (2000) Hotspot and NDVI differencing synergy (HANDS): a new technique for burned area mapping over boreal forest. Rem Sens Environ 74:362–376
French NHF, Kasischke ES, Bourgeau-Chavez LL, Berry D (1995) Mapping the location of wildfires in Alaskan boreal forests using AVHRR imagery. Int J Wildl Fire 5(2):55–62
Gillis MD, Leckie DG (1996) Forest inventory update in Canada. For Chron 72(2):138–156
Gonzalez RC, Woods RE (2008) Digital image processing, 3rd edn. Pearson/Prentice Hall, Upper Saddle River, NJ
Hall KR, Maruca SL (2001) Mapping a forest mosaic—a comparison of vegetation and bird distributions using geographic boundary analysis. Plant Ecol 156(1):105–120
Heath M, Sarkar S, Sanocki T, Bowyer K (1998) Comparison of edge detectors: a methodology and initial study. Comp Vision Image Understand 69(1):38–54
Hessl A, Miller J, Kernan J, Keenum D, McKenzie D (2007) Mapping paleo-fire boundaries from binary point data: comparing interpolation methods. Prof Geogr 59:87–104
Hirata Y, Takahashi T (2011) Image segmentation and classification of Landsat Thematic Mapper data using a sampling approach for forest cover assessment. Can J For Res 41(1):35–43
Holden ZA, Smith AMS, Morgan P, Rollins MG, Gessler PE (2005) Evaluation of novel thermally enhanced spectral indices for mapping fire perimeters and comparisons with fire atlas data. Int J Rem Sens 26:4801–4808
Jakubowski MK, Li W, Guo Q, Kelly M (2013) Delineating individual trees from Lidar data: a comparison of vector- and raster-based segmentation approaches. Rem Sens 5:4163–4186
James PMA, Fortin MJ, Fall A, Kneeshaw DD, Messier C (2007) The effects of spatial legacies following shifting management practices and fire on boreal forest age structure. Ecosystems 10:1261–1277
Johnson EA (1992) Fire and vegetation dynamics: studies from the North American boreal forest. Cambridge University Press, Cambridge
Johnson EA, Miyanishi K (eds) (2001) Forest fires: behavior and ecological effects. Academic Press, San Diego, CA
Kasischke ES, French NHF (1995) Locating and estimating the areal extent of wildfires in Alaskan boreal forests using multiple-season AVHRR NDVI composite data. Rem Sens Environ 51(2):263–275
Kasischke ES, French NHF, Harrell P, Christensen NL, Ustin SL, Barry D (1993) Monitoring of wildfires in boreal forests using large-area AVHRR NDVI composite image data. Rem Sens Environ 45(1):61–71
Keeley JE (2009) Fire intensity, fire severity and burn severity: a brief review and suggested usage. Int J Wildl Fire 18:116–126
Kettig RL, Landgrebe DA (1976) Classification of multispectral image data by extraction and classification of homogeneous objects. IEEE Trans Geosci Electron 14:19–26

Kim JS, Cho HS (1994) A fuzzy logic and neural network approach to boundary detection for noisy imagery. Fuzzy Sets Syst 65:141–159
Knight JF, Lunetta RS (2003) An experimental assessment of minimum mapping unit size. IEEE Trans Geosci Rem Sens 41(9):2132–2134
Ko C, Sohn G, Remmel TK (2013) Tree genera classification with geometric features from high-density airborne LiDAR. Can J Rem Sens 39(s1):S73–S85
Leckie DG, Gillis MD (1995) Forest inventory in Canada with emphasis on map production. For Chron 71(1):74–88
Lim K, Treitz P, Baldwin K, Morrison I, Green J (2003) Lidar remote sensing of biophysical properties of tolerant northern hardwood forests. Can J Rem Sens 29(5):658–678
Liu Z, Yang J, He HS (2013) Studying the effects of fuel treatment based on burn probability on a boreal forest landscape. J Environ Manage 115:42–52
Mancini F, Dubbini M, Gattelli M, Stecchi F, Fabbri S, Gabbianelli G (2013) Using unmanned aerial vehicles (UAV) for high-resolution reconstruction of topography: the structure from motion approach on coastal environments. Rem Sens 5:6880–6898
Mandelbrot B (1967) How long is the coast of Britain? Statistical self-similarity and fractional dimension. Science 156(3775):636–638
McAlpine RS, Wotton BM (1993) The use of fractal dimension to improve wildland fire perimeter predictions. Can J For Res 23(6):1073–1077
McGarigal K, Marks BJ (1995) FRAGSTATS: spatial pattern analysis program for quantifying landscape structure. Gen Tech Rep PNW-GTR-351. USDA For Serv, Pacific Northwest Res Stn, Portland, OR
Merino-de-Miguel S, Huesca M, Gonzalez-Alonso F (2010) MODIS reflectance and active fire data for burn mapping and assessment at regional level. Ecol Model 221:67–74
Mitchell SW, Remmel TK, Csillag F, Wulder M (2008) Distance to second cluster as a measure of classification confidence. Rem Sens Environ 112(5):2615–2626
Morgan P, Hardy CC, Swetnam TW, Rollins MG, Long DG (2001) Mapping fire regimes across time and space: understanding coarse and fine-scale fire patterns. Int J Wildl Fire 10: 329–342
Murphy KA, Reynolds JH, Koltun JM (2008) Evaluating the ability of the differenced normalized burn ratio (dNBR) to predict ecologically significant burn severity in Alaskan boreal forests. Int J Wildl Fire 17(4):490–499
Nadeau LB, Englefield P (2006) Fine-resolution mapping of wildfire fuel types for Canada: fuzzy logic modeling for an Alberta pilot area. Environ Monit Assess 120:127–152
Oden NL, Sokal RR, Fortin MJ, Goebl H (2010) Categorical wombling: detecting regions of significant change in spatially located categorical variables. Geogr Anal 25(4):315–336
Oliva P, Martin P, Chuvieco E (2011) Burned area mapping with MERIS post-fire image. Int J Rem Sens 32:4175–4201
O'Neill RV, Krummel JR, Gardner RH, Sugihara G, Jackson B, DeAngelis DL, Milne BT, Turner MG, Zygmunt B, Christensen SW, Dale VH, Graham RL (1988) Indices of landscape pattern. Landsc Ecol 1(3):153–162
Ostapowicz K, Vogt P, Riitters KH, Kozak J, Estreguil C (2008) Impact of scale on morphological spatial pattern of forest. Landsc Ecol 23:1107–1117
Parisien MA, Parks SA, Miller C, Krawchuk MA, Heathcott M, Moritz MA (2011) Contributions of ignitions, fuels and weather to the spatial patterns of burn probability of a boreal landscape. Ecosystems 14:1141–1155
Park W, Yu K (2011) Hybrid line simplification for cartographic generalization. Pattern Recogn Lett 32(9):1267–1273
Perera AH, Buse LJ (2014) Ecology of wildfire residuals in boreal forests. Wiley, Hoboken, NJ
Perera AH, Buse LJ, Weber MG (eds) (2004) Emulating natural forest landscape disturbances. Columbia University Press, New York, NY
Peters DPC, Groffman PM, Nadelhoffer KJ, Grimm NB, Collins SL, Michener WK, Huston MA (2008) Living in an increasingly connected world: a framework for continental-scale environmental science. Front Ecol Environ 6(5):229–237

Philibert MD, Fortin MJ, Csillag F (2008) Spatial structure effects on the detection of patches boundaries using local operators. Environ Ecol Stat 15:447–467

Pontius RG, Shusas E, McEachern M (2004) Detecting important categorical land changes while accounting for persistence. Agric Ecosyst Environ 101(2–3):251–268

Pourghasemi HR, Beheshtirad M, Pradhan B (2016) A comparative assessment of prediction capabilities of modified analytical hierarchy process (M-AHP) and Mamdani fuzzy logic models using Netcad-GIS for forest fire susceptibility mapping. Geomatics Nat Hazard Risk 7(2):861–885

Pozo D, Olmo FJ, Aladosarboledas L (1997) Fire detection and growth monitoring using a multitemporal technique on AVHRR mid-infrared and thermal channels. Rem Sens Environ 60:111–120

Prusinkiewicz P, Lindenmayer A (1990) The algorithmic beauty of plants. Springer-Verlag, New York, NY

Quattrochi DA, Goodchild MF (1997) Scale in remote sensing and GIS. CRC Press, Boca Raton, FL

Rayfield BR, James PMA, Fall A, Fortin MJ (2007) Comparing static versus dynamic protected areas in dynamic boreal ecosystems. Biol Conserv 141(2):428–449

Remmel TK (2009) Investigating global and local categorical map configuration comparisons based on coincidence matrices. Geogr Anal 41(1):113–126

Remmel TK (2015) ShrinkShape2: a FOSS toolbox for computing rotation-invariant shape spectra for characterizing and comparing polygons. Can Geogr 59(4):532–547

Remmel TK, Csillag F (2003) When are two landscape pattern indices significantly different? J Geogr Syst 5(4):331–351

Remmel TK, Csillag F (2006) Mutual information spectra for comparing categorical maps. Int J Rem Sens 27(7):1425–1452

Remmel TK, Fortin MJ (2013) Categorical, class-focused map patterns: characterization and comparison. Landsc Ecol 28(8):1587–1599

Remmel TK, Perera AH (2001) Fire mapping in a northern boreal forest: assessing AVHRR/NDVI methods of change detection. For Ecol Manage 152(1–3):119–129

Remmel TK, Perera AH (2002) Accuracy of discontinuous surfaces: a case study of boreal forest fires. Int J Geogr Inf Sci 16(3):289–300

Remmel TK, Perera AH (2009) Mapping natural phenomena: boreal forest fires with non-discrete boundaries. Cartographica 44:274–288

Remmel TK, Csillag F, Mitchell S, Wulder MA (2005) Integration of forest inventory and satellite imagery: a Canadian status assessment and research issues. For Ecol Manage 207:405–428

Richardson LF (1961) The problem of contiguity: an appendix to statistics of deadly quarrels. In: General systems: yearbook of the society for the advancement of general systems theory, vol 61. Ann Arbor, Michigan, pp 139–187

Riitters KH, O'Neill RV, Hunsaker CT, Wickham JD, Yankee DH, Timmins SP, Jones KB, Jackson BL (1995) A factor analysis of landscape pattern and structure metrics. Landsc Ecol 10(1):23–39

Rothermel RC (1972) A matpredicting fire spread in wildland fuels. USDA For Serv, Intermountain For Range Exp Stn, Ogden, UT. Research Paper INT-116

Rouget M, Cowling RM, Lombard AT, Knight AT, Kerley GIH (2006) Designing large-scale conservation corridors for pattern and process. Conserv Biol 20(2):549–561

Ruiz M, Lopez F, Paez A (2012a) Comparison of thematic maps using symbolic entropy. Int J Geogr Inf Sci 26:413–439

Ruiz JAM, Riano D, Arbelo M, French NHF, Ustin SL, Whiting ML (2012b) Burned area mapping time series in Canada (1984–1999) from NOAA-AVHRR LTDR: a comparison with other remote sensing products and fire perimeters. Rem Sens Environ 117:407–414

Saura S (2002) Effects of minimum mapping unit on land cover data spatial configuration and composition. Int J Rem Sens 23(22):4853–4880

Schroeder TA, Wulder MA, Healey SP, Moisen GG (2011) Mapping wildfire and clearcut harvest disturbances in boreal forests with Landsat time series data. Rem Sens Environ 115:1421–1433

Shi WZ, Cheung CK (2006) Performance evaluation of line simplification algorithms for vector generalization. Cartogr J 43(1):27–44

Siegert F, Hoffmann AA (2000) The 1998 forest fires in East Kalimantan (Indonesia): a quantitative evaluation using high resolution, multitemporal ERS-2 SAR images and NOAA-AVHRR hotspot data. Rem Sens Environ 72:64–77

Smith A (2010) Image segmentation scale parameter optimization and land cover classification using the Random Forest algorithm. J Spatial Sci 55(1):69–79

Spichtinger N, Damoah R, Eckhardt S, Forster C, James P, Beirle S, Marbach T, Wagner T, Novelli PC, Stohl A (2004) Boreal forest fires in 1997 and 1998: a seasonal comparison using transport model simulations and measurement data. Atmos Chem Phys 4(7):1857–1868

Sullivan AL (2009) Wildland surface fire spread modelling, 1990–2007. 3: Simulation and mathematical analogue modes. Int J Wildl Fire 18:387–403

Sun J, Southworth J (2013) Remote sensing-based fractal analysis and scale dependence associated with forest fragmentation in an Amazon tri-national frontier. Rem Sens 5:454–472

Turner MG (2005) Landscape ecology: what is the state of the science? Ann Rev Ecol Evol Syst 36:319–344

Vogt P, Riitters KH, Estreguil C, Kozak J, Wade TG, Wickham JD (2007) Mapping spatial patterns with morphological image processing. Landsc Ecol 22:171–177

van Wagner CE (1969) A simple fire-growth model. For Chron 45:103–104

Wang F, Hall GB (1996) Fuzzy representation of geographical boundaries in GIS. Int J Geogr Inf Syst 10(5):573–590

Wang G, Oyana T, Zhang M, Adu-Prah S, Zeng S, Lin H, Se J (2009) Mapping and spatial uncertainty analysis of forest vegetation carbon by combining national forest inventory data and satellite images. For Ecol Manage 258:1275–1283

Whelan RJ (1995) The ecology of fire. Cambridge Studies in Ecology, Cambridge Univ. Press, Cambridge

Wu ST, Marquez MRG (2003) A non-self-intersection Douglas-Peucker algorithm. In: Proceedings of the XVI Brazilian symposium on computer graphics and image processing (SIBGRAPI'03) IEEE computer society. Unpaginated http://ieeexplore.ieee.org/document/1240992/authors

Wu B, Yu B, Yue W, Shu S, Tan W, Hu C, Huang Y, Wu J, Liu H (2013) A voxel-based method for automated identification and morphological parameters estimation of individual street trees from mobile laser scanning data. Rem Sens 5:584–611

Yu J, Chen D, Lin Y, Ye S (2017) Comparison of linear and nonlinear spectral unmixing approaches: a case study with multispectral TM imagery. Int J Remote Sens 38(3):773–795

Zhang J, Stuart N (2001) Fuzzy methods for categorical mapping with image-based land cover data. Int J Geogr Inf Sci 15(2):175–195

Zipperer WC (1993) Deforestation patterns and their effects on forest patches. Landsc Ecol 8(3):177–184

Airborne LiDAR Applications in Forest Landscapes

Connie Ko and Tarmo K. Remmel

Abstract This chapter provides an introduction and overview of using light detection and ranging (LiDAR) in forest applications. The first section explains the principles and basic terminology for LiDAR and introduces the use of LiDAR on three different platforms (spaceborne, airborne, and terrestrial) for forest applications. The second section discusses applications in relation to the primary measurements from a LiDAR point cloud, primarily information derived from distance (from the aircraft to the target). We cover concepts related to different representations of surfaces (e.g., digital surface model, digital terrain model, digital elevation model, and canopy height model). Typically, single trees can be identified from the canopy height model and there are two different ways to assign LiDAR points to individual trees, the surface-based method and the point-based method. The third section discusses forest applications in relation to secondary measurement from a LiDAR point cloud, information derived from point cloud geometry rather than direct distance measurements. This section covers tree genera classification; the use of allometric equations for deriving DBH, biomass, and other forest attributes; and the classification of vegetation types. Three ways of getting genera information are discussed, including the vertical profile method, methods relying on geometry derived from individual tree point clouds, and methods that incorporate spectral information. The fourth section provides a case study for identifying potential tree hazards along a powerline corridor in Ontario, Canada. We conclude by discussing the future of this technology.

Abbreviations

2D	Two-dimensional
3D	Three-dimensional
ALS	Airborne laser scanning
BA	Basal area

C. Ko (✉) • T.K. Remmel
Department of Geography, York University,
4700 Keele Street, Toronto, ON, Canada, M3J 1P3
e-mail: cko@yorku.ca; remmelt@yorku.ca

T.K. Remmel, A.H. Perera (eds.), *Mapping Forest Landscape Patterns*,
DOI 10.1007/978-1-4939-7331-6_4

CHM	Canopy height model
DBH	Diameter at breast height
DEM	Digital elevation model
DSM	Digital surface model
DTM	Digital terrain model
GIS	Geographical information system
GLAS	Geoscience laser altimeter system
GPS	Global positioning system
ICESat	Ice, cloud and land elevation satellite
IMU	Inertial measurement unit
L_t	The travel time of a light pulse
LiDAR	Light detection and ranging
MODIS	Moderate resolution imaging spectroradiometer
MVCD	Minimum vegetation clearance distance
p	Number of points projected into a horizontal area (i.e., the point density)
R	Range for LiDAR
SLS	Spaceborne laser scanning
t_L	The total travel time of a single energy pulse
t_{rise}	Rise time of an energy pulse
TLS	Terrestrial laser scanning
UAV	Unmanned aerial vehicle

Introduction

During the past few decades, the use of light detection and ranging (LiDAR) systems for mapping forests and their characteristics has become highly popular. Specifying the platform for a LiDAR system identifies whether it is mounted on a spacecraft (*spaceborne laser scanning, SLS*), aircraft (*airborne laser scanning, ALS*), or a tripod at ground level (*terrestrial laser scanning, TLS*), each of which provides slightly different data characteristics. Each of these types of LiDAR system provide precise distance measurements between the sensor and targets by measuring the total transmission and return time of laser pulses with known velocity (i.e., distance = velocity × time) and then halving that distance to compensate for the fact that the signal travels the same path once in each direction. When measuring forests from the ground (TLS), but more typically from above the canopy (ALS), height metrics for vegetation and landscape features can be attained if the sensor's location and attitude are accurately known based on data streams from the instrument's inertial measurement unit (IMU) and global positioning system (GPS) receiver. This arrangement makes it possible to accurately describe the three-dimensional (3D) characteristics of terrain, of a vegetated canopy, or of another surface that is capable of reflecting the signal.

Because lasers often pass partially through their targets and return a pulse from one or more subsequent targets, LiDAR systems generally provide more than a single return pulse of energy for any outgoing laser pulse, so processing of the multiple return signals can extend the mapping capability to penetrate vegetated canopies and can often reach all the way to ground level, especially for ALS systems. Therefore, LiDAR has the potential to provide detailed spatial and vertical characterizations of landscapes, even when used over vegetated land cover classes.

Detailed, accurate estimation of forest attributes is important for producing and updating forest inventories, monitoring disturbances or other dynamic processes, and supporting the management of natural and commercial forests. The 3D characterization of landscapes with LiDAR provides insights about forest structures that go beyond the typically available surface reflectance imagery (e.g., airphotos, passive satellite sensors), and this allows the measurement and estimation of supplementary forest characteristics, including vegetation height, diameter at breast height (DBH), biomass, and canopy volume. However, the problem of species identification has not yet been solved, and remains an active area of research.

LiDAR produces a point cloud rather than a flat image, and each point has a 3D coordinate that represents the location where a reflection occurred within a particular frame of reference. A point cloud scene can be generated from systems that record discrete points with (x, y, z) position triplets in space; however, point clouds can also be derived from full-waveform systems that record a continuum of received power signals from the reflected targets, and this approach is becoming more common. (We will discuss discrete and waveform systems in section "Defining ALS LiDAR" of this chapter.) Most LiDAR systems can also provide additional attributes pertaining to these reflections, including the scan angle, intensity of the reflection, and number of reflections; however, it is the position of the points in space that has thus far received the greatest attention. The points in the cloud represent reflections from *footprints* that represent the areas on a surface that have intercepted the laser beam, perpendicular to the incidence angle; that is, they represent the area "painted" by the laser, and that area depends on the angle of the beam and its orientation with respect to the surface. Footprints are circular to elliptical and their size is a function of both the beam divergence (i.e., how much the beam spreads during its travel) and sensor altitude (i.e., the distance to the target).

The footprint area (and thus the cost of data acquisition for a given area) is generally greatest for SLS LiDAR, followed by ALS and then TLS due to the decreasing distance from the scanner to the target. Small-footprint systems (with footprints approximately 0.1–0.3 m in diameter) and large-footprint systems (approximately 8–70 m in diameter) can both be used to estimate and map forest inventory attributes (Bouvier et al. 2015). However, small-footprint systems are often preferred because they provide a more detailed characterization of the forest canopy and therefore improve analyses at the levels of individual trees, plots, or stands (Maltamo et al. 2014). Furthermore, ALS allows better estimates of ground elevation than TLS or SLS due to its combination of a close vantage point with a small footprint, thereby allowing more accurate measurement of tree heights. In this chapter, we discuss LiDAR in general, but focus our review, examples, and discussion on the use of small-footprint ALS to retrieve forest attributes for landscape-scale mapping.

Defining ALS LiDAR

LiDAR describes a family of methods that use active remote sensing technology. *Active* means that these sensors generate and emit energy pulses and then record the timing and quantity of that energy that is reflected (returned) back to the incorporated detector. This approach differs substantially from passive imaging systems such as most satellite sensors, which record reflected or emitted energy that originated from other sources, such as sunlight. The primary data acquired by LiDAR systems are range (distance) measurements that provide high-precision measurements of the distance between the LiDAR sensor and the target that reflected the energy pulse. When these systems are mounted on airplanes, helicopters, or other airborne platforms, they are referred to as ALS, primarily to clarify the expectations for footprint size and swath width, which differ substantially among SLS, ALS, and TLS LiDAR.

Single-channel ALS systems emit energy pulses (signals) at a particular wavelength and record the reflection time and amplitude. The nature of the interaction of the energy pulse with a target will be determined by a combination of the target material, the incidence angle, and the wavelength of the energy pulse. Some newer multichannel LiDAR systems are capable of multispectral scanning; they emit signals at multiple wavelengths and record the returned characteristics independently for each wavelength, thereby providing a 3D structural point cloud with multispectral attributes. Although the travel time is identical for all of the wavelengths, the wavelengths differ in their ability to penetrate or reflect from targets with different characteristics. Thus, one wavelength may provide reflections from surfaces that completely absorb or completely reflect a different wavelength.

The range (R) for LiDAR is computed using Eq. (1), in which c is the constant speed of light (approximately 3×10^8 m s^{-1}), t_L is the total travel time of a single energy pulse, and dividing by 2 compensates for the fact that t_L actually records the time for a pulse to travel both to the target and back again (i.e., twice the actual distance). Thus, the range between the sensor and the target is only half the distance recorded by the time variable.

$$R = \left(c \times t_{\mathrm{L}}\right)/2 \tag{1}$$

The process of emitting pulses and recording their returns, converting this data into distances, attaching the scanning attributes, and obtaining the coordinates of the actual reflection point (x, y, z) by corrections based on the platform's position is repeated while flying over an area. The number of points recorded will depend on the scanning rate (in Hz, which represents the number of energy pulses per second). The collection of points, which is called a point cloud scene, is generated when all points from a given survey are mapped collectively in the same 3D space.

Figure 1 provides an example of a point cloud scene created from nearly 500,000 points, representing approximately 1 ha of a forested area. The image is shaded such that darker points represent lower elevations (i.e., positions closer to the ground).

Fig. 1 Point cloud scene with approximately 500,000 points, covering an area of approximately 1 ha. *Darker shades* represent lower elevations. *White* represents areas from which no signal was received

Although the point cloud appears dense, it is not continuous and the sense of continuity is a function of the small size of the image and the large number of points covering the targeted area. The number of points (p) projected into a horizontal area is referred to as the point density (points per m^2), and is often interpreted similarly to the spatial resolution of an image.

The point density can vary greatly among studies, and although a higher point density can provide greater detail about the study area, the amount of data, the processing time, and the computational complexity also increase. Conversely, if the point density is low, objects may not be portrayed sufficiently accurately, and smaller objects may not even be discernable; thus, a balance between point density and analytical needs is required. The optimal point density is affected not only by a user's needs (or desires) in terms of the required detail, but also by the reality of a landscape's complexity and by budgetary constraints. Flight altitude also imposes trade-offs: a higher altitude provides a wider data collection swath width in exchange for a decreased point density and a larger footprint, whereas lower altitude provides increased point density and reduced footprint size, but at the cost of a narrower swath width. Unless local flight regulations constrain the permitted altitude, the flight altitude is a subjective choice that represents the researcher's compromise among the swath width, point density, and footprint area. For example, Persson et al. (2002) used a point density of about 4 points per m^2 to derive a tree height model. Among studies with a higher point density, Persson et al. (2006) used LiDAR data with a density of about 50 points per m^2 to identify tree species and Vauhkonen et al. (2008, 2009) used LiDAR data with a density of 40 points per m^2. In contrast,

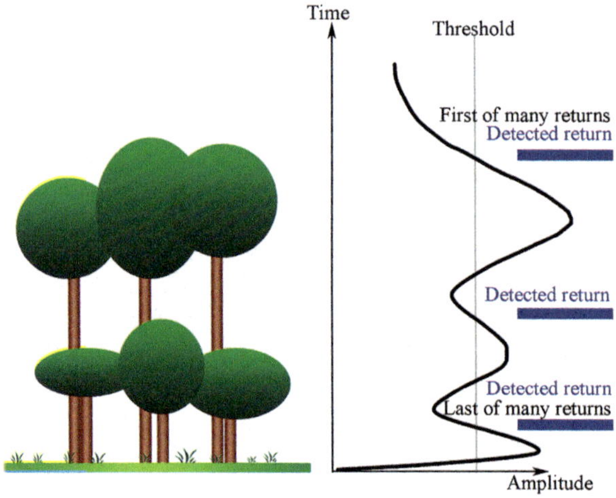

Fig. 2 Comparison of a waveform-digitization system's data (the *black* line) with data from a discrete-return system (*blue* bars, based on the threshold identified by the *grey* line). Signals refer to the hypothetical forest depicted at the *left*

Woods et al. (2010) successfully predicted forest attributes using a LiDAR point density of 0.5 points per m^2.

LiDAR systems can be categorized into two subtypes: (1) waveform digitization systems that record a continuous wave of reflection intensity, and (2) discrete-return systems that retain only intensities greater than a specified threshold value. Waveform digitization produces a continuous waveform (a time series) that fully describes the vertical structure (along the scan axis) of targets on the ground, whereas a discrete-return system detects peaks when the return signal exceeds a detection threshold (Fig. 2). The figure illustrates the *intensity* (amplitude) recorded by the sensor with respect to time; these recorded signal profiles result from the interaction between the emitted pulse and the target. The forms of these waveforms (described by descriptive statistics such as the peak width, amplitude, and skewness) depend on a target's properties (e.g., reflectance, roughness) and on the scanning incidence angle. Figure 2 also shows that the intensity increases when the LiDAR signals interact with the relatively flat reflective surface of the canopy and then decreases at the gap between the bottom of the canopy and the understory (or a lower, immature canopy), where the vertical space is dominated primarily by woody stems. Once the energy pulse reaches the understory, the intensity increases again as it reflects from either the understory or a secondary canopy. Finally, the intensity increases again when the signal reflects from the ground.

To obtain discrete returns from the waveform, a threshold is introduced, and each time the intensity exceeds the specified threshold, a return is generated (e.g., the grey line in Fig. 2 results in three returns). Depending on the nature and geometry of the target, one LiDAR pulse can generate a single return or multiple returns. For example, pulses that reflect from a flat surface, such as exposed bedrock, may only generate a single return, whereas pulses reflected from vegetated targets may gener-

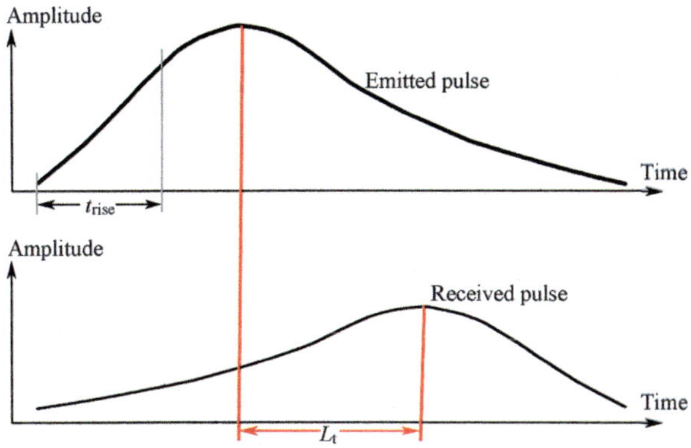

Fig. 3 Travel time of a single energy pulse in a waveform system. L_t represents the travel time of the LiDAR pulse and t_{rise} represents the time required for the system to amplify the signal sufficiently to generate a detectable return signal. The amplitude of the received pulse is lower than that of the emitted pulse due to a loss of energy to the environment

ate multiple returns, often distributed throughout the vertical canopy structure. In theory, the first return represents the top of the canopy and the last return represents the ground, but this assumes that the LiDAR pulse had sufficient energy to penetrate obstacles such as leaves and reach the ground or was not prevented from reaching the ground by leaves, branches, or other materials. The reality is much more complex, as we will see shortly.

To further conceptualize the idea of waveform pulses, Fig. 3 illustrates a single emitted pulse and the received (reflected) pulse. For a waveform system, the range is measured based on the phase difference between the transmitted and received signal (Mallet and Bretar 2009). Here, the phase difference represents the time difference between the peaks of the emitted and received pulses. The distance between the two vertical *red* lines represents the travel time of a light pulse (L_t), which is used to calculate the distance between the scanner and the target. The rise time (t_{rise}) represents the time needed for the system to amplify the signal sufficiently to generate a detectable return signal. The amplitude of the received pulse is always lower than that of the emitted pulse due to a loss of energy to the environment (e.g., atmospheric absorption, energy absorption by structures contacted by the beam). For all LiDAR systems, a delay is added between pulses to allow a sufficient time gap that allows data from one pulse to be distinguished from data from the next one.

Obtaining the precise location of a point and using this data to generate a point cloud scene requires more than just a LiDAR scanner. For ALS, it's necessary to obtain a precise measurement of the flight path and sensor location while each measurement was made, along with detailed sensor orientation information. Flight path data is easily recorded by means of an onboard GPS receiver, but the position of the scanner relative to the aircraft's true position is documented by the IMU that is mounted on the aircraft to measure its yaw, pitch, and roll attributes (Fig. 4). As the LiDAR sensor emits energy pulses, the return reflections from targets are accurately measured and associated with the corresponding IMU measurements, which are

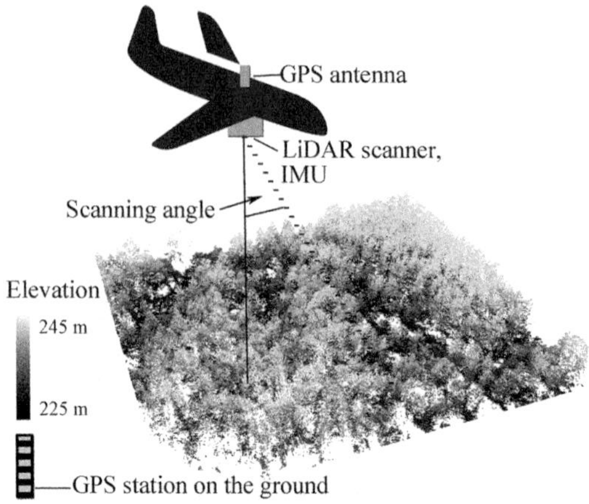

Fig. 4 Illustration of an ALS system, showing the vertical (nadir) direction (*black* vertical line), the scanning angle, flight platform, GPS, IMU, and LiDAR components as they collect data over a forested surface. White represents areas from which no signal was received

used to compute the correct range and absolute position. Because the accuracy of a height measurement depends critically on the accuracy of the platform's measured height, the measurement error ranges from a few centimeters to a few decimeters, depending on the platform and the quality of its instrumentation.

For any LiDAR system, the selection of a signal wavelength suitable for the purpose of the study is important. For vegetation studies, common wavelengths include 900, 1064, 1470, and 1560 nm for single-channel ALS due to the high reflectance in those regions (Petrie and Toth 2009). In the near-infrared region of the electromagnetic spectrum, energy emitted from the sensor is easily refracted and reflected by leaves, resulting in an easily recorded data point. This makes the near-infrared wavelengths popular for studying vegetation. For multispectral LiDAR or multichannel LiDAR (e.g., the Titan system; Teledyne Optech, http://www.teledyneoptech.com), common combinations of wavelengths include 532, 1065, and 1550 nm; the TLS Dual-Wavelength Echidna LiDAR (DWEL) simultaneously scans at 1064 and 1548 nm (Douglas et al. 2012).

Introduction to the Three Common LiDAR Platforms

The three common LiDAR platforms for vegetation applications are SLS, ALS, and TLS. SLS LiDAR offers the possibility of mapping a forest's vertical structure globally, albeit with large footprints and a sparse point density. The ice, cloud and land elevation satellite with the geoscience laser altimeter system (ICESat/GLAS) acquired waveform LiDAR data globally from 2003 to 2009 (Abshire et al. 2005; Simard et al. 2011). Although GLAS did not provide complete global coverage, it was the only LiDAR platform that came close (Los et al. 2012). GLAS scanned at a wavelength of 1065 nm, recording reflected energy from footprints spaced 172 m apart (Harding and Carabajal 2005). This dataset allowed Lefsky (2010) to generate global forest height maps by combining the LiDAR data with other variables

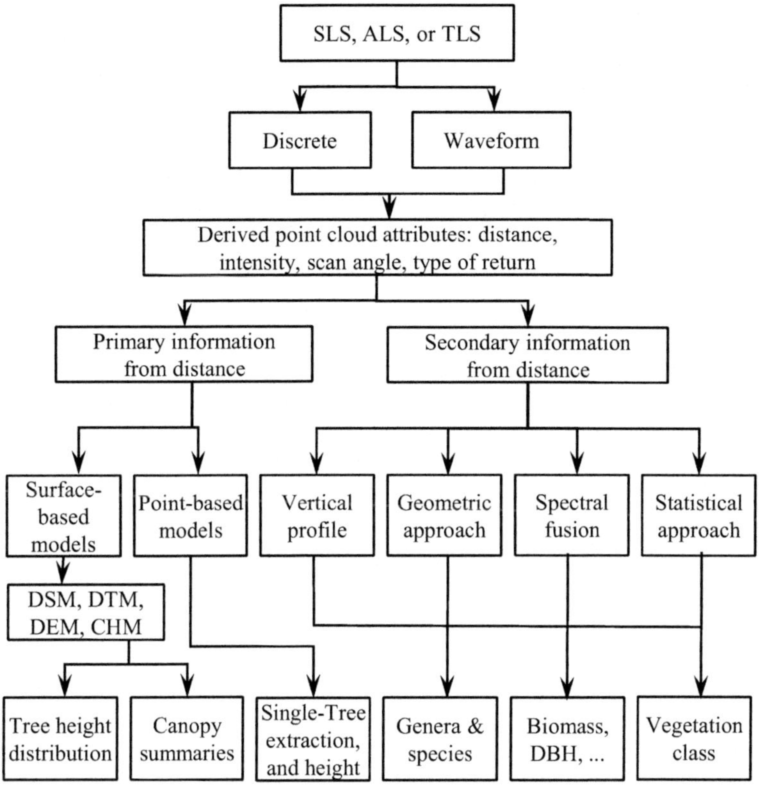

Fig. 5 Schema for organizing LiDAR analysis methodologies. *CHM* canopy height model, *DBH* diameter at breast height, *DEM* digital elevation model, *DSM* digital surface model, *DTM* digital terrain model

obtained from the moderate resolution imaging spectroradiometer (MODIS). This approach permitted values for gaps in the LiDAR coverage to be estimated from the ancillary data. Figure 5 presents a top-down organization of LiDAR analytical methods that is discussed in this chapter to organize the terminology and relationships among the methods.

ALS has become a popular tool for forestry applications because of its combination of relatively low cost and high resolution (Nelson 2013), but proposals for this technology date back to the 1960s (e.g., Rempel and Parker 1964), when suggested uses included measuring micro-relief and tree heights with a laser terrain profiler from airborne platforms (Ritchie 1996). The continued commercialization of LiDAR equipment has vastly increased the number of studies that employ this technology. Hyyppä et al. (2008, 2012) reviewed the numerous applications of small-footprint ALS for forest inventory collection and mapping, and concluded that individual tree-based features and inventories are improving and that in addition to the first return the last return can contain important information for estimating tree, stand, or forest characteristics.

Conventionally, sensors are mounted on an airplane or helicopter (the airborne platform), although the use of unmanned aerial vehicles (UAVs) has begun to attract many users due to their increasing availability and relatively low cost. Wallace et al. (2012) describe an example of using UAV-based LiDAR to acquire forest inventory attributes. Another stream of research closely related to the use of UAVs is the derivation of photogrammetric point clouds for forests. Instead of acquiring point clouds from laser scanning, photogrammetric point clouds are created from processing stereo-pairs of aerial photographs. This method of data acquisition is less expensive than LiDAR, as a laser scanning sensor is not required. The recent development of software to support such analyses has initiated research on using this technique in forests, such as the studies by Baltsavias et al. (2008) and by Rosnell and Honkavaara (2012).

To characterize forests or single trees, ALS studies can be categorized into two broad groups: area-based approaches and single-tree approaches. Area-based approaches extract, analyze, and compare statistical characteristics of LiDAR point clouds for representative stands of trees, whereas single-tree approaches require an initial segmentation of LiDAR point clouds into separate clouds that represent individual trees before detailed statistical, structural, and geometric descriptors or summaries can be calculated for the points contained within the point clouds for individual trees.

Typically, for an area-based approach (White et al. 2013), descriptive statistics can be computed to characterize forest canopy heights within a region of interest, and then specific attributes (e.g., timber volume, DBH, or biomass) can be modeled by regression (Woods et al. 2010; Treitz et al. 2012) or other statistical methods. Means et al. (2000) successfully predicted the characteristic stand height, basal area (BA), and volume of a forest by using LiDAR-derived attributes to build relationships in the Western Cascades of Oregon. Lim et al. (2003) derived ten biophysical forest metrics in Ontario: maximum tree height, Lorey's mean tree height, mean DBH, total BA, percent of canopy openness, leaf area index, ellipsoidal crown closure, total aboveground biomass, total wood volume, and stem density (number per unit area). The R^2 values ranged from 0.63 to 0.86. Næsset et al. (2005) used regression analyses to predict tree height, BA, and volume in southern Norway. For surface-based methods, where trees (or small clusters of trees) are represented by a simplified polygon, data representation can be greatly simplified to minimize the volume of data. In these cases, attributes such as tree height, crown width, or DBH can easily be attached to the polygons using a geographical information system (GIS) once the attributes have been determined.

Single-tree approaches begin by segmenting the LiDAR point clouds into groups of data for individual trees. There are generally two ways to achieve this: the surface-based method (see section "Detection and Delineation of Individual Trees: Surface-Based Methods") and the point-based method (see section "Detection and Delineation of Individual Trees: Point-Based Methods"). The surface-based method relies on drawing a polygonal boundary around each observed tree crown on a horizontal (nadir view) two-dimensional (2D) surface and using that boundary as a vertical cookie-cutter to identify all points within the polygonal solid volume as belonging to the same tree. Obvious concerns arise where trees grow in close proximity, since their crowns can touch or overlap, thus causing segmentation assumption errors (i.e., errors that result from classifying a point as belonging to the wrong tree). In contrast, the point-based methods classify points to a single tree using clustering based on properties of the individual points, rather than the brute-force vertical section through the 3D data product that is used in area-based methods.

For forest applications, LiDAR is well recognized for measuring tree height and estimating volume and biomass. However, species recognition is a tantalizing area of continuing research. More often, only the genus is identified, since that coarser hierarchical classification leads to higher classification accuracy. Two approaches have been used extensively to scrutinize the distribution of signal returns for classifying the genus. The first approach considers the vertical distribution of LiDAR returns (points) within a point cloud segmented for each tree (see section "Vertical Profile for a Single Tree"). Visually, these summaries include plots of the frequency of points along the x-axis relative to height above the ground on the y-axis. The full resolution of this distribution can be used to characterize a whole tree, or to summarize the point frequencies grouped into vertical height bins (often by slicing the height axis into deciles); together, these methods form the category of vertical profile approaches (Holmgren and Persson 2004; Brandtberg 2007; Ørka et al. 2007, 2009; Korpela et al. 2010; Kim et al. 2011). The second approach is to derive geometric attributes from the segmented single-tree point clouds (Kato et al. 2009; Vauhkonen et al. 2008, 2009, 2010; Ko et al. 2013) and to use those geometric descriptors to infer the genus or species (see section "Tree Genus and Species Classification").

TLS LiDAR can be mounted on a stationary object (e.g., tripod) or on moving objects (e.g., a vehicle or a person; in the latter case, this is called *mobile laser scanning*). TLS has been used to derive DBH estimates in the Kielder Forest District of Northern England (Watt and Donoghue 2005), whereas Tansey et al. (2009) measured stem densities and tree DBH using an automatic stem recognition model, and used the resulting data to compute BA. These metrics were extended to include stem counts (Liang et al. 2012), aboveground woody biomass (Yao et al. 2011; Yu et al. 2013), and tree canopy volume and BA by means of voxel-based point cloud slicing (Moskal and Zheng 2012). (A voxel is the 3D equivalent of a 2D pixel. We'll discuss this in more detail in section "Detection and Delineation of Individual Trees: Point-Based Methods") Although many studies using TLS for forestry applications relate to attribute extraction and mapping, Doneus et al. (2010) used TLS as a validation data source for full-waveform ALS data.

Intensity, Point Density, and Multispectral LiDAR

The primary measurement obtained from LiDAR is the range to a target, and by extension, tree, or canopy height. However, most LIDAR systems are capable of recording additional information during the scanning process, such as the scan

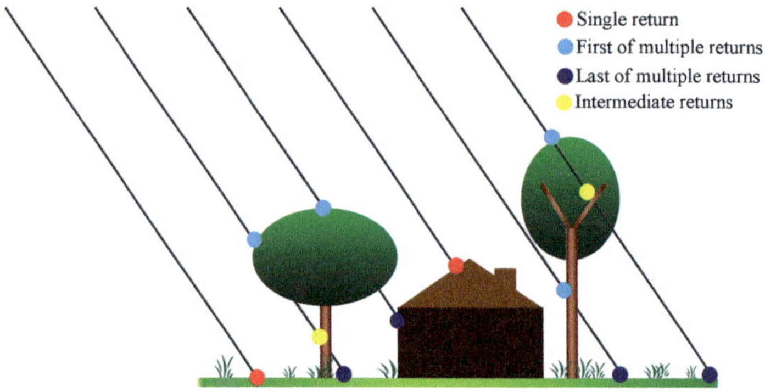

Fig. 6 A hypothetical example depicting the travel vectors for laser pulses (*black* lines) and reflection points from features, color-coded to identify the type of return

angle, return type, and return intensity. The scan angle represents the angle of the laser pulse between the vertical nadir line and the directional vector that points away from the nadir point, originating at the scanner and making contact with a target (Fig. 4). The return type indicates the sequence in which subcomponents of a laser pulse are recorded based on sequence of the returns (e.g., the first return from a pulse vs. the last return). The intensity is the amplitude of the laser radiation returned.

A single laser pulse can generate either a single return or multiple returns (Fig. 2), depending on the target and the characteristics of its interaction with the signal. Knowledge of the type of return can be valuable for classifying objects, for separating points that belong to either terrain or non-terrain features, and for generating digital elevation models (DEMs; Fig. 6). There are four main types of return: (1) a single return, with only one reflected return received from the emitted pulse; (2) the first of multiple returns, with the first reflected return normally interpreted as the top of the canopy for a forested area; (3) the last of multiple returns, with the reflected return normally interpreted as the ground level; and (4) intermediate returns, with the reflected returns arriving between the first and last of multiple returns.

Intensity is the amplitude of the returned energy for a laser pulse. This intensity is affected by the scan angle, flight path, flight altitude, and incidence angle at the point of interaction. Intensity is also affected by the target orientation and material (whose reflectivity determines the amount of energy reflected by the material); thus, unless intensity data is calibrated using known targets, the values recorded from the sensor are relative measurements and should be used with caution. In some studies, attributes derived from the intensity have been used to classify tree species and heights (Ørka et al. 2007, 2009; Korpela et al. 2010; Morsdorf et al. 2010; Vauhkonen et al. 2010; Kim et al. 2011.

It has been suggested that the minimum pulse density for retrieving reliable estimates of forest parameters is 0.1 points per m^2 (Næsset 1997; Holmgren 2004; Vauhkonen et al. 2014). However, to delineate individual trees, at least five and potentially more than 10 points per m^2 are necessary (Hyyppä and Inkinen 1999;

Persson et al. 2002; Vauhkonen et al. 2008), and the accuracy of the estimated tree attribute decreases substantially as the pulse density decreases (Magnusson et al. 2007; Vauhkonen et al. 2008). Strunk et al. (2012) suggested that there was almost no loss in precision when using 0.05 points per m^2 rather than 3 points per m^2 in western Washington state, USA. In contrast, Ko et al. (2014) concluded that the accuracy of classifying boreal tree genera in Ontario, Canada, decreases significantly below a pulse density of 5 points per m^2. Although both Strunk et al. (2012) and Ko et al. (2014) studied forests dominated by conifers, differences in the terrain, tree species, and forest characteristics may explain their different results. Finding the lower limit for point density is important in terms of lowering the overall cost of data acquisition, but this value is often subjective and related more to the extent of the study area and budgetary constraints than to scientifically and statistically determined optimal values.

Although LiDAR data is often used independently, its full potential is realized when it is combined with additional data sources (e.g., aerial photographs, multispectral satellite imagery). Improved tree and canopy height estimates have been obtained by combining segmented Landsat-5 Thematic Mapper data (Wulder and Seemann 2003) and QuickBird images (Hilker et al. 2008) with LiDAR point clouds. Hyperspectral imagery from the HyMap sensor (Hill and Thomson 2005), the multispectral Airborne Thematic Mapper (Koukoulas and Blackburn 2005), and the QuickBird satellite images (Ke et al. 2010) have been fused with LiDAR to produce classified vegetation and species maps.

Another way of fusing information is to combine LiDAR data acquired at different wavelengths. Although this technology is still being developed, Gaulton et al. (2013) used dual-wavelength LiDAR to estimate the vegetation moisture content and Wang et al. (2014) used this tool for land cover classification. Although the former study experimented on leaf samples, the results nonetheless indicate the potential to scale up the approach to a canopy level.

Primary Measurements

Primary measurements generally relate directly to the range values measured by LiDAR and refer in some way to various vegetation canopy height components. Height is the most direct information that can be derived from the recorded locations of points in a 3D space (x, y, z), and is therefore a primary measurement. Most other information is inferred or modeled from the locations of the points within the cloud and from secondary information, including many forestry attributes (e.g., species, BA, biomass); we will discuss this information in section "Secondary Measurements".

Displays of unprocessed point clouds normally color each point based on its height attribute to aid in visualization, but these representations do not necessarily provide much information beyond a cursory visualization of the study area (Fig. 1). Height, however, can provide valuable information and is often a key variable depicted in forest maps, as in the global forest height map generated by combining MODIS data with 500 m spatial resolution and GLAS data to derive canopy heights (Lefsky 2010).

To improve on the results of Lefsky (2010), Simard et al. (2011) used the Random Forests software to predict heights that were not covered by the GLAS LiDAR data, basing their predictions on seven global ancillary variables: annual mean precipitation (mm), precipitation seasonality (the difference, in mm, between the mean growing-season and dormant-season precipitation), annual mean temperature (°C), temperature seasonality (the temperature difference, in °C, between the winter minimum and summer maximum temperature), elevation (m), tree cover (%), and protection status. The modeled canopy height (Fig. 7) had a root-mean-square error of 6.1 m ($R^2 = 0.5$); when seven outliers were removed, the error improved to 4.4 m ($R^2 = 0.7$).

In addition to mapping forest stand heights or individual tree heights, height information can also be used to generate variations of 2D surface (raster) models. The simplification of 3D data into a 2D surface reduces the volume of data and allows algorithms designed for analyzing 2D data to be utilized. Although consensus is lacking in the literature regarding the definitions of the names for some surface models, such as digital surface models (DSMs), DEMs, canopy height models (CHMs), and digital terrain models (DTMs), all of these elevation surfaces can be generated from a LiDAR point cloud. (We will provide our own definitions of these model types in section "Surface Models (DEM, DSM, DTM, CHM)".) CHMs are particularly useful for detecting and delineating individual trees (see section

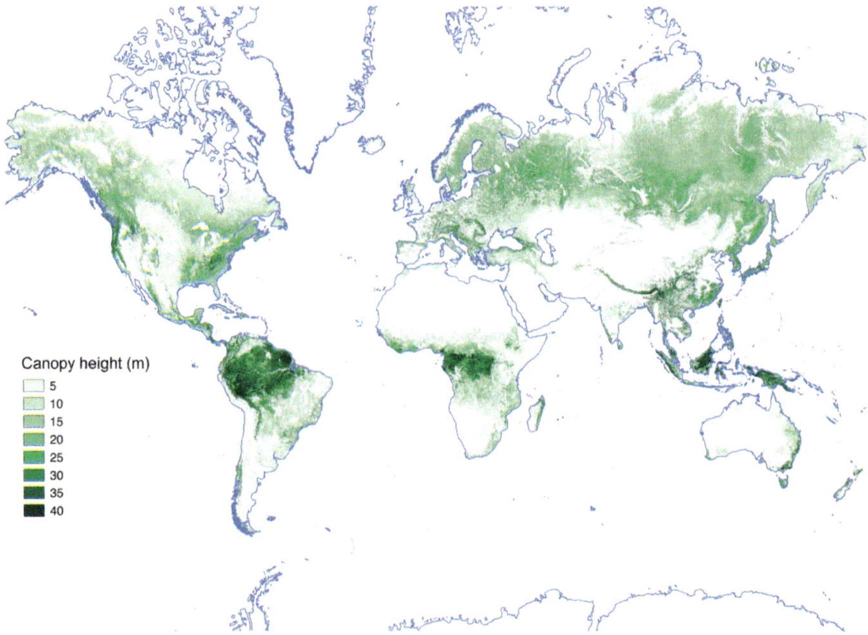

Fig. 7 Global forest height map simulated using predictions from the Random Forests software and seven global ancillary variables for regions where GLAS LiDAR coverage did not exist. Source: Marc Simard, after Simard et al. (2011)

"Canopy Height Models and Detection and Delineation of Individual Trees"), since these models can potentially identify individual curved tree crowns and the local minima (valleys) between adjacent crowns.

Surface Models (DEM, DSM, DTM, CHM)

The term DEM is used to refer to a superset of both DSMs and DTMs. DSMs are surfaces that represent the *upper surface* of any scanned area. In a forested area, the DSM represents the top of the canopy (where vegetation is present) or the top of whatever highest feature exists at a location without a tree that is illuminated by the LiDAR pulse. DSMs are normally generated from the first returns of the laser pulse and represent the features closest to the sensor. Conversely, DTMs represent the *topography* of the bare ground (with all living natural or artificial features removed) and are usually generated from the last returns of the laser pulses.

Both surface types (i.e., DSMs, DTMs) are useful for retrieving forest attributes and in other applications such as hydrological modeling, flood prediction, and urban studies. CHMs are easily obtained by subtracting the DTM for an area from the corresponding DSM, thereby providing the local heights of the trees or canopy. Figure 8 illustrates the difference between a site's DSM (*red* line), DTM (*blue* line), and CHM (*black* line).

The advantage of simplifying the 3D point cloud into 2D raster data is that this allows further processing by using neighborhood operations (i.e., calculations based on data in the area surrounding a particular target such as a tree) in a substantially more efficient manner than if it were necessary to work with 3D data, but results in a loss of precision because interpolation of the raster height in areas with few to no data points may introduce errors. A classic example of this would be where a bridge crosses a chasm; in reality, this location has two elevations (the bridge surface and the bottom of the chasm), but only one of them can be depicted in the surface model (Pfeifer and Mandlburger 2009).

Fig. 8 Differences among three standard surface models for a hypothetical forest stand growing on a slope: DSM, DTM, and CHM

To generate these surfaces, the first step is to classify a point cloud scene into two major groups of points: terrain and non-terrain points. To generate a DTM, the classified terrain points become candidate points for generation of the DTM. The second step is generation of the surface by interpolation. Interpolation is a process of estimating height values for locations where they do not exist (i.e., between pairs of known and measured points obtained from the LiDAR data) to generate a continuous surface. The surfaces can be represented as triangulated irregular networks, in which the surface is represented as a series of triangles with shared sides, or as raster surfaces. The selection of an interpolation method is beyond the scope of this chapter, but some options include inverse distance-weighted interpolation (Shepard 1968), nearest-neighbor interpolation (Sibson 1981; Sambridge et al. 1995), and kriging (Cressie 1988; Caruso and Quarta 1998). Each type of interpolation has advantages and drawbacks (Caruso and Quarta 1998).

Obtaining a precise DTM is crucial for precise estimation of tree heights and therefore affects many other estimates of secondary attributes (e.g., volume) that depend on tree height. There are many algorithms and filters that extract ground points from a LiDAR point cloud and use these points for generating a DTM or DEM; here, we will introduce a few popular methods. Further details are presented by Sithole and Vosselman (2004), who reviewed eight filters and tested their relative performance using 12 datasets. Kraus and Pfeifer (1998) identified terrain points by comparing LiDAR data with an approximation of the terrain surface. Residuals were calculated between the LiDAR data and the surface, with the points assigned weights according to the calculated residuals; a new surface was then constructed according to the assigned weights, and this process was repeated until the minimum error was reached or the maximum number of iterations was attained. Vosselman (2000) identified non-terrain points by comparing the height differences between neighboring points that were above a certain threshold. Another example is given by Tóvári and Pfeifer (2005), who used a segmentation method based on growing the region around randomly selected points based on the similarity of normal vectors to a plane fit through *n* nearby points to group LiDAR points into segments. By assuming homogeneity of the terrain surface, segments are obtained by allowing grouping of adjacent points so that the points are separated by a distance of 2 m or less and the normal vectors at these points differ by 5° or less. Alternatively, Evans and Hudak (2007) used a multiscale curvature algorithm to iteratively separate LiDAR data into ground and non-ground points.

DTMs have many applications, such as to support the development of a hydrological model, to understand the effects of slope on erosion, and in urban planning to determine where structures should be built.

Canopy Height Models and Detection and Delineation of Individual Trees

In forested areas, CHMs provide a raster surface that represents the absolute height of the canopy (Fig. 8). Although a CHM can be used as a map to represent the height of the vegetation within a study area, the most popular use of a CHM is

Fig. 9 An example of a CHM created using inverse distance-weighted interpolation over a forested area consisting mainly of pine (genus *Pinus*), poplar (genus *Populus*), and maple (genus *Acer*). This image represents the same study area depicted in Fig. 1

for detection and delineation of individual trees. However, CHMs have also been used to estimate canopy density and tree height. Figure 9 shows an example of a CHM created using inverse distance-weighted interpolation over a forested area consisting mainly of pine (genus *Pinus*), poplar (genus *Populus*), and maple (genus *Acer*). In another example, St-Onge et al. (2008) produced a hybrid photogrammetric LiDAR CHM by subtracting the DTM from a photogrammetric DSM (derived from archived aerial photographs) and used that to retroactively map forest structures.

Detection and Delineation of Individual Trees: Surface-Based Methods

For the surface-based (raster) method of single-tree detection, it is assumed that the CHM is either available or can be produced. Tree locations are subsequently detected using the local maxima in the CHM. However, this usually results in a large number of false-positive detections because individual tree crowns are often highly variable and may contain multiple local maxima. In this approach, local maxima are minimized within a window that is an arbitrary number of pixels wide by computing the local average, then the window is moved by one pixel, and the process is repeated. The smoothing of the CHM is influenced by the characteristics of the study area (e.g., slope) and the expected areas of the tree crowns; the optimal smoothing window will be large enough to eliminate local variations within a single tree but small enough to be able to detect trees growing in close proximity. A 3 pixel × 3 pixel filter (Hyyppä et al. 2001; Persson et al. 2002; Morsdorf et al. 2004) and larger filters (Popescu and Wynne 2004; Weinacker et al. 2004) have been used successfully.

The tree detection results are usually validated using field data whose position is determined using a GPS Global Navigation Satellite System (GNSS) receiver (Solberg et al. 2006; Ørka et al. 2007). Success rates for tree counts at validated study sites have ranged from 45.4 to 61% for coniferous trees (Heurich 2008) and averaged 44% for deciduous trees (Reitberger et al. 2007). A study by Yu et al. (2004) detected 61/83 trees, for a 73% success rate, whereas Koukoulas and Blackburn (2005) detected 82/103 trees (80%), but this value decreased to 50% at a site with high stem density.

The top of a forest canopy, when inverted, often resembles a collection of small depressions (the tree crowns) separated by ridges (the edges of individual tree crowns). A parallel to watersheds can be drawn, and hence research on watershed identification also provides some insights into surface-based analysis for segmenting LiDAR data into individual tree crown maps (Koch et al. 2006). Watersheds represent all upstream areas drained by a single river or stream and watershed algorithms refer to tools for processing an image or DEM to automatically or semiautomatically identify features that define watershed boundaries based on a predefined threshold that determines their minimum area. Typically, terrain elevation is used as an input based on the assumption that water flows from high elevations to low elevations. Boundaries of watersheds can be delineated by detecting and connecting the areas (pixels) with the highest elevations.

This approach can also be applied to identify tree crown perimeters, since a canopy comprising multiple crowns is similar to an inverted watershed map. For crown delineation, the edges represent the "valleys" between maxima for adjacent tree crowns and trees, and are delineated using methods similar to those used for watershed delineation (Fig. 10), except that the surface is inverted (i.e., the valleys

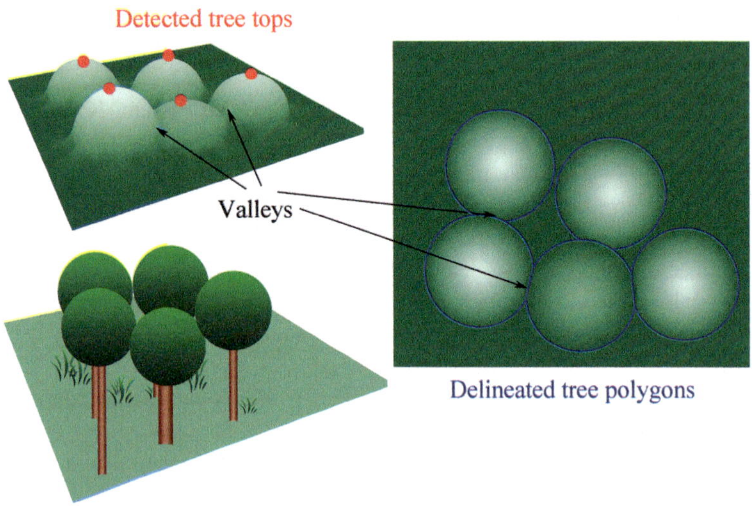

Fig. 10 Individual trees can be detected through local maxima in the derived CHM, and tree crowns can be delineated as polygons by following the valleys in the CHM

represent the boundaries, not the ridges). For example, Hyyppä et al. (2001) used a CHM method similar to delineating an inverted watershed to avoid inappropriate shapes for tree crowns, but included a condition related to the center of gravity to make sure that the segmented area was roughly circular (i.e., that it conformed to a realistic tree crown shape). Similarly, Weinacker et al. (2004) used rules to identify shapes with a boundary length that was more than 2.5 times the crown diameter to ensure realistic tree crown detection, and then further analyzed the delineated crowns to split overly elongated shapes (which probably represented more than one tree) into multiple parts.

In addition to watershed methods, Persson et al. (2002) used an active-contour method to generate a DSM for use in tree crown segmentation, and Holmgren and Persson (2004) used a parabolic surface-fitting method to identify tree crown shape for use in segmentation and identification of stem positions. Researchers have also grown regions outward, starting from seed points, with these points normally representing the local maxima in the CHM. For example, Leckie et al. (2003) employed a valley-following approach to isolate crowns, whereas other researchers have developed methods for segmentation that integrated LiDAR data with satellite imagery or with aerial photos (Brandtberg et al. 2003; Popescu et al. 2003). Brandtberg et al. (2003) converted LiDAR points into a raster image in which brightness levels were used to represent the recorded height. The authors then used a second-derivative filter to detect blob-shaped structures (i.e., tree crowns). Solberg et al. (2006) imposed a restriction on the region-growing algorithm so that the delineation could only grow in a star shape, thereby preventing the growth of long and thin shapes that were unlikely to represent trees. However, because individual trees are detected from 2D surfaces in all surface-based methods, this approach may become problematic when the canopy has multiple strata and when crowns touch or intersect. The delineation also assumes perfectly vertical boundaries at the CHM valley locations, which is unrealistic.

Detection and Delineation of Individual Trees: Point-Based Methods

With point-based methods, LiDAR points are not interpolated onto a surface; instead, all points of interest are taken into consideration for processing. For example, Morsdorf et al. (2003) used 3D *k*-means clustering to create groups of points by using local maxima of the CHM as starting locations for the cluster centroids. Korpela et al. (2007) used a template-matching method in which templates with different crown shapes were created using explicit mathematical functions; the templates were designed to match the LiDAR points so that tree crowns that were enveloped within the templates could be segmented. Popescu and Zhao (2008) used a voxel-based approach to divide single trees into height bins and then fitted a fourth-order polynomial function to the height profile to arrive at height estimates at the base of the tree crown. In this context, "voxels" (3D equivalents of 2D pixels) represent rectangular solids (normally cubes) distributed through the 3D space of a site; each voxel can contain no, one, or multiple LiDAR points. This is analogous to 2D pixels, which can contain either no spectral signature or a pure or mixed signature. Holmgren et al. (2008) defined an "alpha shape" for each tree, which resembles a 3D convex hull but allows inward curvature at some point locations on the hull (in this field of research, the surface is called a *hull* because it resembles the hull of an inverted ship). The amount of allowed curvature is determined by a user-defined parameter called alpha (α), which represents the radius of a circle (in 2D) or sphere (in 3D) at a position where it contains no points. The smaller the value of α, the more curvature is allowed for the alpha shape; therefore, as α approaches ∞, the alpha shape becomes a perfectly convex hull. Since the alpha shape is based on a series of tetrahedrons, Holmgren et al. (2008) calculated the sum of the surface area of the tetrahedrons within a voxel defined with respect to tree height and obtained the base of the tree crown by locating the minimum calculated area.

Reitberger et al. (2009) combined a normalized graph-cut function with watershed segmentation and was able to improve the tree detection rate compared with using the watershed method alone. Graph-cut segmentation is a method for separating objects according to a mathematical model, in which the best separation of objects occurs when the model attains a minimum, since the goal is to identify individual objects (i.e., trees), which have maximum internal similarity, while separating points into different objects based on minimum similarity among canopies. The proportion of the trees detected in this example increased from 47 to 56% when 3D normalized-cut segmentation was used in addition to the 2D watershed segmentation. In another example of 3D detection, Kato et al. (2009) developed a method based on radial basis functions to create an iso-surface that wrapped around surface points that represented tree crowns.

A study comparing six tree detection algorithms for a variety of forests in Brazil, Germany, Norway, and Sweden showed that detection rates decrease when tree density increases (Vauhkonen et al. 2011). None of the compared algorithms exhibited a significantly better detection rate, which suggests that the overall performance depends more on the forest structure than on the actual algorithm that is used.

An advantage of point-based methods is that they make no assumptions about the degree of canopy intersection and therefore are not constrained by prior assumptions, but this limits the ability to detect individual trees in dense stands (where canopy overlaps are common) or where multiple strata of crowns exists. A disadvantage is that the computational cost is much higher for 3D data than for a flattened (2D raster) representation of the study site.

Secondary Measurements

In this section, we discuss the secondary level of information that can be retrieved from LiDAR data, beyond simply using range information to measure tree height. Secondary measurements exploit information from the 3D distribution of points and allow derivation of attributes for vertical columns of points. By studying the characteristics of the spatial distributions of LiDAR points, it becomes possible to infer or model stand attributes such as DBH, biomass, BA, and wood volume (yield). In the following sections, we will discuss three approaches for extracting secondary information from LiDAR point clouds. For studies involving wide extents or for operational forest management, both of which generate vast quantities of data to be processed, area-based approaches are often preferred. The first approach combines LiDAR data with allometric equations and other regression models to compute and map forest characteristics such as biomass, DBH, and wood volume over vast areas. Next, we discuss mapping of vegetation types using approaches similar to land cover classification. Finally, we examine methods for obtaining species information from LiDAR point clouds. We conclude these sections with a case study conducted near Thessalon, Ontario, Canada, where LiDAR and geometric spatial processing were used to detect potentially hazardous trees near a power transmission line corridor to support vegetation management.

Regression Models and Allometric Equations

Allometric equations and other regression models have been useful for many applications of forest description because the concepts are easy to implement and are easily applied to area-based models. Users define a region of interest that contains points to which attributes are attached and are subsequently used in the calculation of metrics. Examples of metrics include the mean, standard deviation, skewness, and kurtosis of height (or height percentiles), and the values can be used as inputs for regression models that typically predict forest inventory attributes such as biomass, BA, DBH, stem density, tree crown size, height of the base of the crown, and stem volume. Using field-validation data, prediction equations can be empirically established and then applied elsewhere. As in the case of regression models and allometric equations, these equations are popular for estimating forest parameters. Allometric equations are mathematical models that relate tree biomass to variables such as DBH or tree height. Allometric equations are empirically derived, normally through field measurements, and are often site specific, but can sometimes be applied locally, regionally, or even more broadly within specific and clearly defined scopes (i.e., under conditions similar to those for which the model was developed).

A major benefit is that as long as a forest's composition and the environmental conditions that influence its composition and structure remain consistent, the equations can be reused. Conversely, allometric equations cannot be applied in areas where the underlying characteristics of the sites differ from those at the sites used to develop the equations. Lefsky et al. (1999) effectively predicted biomass in eastern Maryland, USA, with an allometric equation derived by Monk et al. (1970) that relates biomass per stem to DBH, but estimated the DBH values from LiDAR data. Lefsky et al. (2005) predicted biomass at three field sites (in Oregon and Tennessee, USA, and in Santarém, Brazil) using equations derived from Brown (1997) and Nelson et al. (1999).

These examples show that with different study areas and LiDAR datasets, measuring the same forest parameters can lead to different allometric equations. Similarly, regression equations can be generated for site- or study-specific purposes when ground-truthing data are available. Drake et al. (2003) estimated the quadratic mean stem diameter, BA, and aboveground biomass near the Sarapiquí River in northeastern Costa Rica. In their study, the stem diameter, BA, and biomass were estimated at the footprint level (0.05 ha) and the plot level (0.25–0.50 ha) using two separate sets of allometric equations.

The allometric equations derived by Brown (1997) have been used for estimating oven-dried aboveground biomass for each stem. In another example, from eastern Texas, USA, Zhao et al. (2009) estimated tree height, crown width, and height of the base of the crown for pines and deciduous trees from LiDAR-derived height metrics. Biomass and DBH were estimated using four different models derived by Jenkins et al. (2003), namely two regression equations for DBH and two allometric equations for biomass (each separating pines and deciduous trees). Careful calibration and selection of equations are essential for accurate estimates, since empirically determined coefficients can vary greatly among sites and influence allometric predictions.

The area-based regression method is efficient and simple for mapping large study areas. For example, Maltamo et al. (2016) predicted the aboveground biomass in mountain forests in Norway, in a study area that was 1500 km long and approximately 200 m wide. Næsset (2004) studied forest stand attributes such as mean diameter, number of stems, BA, and volume for a 6500-ha study area. In Ontario, Woods et al. (2010) predicted forest variables for each 400-m^2 tile in a 630,000-ha forest using data provided by the local forestry company (Tembec Inc.). These large-area studies have proven the effectiveness of using LiDAR to support forest operations.

Vertical Profile for a Single Tree

The approach described in section "Regression Models and Allometric Equations" can also let researchers calculate statistical parameters for individual trees. Attributes can be derived from the vertical point distribution of a group of points (such as a single tree or single tree crown) for classification purposes. In many cases, species have characteristic crown forms and the vertical frequency distribution of points differs among species, allowing them to be classified. To visually illustrate how different species may lead to a different vertical profile of LiDAR points, Fig. 11 shows an example of vertical profiles for a pine, poplar, and maple tree based on the first return, single (i.e., only) return, last return, and all return types combined.

There are three primary families of features that can be extracted from vertical profiles. The first considers the distribution of returns in the entire column or with that column sliced into vertical groups as a percentile (or other intervals such as deciles) of the height: the total count or proportion of the return types, height summaries, and descriptive statistics such as means and standard deviations. Since a single LiDAR pulse can reflect multiple times and produce multiple returns, it is possible to calculate the proportions of first, single (i.e., only), intermediate, and last returns for each height

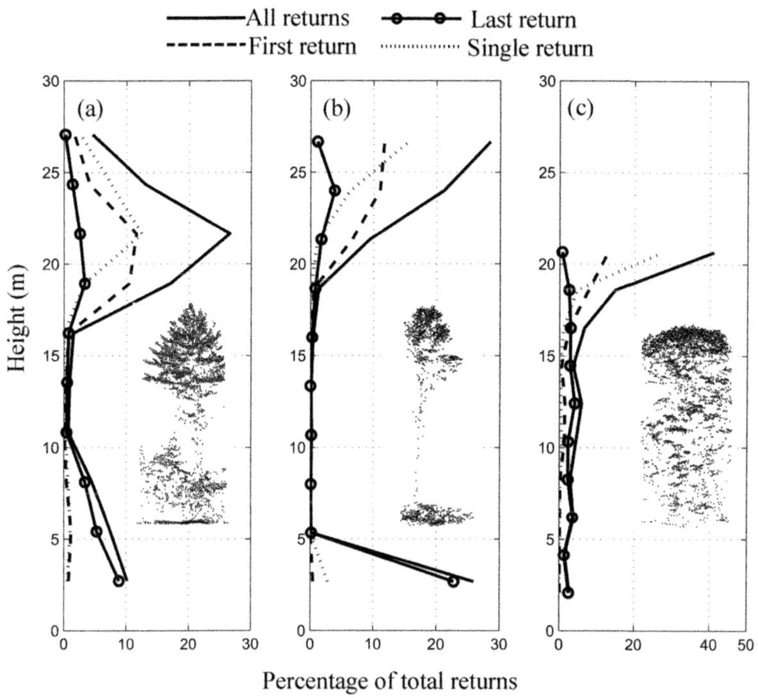

Fig. 11 Frequency distributions of LiDAR points as a function of height for three tree species. The frequency distributions are provided for all returns, the first return, the last return, and the single return for (**a**) pine, (**b**) poplar, and (**c**) maple

interval within the point cloud column. In the second family, the relationship between height and return frequency can be used to characterize the abundance of returns within a vertical column through the forest canopy. The third family of features can represent any summary or descriptive statistic (e.g., mean, standard deviation, coefficient of variation, kurtosis, skewness) for the height or intensity measurements in each interval. Calculating vertical profile features based on the entire tree emphasizes the importance of analyzing a tree as a whole object, whereas analyzing the features derived from intervals can identify and exploit the internal variations of returns relative to their position within the tree. Both groups of features can be used as inputs for classifiers or models for estimating or inferring secondary features of a tree or stand.

Classification of Vegetation Types

The application of LiDAR to classify land types or land covers is a complex topic, and is beyond the scope of this chapter. In summary, LiDAR data can be used to classify forest types or vegetation cover types with characteristically different LiDAR profiles. For example, Brennan and Webster (2006) generated surfaces

(DSM, DEM, intensity, multiple echoes, and normalized heights) from LiDAR data and classified their study area into ten classes, including four vegetation classes: arid low vegetation, and coniferous, deciduous, and low vegetation growing in wet soils. Similarly, Antonarakis et al. (2009) generated eight surfaces from LiDAR data in 10 m × 10 m cells: the canopy surface, terrain, vegetation height, intensity, intensity difference, skewness, kurtosis, and percentage canopy model (the percentage of LiDAR hits that were reflected from the canopy compared to the total LiDAR hits in a 10 m × 10 m cells). They used two methods to classify the area into nine categories, the first included the influence of ground and the other had ground points removed before classification. The result was six vegetation classes, including short vegetation, young planted forest, intermediate planted forest, mature planted forest, young natural forest, and mature natural forest.

Instead of using LiDAR data to generate multiple surfaces, as in the two previous methods, vertical profile metrics (statistical attributes derived from height percentiles) can also be used as features for classification. For example, Falkowski et al. (2009) used 34 features derived from a vertical column to classify different successional stages, including open, stand initiation, young multi-stratum, mature multi-stratum, and old multi-stratum stages. Although the classification of vegetation is not forest specific (classification classes often contain non-forest vegetation types), the classified maps are useful in many forest-related studies.

Tree Genus and Species Classification

It is common to further classify points in the LiDAR point cloud that have been identified as non-terrain points into subclasses that are related to the types of objects being scanned. In urban areas, possible subclasses include buildings, roads, and vegetation, whereas in natural areas, possible subclasses include vegetated areas, open water, and exposed rock. Classification of vegetation to the genus or species level can provide valuable information for forest management, particularly for predicting the growth and yield rates used to update a forest resource inventory or for scheduling harvests.

To classify LiDAR points belonging to a single tree, features (attributes) must be extracted to describe the associated point cloud. To classify LiDAR points into individual tree genera or species, it is often easiest to base the classification on features obtained from the whole column of LiDAR points that represent a single tree. These features can be calculated from either the entire column of LiDAR points or the summarized aggregates of the points after they have been grouped into percentiles, deciles, or other statistical bins. Feature descriptors are then computed from each vertical bin of LiDAR points to produce a vertical profile. A second way to classify tree species from LiDAR data is by exploiting the geometry of the tree, since geometric descriptors can be constructed or extracted from points in the LiDAR point cloud (e.g., convex hull attributes, reconstruction of the branching pattern). Third, classification can involve the combination or fusion of spectral information with the LiDAR point cloud data. Each of the three methods has shown promising results.

Holmgren and Persson (2004) used features of the vertical profile to identify Norway spruce (*Picea abies*) and Scots pine (*Pinus sylvestris*) individuals by utilizing the proportion of return types (i.e., the proportion of single returns and the proportion of the first returns), with an accuracy of 95%. Furthermore, consideration of the height attributes from the 90th height percentile divided by the estimated tree height and the relative standard deviation of heights helped to separate the species. The authors also used attributes related to the signal intensity from the surface returns, such as its standard deviation or mean, for the classification. Additional attributes, such as parameters of the fitting of parabolic surfaces to the top of each tree, described the tree crown geometry; these concepts are described in more detail in the next section.

In another example, Moffiet et al. (2005) used the proportions of return types, heights, and intensity attributes of the canopy to identify cypress pine (*Callitris columellaris*), poplar box (*Eucalyptus populnea*), silver-leaved ironbark (*Eucalyptus melanophloia*), smooth-barked apple (*Angophora costata*), and brigalow (*Acacia harpophylla*). LiDAR data are often acquired during the summer (leaf-on conditions), but vegetation classification by Kim et al. (2009) used both leaf-on and leaf-off data separately and jointly, along with attributes such as the proportion of return types, heights, and intensities to identify eight broadleaved species and seven coniferous species with up to 74.9% classification accuracy.

Many other studies have used similar approaches. For example, Suratno et al. (2009) used the proportion of return types and the mean and standard deviations of heights and intensity to identify ponderosa pine (*Pinus ponderosa*), Douglas-fir (*Pseudotsuga menziesii*), western larch (*Larix occidentalis*), and lodgepole pine (*Pinus contorta*). Using attributes derived from the 10th, 50th, and 90th height percentiles for return types, heights, and intensities, Ørka et al. (2009) classified large Norway spruce and birch (*Betula* spp.) successfully, with 74% accuracy. Using two LiDAR sensors, Korpela et al. (2010) achieved an accuracy of up to 90% by identifying Scots pine, Norway spruce, and birch based on the proportion of return types, height, and intensity attributes calculated for height deciles.

The vertical-profile approach leverages variations of the point distributions within the vertical column, and whether or not the attributes are related to the entire tree or to a particular percentile, these attributes take advantage of LiDAR's ability to penetrate the canopy and see past the upper surface. The attributes derived from the point cloud *relate* to the geometry of the tree crown, but by not explicitly deriving geometric information; another set of attribute options can be derived that unambiguously capture a tree's geometric characteristics.

Trees grow in a variety of shapes and sizes, but trees of the same species tend to have general similarity with respect to the positioning and shape of the crown and the arrangement of branches; together, these similarities define the form of the tree. The general form that a species exhibits in the vertical plane facilitates tree identification from silhouettes. This property has been exploited for species classification in the field. Since LiDAR provides 3D information about trees (i.e., in essence, their form), research into incorporating the geometric measurements that define tree form into the classification of point clouds has become increasingly popular. Holmgren and Persson (2004) used vertical profile attributes and attributes derived from fitting a parabolic surface to LiDAR points at tree tops, and therefore included the extent of the curvature of the parabola in their analysis. Yao et al. (2012) also used attributes related to the curvature of a parabolic surface along with attributes of the vertical profile to classify individuals as coniferous and deciduous trees and to further discriminate between spruce (*Picea* spp.) and fir (*Abies* spp.).

Alpha shapes include a set of 2D or 3D points, and are similar to convex hulls that allow internal curvature; to understand this, imagine shrink-wrapping a tree's crown. Vauhkonen et al. (2008, 2009, 2010) used attributes from the alpha shapes of individual trees to estimate tree parameters such as DBH, volume, age, and crown diameter. Vauhkonen et al. (2008) used alpha shape metrics to classify Scandinavian commercial species, including pine, spruce, and deciduous trees. Ko et al. (2013) discussed and derived geometric attributes related to linear features derived within tree crowns (i.e., stems and branches), convex hulls, point clustering within crowns, and ratios between tree height and relative canopy vertical thickness, and used this approach to identify pine, poplar (*Populus* spp.), and maple (*Acer* spp.) genera. Subsequently they combined these geometric attributes with vertical profile metrics to further improve the classification accuracy (Ko et al. 2014).

Ensemble classification is a strategy that is being used to improve classification accuracy that works by combining results from more than one base classifier to solve a common classification problem. The advantage of using ensemble classification is that the training can be conducted independently for each base classifier. Because the base classifiers are trained independently, each base classifier makes an independent decision which may be the same or different among each base classifier. A vote among the base classifier results yields the final decision and this is seen as more robust than using only a single classifier. Combining information from multiple sources by means of ensemble classification can solve certain problems. In this approach, outputs from two or more related models are combined to provide insights that neither model could provide by itself. This is particularly useful when the information sources or types are mutually exclusive, such as using the combination of geometric and vertical profiles to classify tree species (Holmgren and Persson 2004; Ko et al. 2014).

Spectral information obtained from the top of the canopy can be combined with LiDAR data to further improve the ability to classify features. Typically, the simplest fusion method is to associate a pixel's spectral values with the nearest LiDAR point to produce combinations of positional, geometric, and spectral features. Koukoulas and Blackburn (2005) used LiDAR data to generate a CHM; they then obtained tree top and crown delineation polygons from a combination of LiDAR data and Airborne Thematic Mapper (ATM, type 1268) imagery and used this data to estimate the location, height, and species of individual trees. Holmgren et al. (2008) combined LiDAR data with airborne spectral data to consider the mean pixel values from each delineated tree polygon using three spectral bands obtained from the Digital Mapping Camera (Hexagon Safety & Infrastructure) at 500–650 nm, 590–675 nm, and 675–850 nm and LiDAR-derived attributes such as height distributions, canopy shape, proportion of pulse types, and intensity of returns.

In the previous examples, spectral information was obtained from another platform and subsequently combined with the LiDAR data, which differs from the use of multispectral LiDAR; in the latter, data from multiple LiDAR bands is obtained simultaneously during the same survey. Generally, spectral information obtained from imagery does not form a one-to-one relationship with the acquired LiDAR points, and in this case, there may be multiple LiDAR points within the 2D extent of the given pixels of the multispectral image. In this case, the spectral information is normally associated with the LiDAR point that has the highest elevation, but various averaging options could be used, including weighted averaging or linear combinations spanning multiple spectral channels. To overcome this problem, newer multispectral LiDAR scanners are increasingly capable of recording multiple channels simultaneously, and thus, each LiDAR point contains positional information and multispectral data acquired simultaneously for the same point.

Case Study: Identifying Potentially Hazardous Trees

In this section, we provide details of a case study (Ko et al. 2012, 2013, 2014) that used high-density ALS LiDAR to identify and map potentially hazardous trees along a hydroelectric power transmission corridor near Thessalon, Ontario, Canada. This study sought to identify trees that had the potential to endanger the hydroelectric transmission lines by either growing into the lines or falling onto the lines. The assessment was based on interpreting LiDAR point clouds to identify individual trees and assess the angle at which the trees were leaning from interpreted geometric features; the information was then combined with the genus of each tree to map the trees that represented potential hazards.

LiDAR data was acquired on 7 August 2009 from an altitude of 140 m above the local ground level with a pulse density of approximately 40 points per m^2, and up to five returns per pulse. Individual trees ($n = 186$) were surveyed in the field to form a validation dataset; for each tree, their location, species, and DBH were recorded. Species identified in the field included white birch (*Betula papyrifera*), sugar maple (*Acer saccharum*), northern red oak (*Quercus rubra*), jack pine (*Pinus banksiana*), trembling aspen (*Populus tremuloides*), white pine (*Pinus strobus*), white spruce (*Picea glauca*), and eastern larch or tamarack (*Larix laricina*). Of the validation samples, 67 were pines, 59 poplars, and 34 maples; the remaining 26 trees formed a mixture of minor species and are not part of this case study. Pre-processing required manual segmentation of the LiDAR point cloud to extract individual trees; this step was not automated because it was not the focus of the geometric analysis.

For each tree, 24 geometric features were derived to characterize the form of the tree (Ko et al. 2013). To reduce the complexity of subsequent data processing, an automated feature reduction method was implemented (Ko et al. 2014), which decreased the number of geometric features that were considered to be six. First, each tree crown was segmented into different point clusters, with each cluster representing a branch. The best-fit lines, planes, and volume metrics were calculated for each cluster and used to identify the genus. The six geometric features were the average derived best-fit line segment lengths divided by tree height; the average line segment lengths multiplied by the ratio of crown height to tree height; the volume of the tree crown's convex hull divided by the number of points in the tree crown; the average distance from each LiDAR point to the closest facet of the derived convex hull; and the tree's crown height divided by the tree height. To calculate the 6th metric, a buffer was extended from each LiDAR point outward to a radius of 5% of the tree height, then the overlap volume of that sphere with all adjacent spheres was calculated for each point and divided by the number of points in the tree's crown.

Genus classes were determined using the Random Forests software (Breiman 2001; Liaw and Wiener 2002), which can perform nonparametric ensemble classification to classify tree genus based on training from a subset of the data. The method uses a ranking system of numerous randomized classifications to ultimately select the most likely prediction.

To detect and map potentially hazardous trees, we first identified the location of the transmission line, then we identified the individual trees near the transmission line, and delineated the trees using planar polygons. According to the North American Electric Reliability Corporation FAC-003-2 standard (http://www.nerc.com/filingsorders/us/fercordersrules/e-5_order_fac-003-2_2013.3.21.pdf), the transmission line at the field site requires 15.24 m (50 ft) of clearance. Therefore, we constructed a buffer zone that extended outwards from both sides of the transmission line to create a minimum vegetation clearance distance (MVCD) zone. Next, we delineated trees that intersected the MVCD zone, even if only partially. These marked trees form the first subset of trees that represent potential hazards. We then identified trees taller than 15.24 m as having the potential to contact the transmission line if they fell in that direction. To identify the angle of each potentially hazardous tree from the vertical, we connected the centroids of a vertically moving voxel for each individual tree's LiDAR point cloud to obtain the best estimate for the location of the main stem. By fitting a best-fit line through these points, we approximated the general form and lean of the main stem. To determine whether the tree was leaning towards or away from the transmission lines, we calculated the perpendicular distances from each tree's top and base to the transmission line. If the distance was smaller for the top, then the tree was leaning towards the transmission line. Finally, we flagged trees that could potentially contact the power lines if they fell in the direction of their lean and linked the genus classification with the lean angle for mapping in GIS software.

Figure 12 shows an overview of the LiDAR data for the study site. Site 1 (Fig. 12a) is illustrated in Figs. 12b–d. To identify trees to the genus level, tree samples were taken from all eight sites (Fig. 12a) to capture some of the growth variability caused by environmental variation. Figure 13 shows a plan view of Site 1, with the MVCD zone (*grey*) and trees that had been surveyed in the field in color. The crowns of trees that overlapped this zone are shown in *yellow* or *red*. *Red* trees indicate trees that are in the MVCD zone and taller than 15.24 m and *yellow* trees indicate trees that are in the MVCD zone and shorter than 15.24 m. *Green* trees are

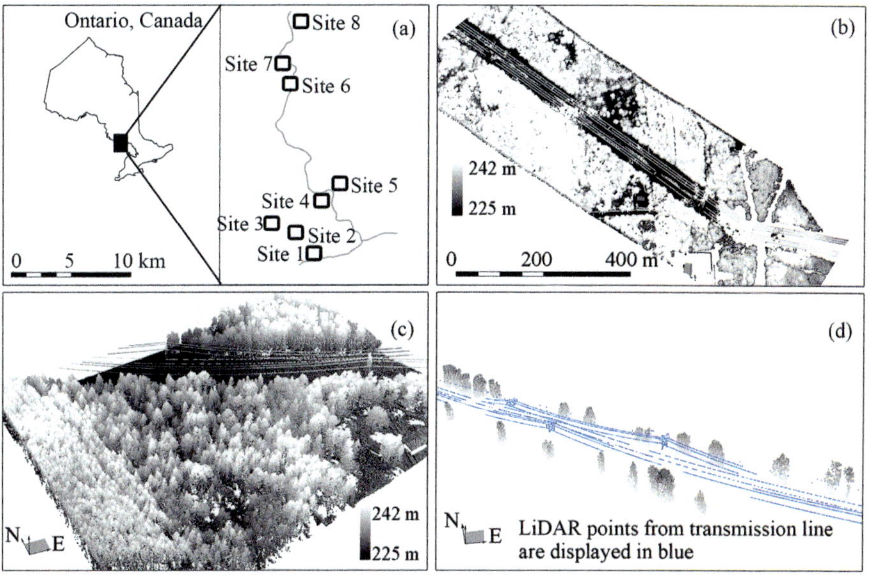

Fig. 12 (**a**) A plan view of the study site, showing the locations of the hydroelectric transmission corridor and of the eight sampling sites in the adjacent forest. (**b**) Site 1 is shown from an oblique perspective, with LiDAR points shaded based on reflection height (lighter = taller). (**c**) 3D view of the transmission line and surrounding forest. (**d**) 3D perspective, indicating individual trees that were identified within the MVCD, with the transmission lines shown in *blue*

the trees that were surveyed in the field but that do not intersect the MVCD zone. Therefore, *red* trees are tall enough to potentially contact the transmission lines if they fall. Figure 14 displays both the trees and their angle from the vertical, shown as a line through each tree's point cloud based on the estimated position of the main stem. Treetops marked with a cross are leaning towards the transmission lines and pose a potential hazard. Figure 15 combines the information obtained from the genus classification with the MVCD zone hazard assessment.

This case study demonstrates LiDAR's ability to monitor trees over a large area and to support an assessment of the potential hazards they pose to the power lines in a hydroelectric transmission corridor. Since trees can grow toward and potentially fall on critical infrastructure, the same approach could be used in other contexts to detect, map, and react to potential hazards in a timely manner. The ability to identify trees to the genus level based on geometric summary metrics extracted from LiDAR point clouds and to integrate this data with GIS buffering, voxel analysis, and measurements of distances and angles demonstrates the utility of LiDAR data. The distance parameters, species identities, and any additional filters that might be deemed necessary to identify hazards can be chosen by the user of the data based on their specific needs. The ability to integrate geometric and 3D data into a unified analytical framework underscores the potential of LiDAR technology.

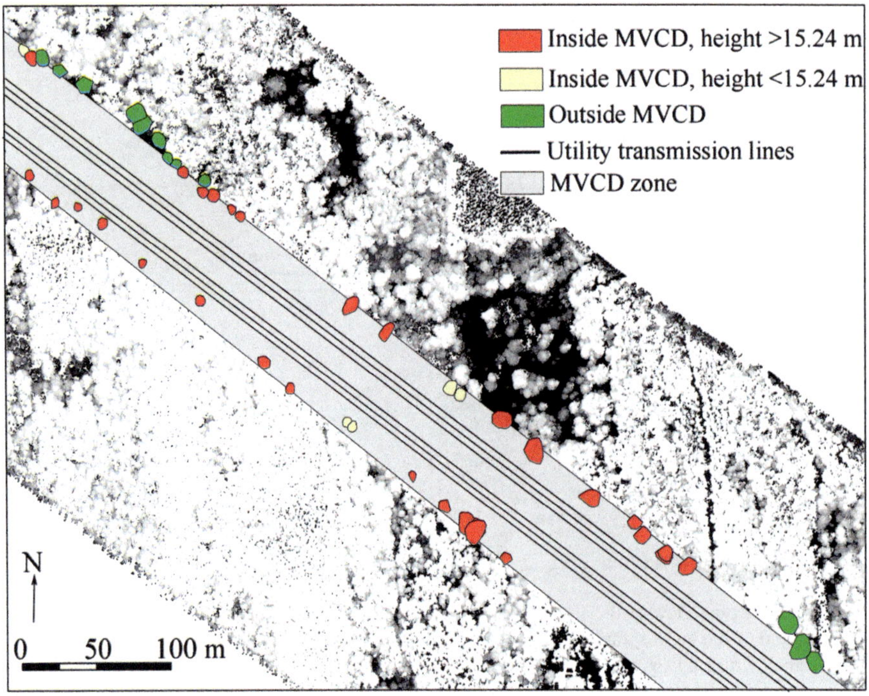

Fig. 13 Map of tree crowns sampled along the hydroelectric transmission corridor (*grey* area). Color codes identify trees that were validated in the field, the subset of trees that overlap the MVCD buffer zone (*red* and *yellow*), and the subset of the latter trees that are taller than 15.24 m and that could potentially fall on the lines (*red*). The forest outside the *grey* area represents the LiDAR point cloud flattened into a 2D plan view

The Future of LiDAR

The use of LiDAR technology in forest applications is still developing rapidly to take advantage of improved (and cheaper) laser scanning technology, multispectral scanners, faster computers, vastly expanded data storage capacities, improved analytical algorithms, and integration with other geospatial sectors. With greater numbers of users, there is also more incentive to build more tools and write more software to support processing, handling, interpretation, and manipulation of the LiDAR data; thus, opportunities for using these data sources will become increasingly abundant, but the tools will also become increasingly user friendly and accessible. Although the costs associated with SLS are still prohibitive, the costs associated with ALS are reasonable and users of the technology now regularly mount scanners on airplanes, helicopters, and unmanned aerial vehicles to perform their surveys. This flexibility is leading to new and innovative uses of the technology.

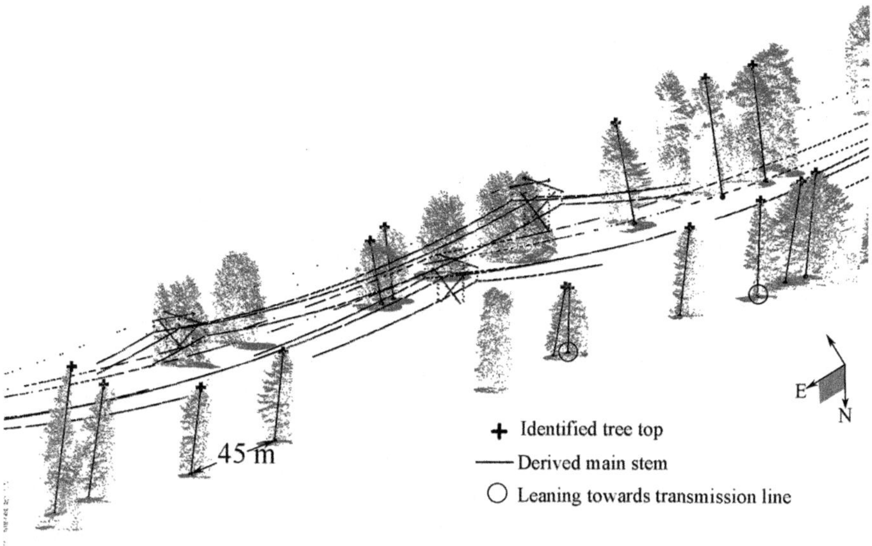

Fig. 14 Example of results showing the main growth direction of the subsample of trees that could potentially fall on the transmission lines. The straight vertical lines are drawn through the centers of the point cloud and connect the tip of the crown with the stem's contact with the ground. Only trees that are within the MVCD zone and that are taller than 15.24 m and leaning towards the lines (circled trees) are considered hazardous

In contrast with the past, software is becoming less expensive or even available at no cost. Although "open-source" software is a relatively recent development, it has quickly become popular among LiDAR users. Open-source software is developed for the benefit of a community, and is made available freely for researchers and other users to use, modify, and distribute. Because of the large benefits this approach offers to the community of LiDAR users and the growing pool of talented users, tools for viewing and processing massive datasets are rapidly advancing. The increased availability of free or inexpensive data and software is leading to rapid advances in the acquisition, processing, and analysis of LiDAR data. Such developments, in turn, attract investment from the business sector, creating a self-perpetuating cycle of improvement.

Hardware advances such as more powerful computers with better storage have also advanced LiDAR development. For example, storage capacity and the associated cost have decreased substantially, and the speed and memory of video cards have increased to the point where real-time manipulation and viewing of 3D data is a reality.

GIS software increasingly includes the ability to process and display 3D data. For example, ESRI's ArcGIS software (http://www.esri.com/software/arcgis) includes an integrated application (ArcScene) for processing 3D data, with specific tools provided for importing, exporting, processing, and analyzing LiDAR point cloud data. Such developments have encouraged the GIS community to incorporate LiDAR data

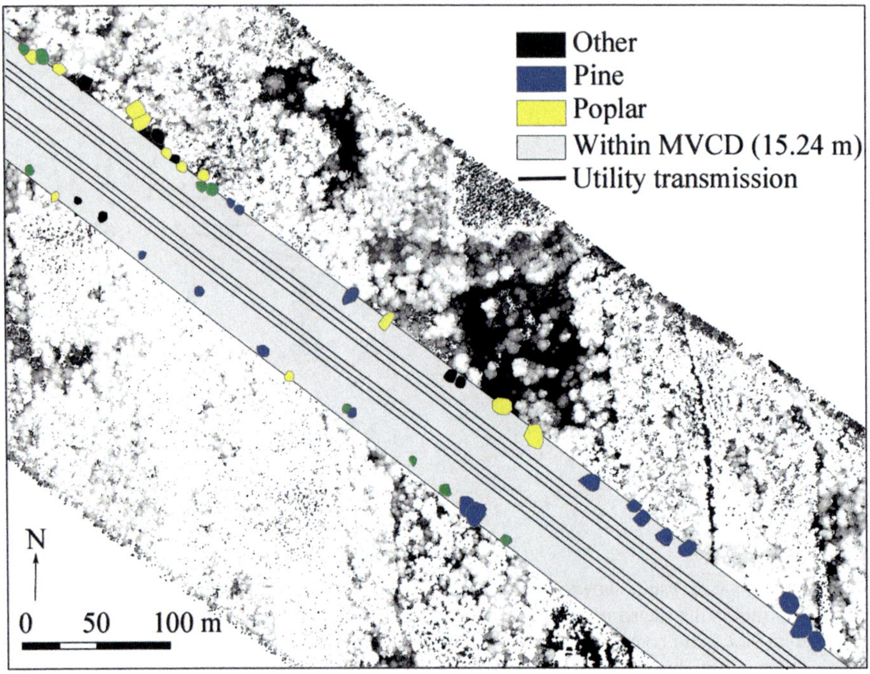

Fig. 15 Map of the hydroelectric transmission corridor (*grey*), showing the trees that were validated in the field, those that overlap the MVCD, and those that pose a leaning hazard. Areas outside the transmission corridor are the flattened LiDAR point cloud for the surrounding forest

into their projects. File format standards for LiDAR data have (nearly) been standardized, with the American Society for Photogrammetry and Remote Sensing (https://www.asprs.org) developing the *.las* file format to replace older formats such as *.xyz*, *.ascii*, or *.txt*. This will greatly simplify the process of data sharing.

Although the area of coverage is smaller when the scanner is mounted on a platform that flies at a lower altitude, the flexibility increases, and although the time required to cover a given area increases, the frequency and density of the coverage can be much greater than with SLS or TLS LiDAR. Miniaturization of the technology is reducing payload weight, increasingly allowing the use of LiDAR technology on lightweight platforms such as UAVs, and increasing the potential for remote and mobile data collection. Once it becomes possible to design multiple sensors that can be carried by platforms that can currently only carry a single sensor, the opportunities for increased multi- and hyperspectral data acquisition and real-time data fusion will improve greatly.

Coupling these improved data collection options with truly mobile data acquisition platforms (e.g., UAVs, automobiles, humans) will allow changes in the perspectives for data collection (e.g., above, below, and within a canopy) and will allow data collection for specific purposes (e.g., mapping hiking trails, tracking large-mammal movement through a dense understory, nest selection by migratory birds, access to an area using all-terrain vehicles). It is foreseeable that small, rugged technology such as that which permits the now-ubiquitous GoPro camera (https://gopro.com) and its cinema-quality videography will become available to allow equally portable remote sensing, ranging, and detection tools that will revolutionize LiDAR data collection. In the not too distant future, animals may be outfitted with LiDAR-enabled GPS collars that provide information on both their movements and their habitat characteristics.

GIS software has only recently started to harness the third and fourth dimensions. Ongoing development from technical, data handling, processing, display, and analytical perspectives will improve opportunities to benefit from the ample data streams that an expanding multispectral LiDAR data collection network will provide. Multispectral LiDAR that provides information such as red, green, blue, infrared, or near-infrared bands will allow the processing of environmental objects, and especially living objects such as vegetation, which is important in the context of this chapter (i.e., forestry applications). Multispectral LiDAR information, combined with precise location information for individual points, has great potential to improve the identification and classification of individual trees, species, characteristics of the forest structure, and environmental conditions.

Spectral information can provide additional classification of features, particularly when used in conjunction with the geometric features and vertical profile characteristics that we have discussed in this chapter. Classification features derived from spectral information have great potential for extraction of new information because the features are often independent of the vertical profile features or geometric features; thus, the additional features derived from the spectral information are less likely to be correlated with traditional LiDAR metrics. The intensity values provided by either single-channel or multiple-channel LiDAR systems require careful geometric and radiometric calibration before the values can be used effectively. Moving forward, research will continue in this domain to find ways to use the intensity values measured by LiDAR scanners as true measured quantities rather than simple indicators.

As LiDAR data becomes cheaper and more readily available, and as data collection becomes easier, data acquisition will become increasingly frequent. A key hurdle to overcome will then be the development of new algorithms to handle the increasing variety, volume, and frequency of the data streams being collected. Newer algorithms and methods for handling data are needed that will permit the handling and analysis of *big data*; related challenges include the storage, processing, sharing, and extraction of tremendous volumes of multidimensional data. There are still some challenges that must be overcome to meet the high demand for onboard LiDAR data processing (e.g., onboard classification, onboard object recognition) to accelerate decision-making processes and final product generation to support environmental, political, planning, and other purposes without requiring time-consuming post-processing of the data.

In terms of classification, ensemble theory has become a popular tool for handling complex cases; efforts to combine multiple classifiers to support better decisions (e.g., to obtain higher classification accuracy) is an area of current emphasis.

As the data volumes increase, ensemble classification can also become an effective method to process the data by grouping it into smaller chunks. The final classification decisions can then be made by combining the decisions from multiple classifiers both to increase classification accuracy and to reduce the processing time (e.g., by taking advantage of parallel rather than serial computing). In ensemble theory, the best scenario is the use of multiple classifiers simultaneously, working with highly diverse classification feature sets, so that the classification features are not redundant; as a result, the development of multiple-channel LiDAR is well suited for these types of methods.

Another popular LiDAR-related project for studies of forested landscapes is the single-photon LiDAR space mission (https://spinoff.nasa.gov/Spinoff2016/ps_6.html) funded by the National Aeronautics and Space Administration (NASA) as part of the ICESat-2 project. In this project, the LiDAR scanner will record every photon that is reflected from targets (at a density of about 12 points per m^2). This mission is scheduled to launch in 2017 and the scanner is expected to generate 32,000 pulses per second within each of its 100 channels (arranged in a 10-by-10 array), thereby providing 3.2 million measurements per second. Since this will be a spaceborne LiDAR system, with a very high altitude and a swath width of several kilometers, the areal coverage will be vast and coverage will be rapid, which should result in lower costs per unit area than with airborne data acquisition. Once this data becomes available, it is likely to become one of the major foci for mapping forest landscapes.

The use of bathymetric LiDAR, which is capable of penetrating shallow water for mapping the bottom of bodies of water, is beyond the scope of this chapter, but a significant amount of research has been completed on this topic. One of the major differences between the LiDAR scanners used for bathymetry and those used for forestry (topographic LiDAR) is the choice of wavelength. Bathymetric LiDAR tends to use a shorter wavelength (e.g., 532 nm) rather than the longer wavelengths (e.g., 1064 nm) that are typically used with vegetation studies such that water penetration can be achieved. The near-infrared part of the electromagnetic spectrum, where vegetation has high reflectivity, is where absorption by water is high, so near-infrared scanners are not suitable for studies involving bodies of water. However, bathymetric LiDAR has many potential uses in terrestrial-aquatic environments such as wetlands, where the land is saturated with water.

To keep up with the accelerating capacity for data collection, researchers will need to improve tools and algorithms for dissecting the 3D LiDAR point clouds. This research will ultimately lead to revolutionary methods for data classification and analysis. The current problems encountered as a result of intersecting tree canopies (i.e., difficulty identifying which data points belong to each canopy) will need to be solved. Similarly, accurate separation of the ground from understory vegetation and overstory vegetation will continue to improve. Although rudimentary tree species classification has been achieved, there is considerable room for improvement of accuracy along with detailed calculations of wood volume (e.g., with branches and leaves removed), rather than relying on regression-type and other allometric relationships. Better estimation of the woody and non-woody biomass fractions will provide precious information to support the forest industry and environmental studies.

There is a push by the forest industry and governments to develop individual-tree inventories rather than the current stand-level products; the reality of this is still distant, but progress is being made. The use of smaller LiDAR instruments has increased mobility and provided us with an opportunity to automatically combine data efficiently from multiple scans with different scanning angles. Individual-tree inventories and improved maps of individual tree crowns are valuable information that will benefit both industrial users and researchers, and will continue to motivate research. Particular challenges will be to decipher the inherent complexities of forest landscapes and automate extraction of this information to support forest management.

References

Abshire JB, Sun X, Riris H et al (2005) Geoscience laser altimeter system (GLAS) on the ICESat mission: on-orbit measurement performance. Geophys Res Lett 32:L21S02

Antonarakis AS, Richards KS, Brasington J (2009) Object-based land cover classification using airborne LiDAR. Remote Sens Environ 112(6):2988–2998

Baltsavias E, Gruen A, Eisenbeiss H et al (2008) High-quality image matching and automated generation of three-dimensional tree models. Int J Remote Sens 29:1243–1259

Bouvier M, Durrieu S, Richard AF et al (2015) Generalizing predictive models of forest inventory attributes using an area-based approach with airborne LiDAR data. Remote Sens Environ 156:322–334

Brandtberg T (2007) Classifying individual tree species under leaf-off and leaf-on conditions using airborne LIDAR. ISPRS J Photogramm Remote Sens 61(5):325–340

Brandtberg T, Warner TA, Landenberger RE et al (2003) Detection and analysis of individual leaf-off tree crowns in small footprint, high sampling density lidar data from the eastern deciduous forest in North America. Remote Sens Environ 85(3):290–303

Breiman L (2001) Random forests. Mach Learn 45(1):5–32

Brennan R, Webster TL (2006) Object-oriented land cover classification of LiDAR-derived surfaces. Can J Remote Sens 32(2):162–172

Brown S (1997) Estimating biomass and biomass change of tropical forests: a primer. UN Food and Agric Org, Rome

Caruso C, Quarta F (1998) Interpolation methods comparison. Comput Math Appl 35(12):109–126

Cressie N (1988) Spatial prediction and ordinary kriging. Math Geol 20(4):405–421

Doneus M, Briese C, Studnicka N (2010) Analysis of full-waveform ALS data by simultaneously acquired TLS data: towards an advanced DTM generation in wooded areas. Int Arch Photogramm Remote Sens Spat Inf Sci 38(Part 7B):193–198

Douglas ES, Strahler A, Martel J et al (2012) DWEL: a dual-wavelength echidna lidar for ground-based forest scanning. In: Proceedings of international geoscience and remote sensing symposium 2012. Munich, Germany 22–27 July 2012, pp 1–4

Drake JB, Knox RG, Ralph O et al (2003) Above-ground biomass estimation in closed canopy neotropical forests using lidar remote sensing: factors affecting the generality of relationships. Glob Ecol Biogeogr 12(2):147–159

Evans JS, Hudak AT (2007) A multi-scale curvature algorithm for classifying discrete return LiDAR in forested environments. IEEE Trans Geosci Remote Sensing 45:1029–1038

Falkowski MJ, Evans JS, Martinuzzi S et al (2009) Characterizing forest succession with lidar data: an evaluation for the inland northwest, USA. Remote Sens Environ 113:946–956

Gaulton R, Danson FM, Ramirez FA et al (2013) The potential of dual-wavelength laser scanning for estimating vegetation moisture content. Remote Sens Environ 132:32–39

Harding DJ, Carabajal CC (2005) ICESat waveform measurements of within-footprint topographic relief and vegetation vertical structure. Geophys Res Lett 32:L21S10

Heurich M (2008) Automatic recognition and measurement of single trees based on data from airborne laser scanning over the richly structured natural forests of the Bavarian Forest National Park. For Ecol Manage 255(7):2416–2433

Hilker T, Wulder MA, Coops NC (2008) Update of forest inventory data with lidar and high spatial resolution satellite imagery. Can J Remote Sens 34(1):5–12

Hill RA, Thomson AG (2005) Mapping woodland species composition and structure using airborne spectral and LiDAR data. Int J Remote Sens 26:3763–3779

Holmgren J (2004) Prediction of tree height, basal area and stem volume using airborne laser scanning. Scand J For Res 19:543–553

Holmgren J, Persson Å (2004) Identifying species of individual trees using airborne laser scanning. Remote Sens Environ 90(4):415–423

Holmgren J, Persson Å, Söderman U (2008) Species identification of individual trees by combining high resolution LiDAR data with multi-spectral images. Int J Remote Sens 29(5):1537–1552

Hyyppä J, Inkinen M (1999) Detecting and estimating attributes for single trees using laser scanner. Photogramm J Finland 16(2):27–42

Hyyppä J, Kelle O, Lehikoinen M et al (2001) A segmentation-based method to retrieve stem volume estimates from 3-D tree height models produced by laser scanners. IEEE Trans Geosci Remote Sens 39:969–975

Hyyppä J, Hyyppä H, Leckie D et al (2008) Review of methods of small-footprint airborne laser scanning for extracting forest inventory data in boreal forests. Int J Remote Sens 29:1339–1366

Hyyppä J, Yu X, Hyyppä H et al (2012) Advances in forest inventory using airborne laser scanning. Remote Sens (Basel) 4:1190–1207

Jenkins C, Chojnacky DC, Heath LS et al (2003) National-scale biomass estimators for United States tree species. For Sci 49(1):12–35

Kato A, Moskal LM, Schiess P et al (2009) Capturing tree crown formation through implicit surface reconstruction using airborne lidar data. Remote Sens Environ 113(6):1148–1162

Ke Y, Lindi JQ, Im J (2010) Synergistic use of QuickBird multispectral imagery and LIDAR data for object-based forest species classification. Remote Sens Environ 114(6):1141–1154

Kim S, McGaughey RJ, Andersen HE et al (2009) Tree species differentiation using intensity data derived from leaf-on and leaf-off airborne laser scanner data. Remote Sens Environ 113(8):1575–1586

Kim S, Hinckley T, Briggs D (2011) Classifying individual tree genera using stepwise cluster analysis based on height and intensity metrics derived from airborne laser scanner data. Remote Sens Environ 115(12):3329–3342

Ko C, Remmel TK, Sohn G (2012) Mapping tree genera using discrete LiDAR and geometric tree metrics. BOSQUE 33(3):313–319

Ko C, Sohn G, Remmel TK (2013) Tree genera classification with geometric features from high-density airborne LiDAR. Can J Remote Sens 39(S1):S1–S13

Ko C, Sohn G, Remmel TK et al (2014) Hybrid ensemble classification of tree genera using airborne LiDAR data. Remote Sens (Basel) 6:11225–11243

Koch B, Heyder U, Weinacker H (2006) Detection of individual tree crowns in airborne lidar data. Photogramm Eng Remote Sens 72(4):357–363

Korpela I, Dahlin B, Schäfer H et al (2007) Single-tree forest inventory using lidar and aerial images for 3D treetop positioning, species recognition, height and crown width estimation. In: Rönnholm P, Hyyppä H, Hyyppä J. (eds) Proceedings of ISPRS workshop on laser scanning, Espoo, Finland, 12–14 September 2007, pp 227–233

Korpela I, Ørka HO, Maltamo M et al (2010) Tree species classification using airborne LiDAR—effects of stand and tree parameters, downsizing of training set, intensity normalization and sensor type. Silva Fennica 44(2):319–339

Koukoulas S, Blackburn GA (2005) Mapping individual tree location, height and species in broadleaved deciduous forest using airborne LIDAR and multi-spectral remotely sensed data. Int J Remote Sens 26(3):431–455

Kraus K, Pfeifer N (1998) Determination of terrain models in wooded areas with aerial laser scanner data. ISPRS J Photogramm 53(4):193–203
Leckie DG, Gougeion FA, Hill DA et al (2003) Combined high-density lidar and multispectral imagery for individual tree crown analysis. Can J Remote Sens 29(5):633–649
Lefsky MA (2010) A global forest canopy height map from the Moderate Resolution Imaging Spectroradiometer and the Geoscience Laser Altimeter System. Geophys Res Lett 37:L15401
Lefsky MA, Harding DJ, Cohen WB et al (1999) Surface lidar remote sensing of basal area and biomass in deciduous forests of eastern Maryland, USA. Remote Sens Environ 67(1):83–98
Lefsky MA, Harding DJ, Keller M et al (2005) Estimates of forest canopy height and aboveground biomass using ICESat. Geophys Res Lett 32:L22S02
Liang X, Litkey P, Hyyppä J et al (2012) Automatic stem mapping using single-scan terrestrial laser scanning. IEEE Trans Geosci Remote Sens 50:661–670
Liaw A, Wiener M (2002) Classification and regression by random forest. R News 2(3):18–22
Lim K, Treitz P, Baldwin K et al (2003) Lidar remote sensing of biophysical properties of tolerant northern hardwood forests. Can J Remote Sens 29:658–678
Los SO, Rosette JAB, Kljun N et al (2012) Vegetation height and cover fraction between 60°S and 60°N from ICESat GLAS data. Geosci Model Dev 5:413–432
Magnusson M, Fransson JES, Holmgren J (2007) Effects on estimation accuracy of forest variables using different pulse density of laser data. Forest Sci 53:619–626
Mallet C, Bretar F (2009) Full-waveform topographic lidar: state-of-the-art. ISPRS J Photogramm 64(1):1–16
Maltamo M, Næsset E, Vauhkonen J (2014) Forestry applications of airborne laser scanning: concepts and case studies. Springer, Dordrecht
Maltamo M, Bollandsås OM, Gobakken T, Næsset E (2016) Large-scale prediction of aboveground biomass in heterogeneous mountain forests by means of airborne laser scanning. Can J For Res 46(9):1138–1144
Means JE, Acker SA, Fitt BJ et al (2000) Predicting forest stand characteristics with airborne scanning lidar. Photogramm Eng Remote Sens 66(11):1367–1371
Moffiet T, Mengersen K, Witte C et al (2005) Airborne laser scanning: exploratory data analysis indicates potential variables for classification of individual trees or forest stands according to species. ISPRS J Photogramm 59(5):289–309
Monk CD, Child GI, Nicholson SA (1970) Biomass, litter and leaf surface area estimates of an oak-hickory forest. Oikos 21:134–141
Morsdorf F, Meier E, Allgöwer B et al (2003) Clustering in airborne laser scanning raw data for segmentation of single trees. Int Arch Photogramm Remote Sens Spatial Info Sci 34(3):W13
Morsdorf F, Meier E, Kötz B et al (2004) LIDAR-based geometric reconstruction of boreal type forest stands at single tree level for forest and wildland fire management. Remote Sens Environ 92(3):353–362
Morsdorf F, Marell A, Koetz B et al (2010) Discrimination of vegetation strata in a multi-layered Mediterranean forest ecosystem using height and intensity information derived from airborne laser scanning. Remote Sens Environ 114(7):1403–1415
Moskal LM, Zheng G (2012) Retrieving forest inventory variables with terrestrial laser scanning (TLS) in urban heterogeneous forest. Remote Sens (Basel) 4(1):1–20
Næsset E (1997) Estimating timber volume of forest stands using airborne laser scanner data. Remote Sens Environ 61(2):246–253
Næsset E (2004) Practical large-scale forest stand inventory using a small-footprint airborne scanning laser. Scand J For Res 19(2):164–179
Næsset E, Bollandsås OM, Gobakken T (2005) Comparing regression methods in estimation of biophysical properties of forest stands from two different inventories using laser scanner data. Remote Sens Environ 94(4):541–553
Nelson R (2013) How did we get here? An early history of forestry LiDAR. Can J Remote Sens 39:6–17
Nelson BW, Mesquita R, Pereira JLG et al (1999) Allometric regressions for improved estimate of secondary forest biomass in the central Amazon. For Ecol Manage 117(1–3):149–167

Ørka HO, Næsset E, Bollandsås OM (2007) Utilizing airborne laser intensity for tree species classification. Int Arch Photogramm Remote Sens Spatial Inf Sci 36(Part 3):300–304
Ørka HO, Næsset E, Bollandsås OM (2009) Classifying species of individual trees by intensity and structure features derived from airborne laser scanner data. Remote Sens Environ 113(6):1163–1174
Persson A, Holmgren J, Söderman U (2002) Detecting and measuring individual trees using an airborne laser scanner. Photogramm Eng Remote Sens 68(9):925–932
Persson A, Holmgren J, Söderman U (2006) Identification of individual trees by combining very high resolution laser data with multi-spectral imagines. Workshop on 3D Remote Sensing in Forestry, Vienna, Austria
Petrie G, Toth CK (2009) Introduction to laser ranging, profiling and scanning. In: Shan J, Toth CK (eds) Topographic laser ranging and scanning: principles and processing. CRC Press, Boca Raton, FL, pp 1–27
Pfeifer N, Mandlburger G (2009) LiDAR data filtering and DTM generation. In: Shan J, Toth CK (eds) Topographic laser scanning and ranging. Principles and processing. CRC Press, Boca Raton, FL, pp 307–334
Popescu SC, Wynne RH (2004) Seeing the trees in the forest: using lidar and multispectral data fusion with local filtering and variable window size for estimating tree height. Photogramm Eng Remote Sens 70(5):589–604
Popescu SC, Zhao K (2008) A voxel-based lidar method for estimating crown base height for deciduous and pine trees. Remote Sens Environ 112(3):767–781
Popescu SC, Wynne RH, Nelson R (2003) Measuring individual tree crown diameter with lidar and assessing its influence on estimating forest volume and biomass. Can J Remote Sens 29:564–577
Reitberger J, Heurich M, Krzystek P et al (2007) Single tree detection in forest areas with high-density LiDAR data. Int Arch Photogramm Remote Sens Spatial Inf Sci 36:139–144
Reitberger J, Cl S, Krzystek P et al (2009) 3D segmentation of single trees exploiting full waveform LIDAR data. ISPRS J Photogramm Remote Sens 64(6):561–574
Rempel RC, Parker AK (1964) An information note on an airborne laser terrain profiler for micro-relief studies. In: Proceedings, 3rd symposium on remote sensing of environment, University of Michigan Institute of Science and Technology, pp 321–337
Ritchie JC (1996) Remote sensing applications to hydrology: airborne laser altimeters. Hydrol Sci J 41(4):625–636
Rosnell T, Honkavaara E (2012) Point cloud generation from aerial image data acquired by a quadrocopter type micro unmanned aerial vehicle and a digital still camera. Sensors 12:453–480
Sambridge M, Braun J, McQueen H (1995) Geophysical parametrization and interpolation of irregular data using natural neighbours. Geophys J Int 122(3):837–857
Shepard D (1968) A two-dimensional interpolation function for irregularly-spaced data. In: Proceedings of the ACM annual conference/annual meeting. Association for Computing Machinery, New York, NY, pp 517–524
Sibson R (1981) A brief description of natural neighbour interpolation. In: Barnett V (ed) Interpreting multivariate data. John Wiley & Sons, New York NY, pp 21–36
Simard M, Pinto N, Fisher JB et al (2011) Mapping forest canopy height globally with spaceborne lidar. J Geophys Res: Biogeosci 116:G04021
Sithole G, Vosselman G (2004) Experimental comparison of filter algorithms for bare-earth extraction from airborne laser scanning point clouds. ISPRS J Photogramm Remote Sens 59(1):85–101
Solberg S, Næsset E, Bollandsås OM (2006) Single tree segmentation using airborne laser scanner data in a structurally heterogeneous spruce forest. Photogramm Eng Remote Sens 72:1369–1378
St-Onge B, Vega C, Fournier RA et al (2008) Mapping canopy height using a combination of digital stereo-photogrammetry and lidar. Int J Remote Sens 29(11):3343–3364
Strunk J, Temesgen H, Andersen HE et al (2012) Effects of pulse density and sample size on a model-assisted approach to estimate forest inventory variables. Can J Remote Sens 38(5):644–654

Suratno A, Seielstad C, Queen L (2009) Tree species identification in mixed coniferous forest using airborne laser scanning. ISPRS J Photogramm Remote Sens 64(6):683–693
Tansey K, Selmes N, Anstee A et al (2009) Estimating tree and stand variables in a Corsican pine woodland from terrestrial laser scanner data. Int J Remote Sens 2009(30):5195–5209
Tóvári D, Pfeifer N (2005) Segmentation based robust interpolation—a new approach to laser data filtering. Int Arch Photogramm Remote Sens Spatial Inf Sci 36(3/W19):79–84
Treitz P, Lim K, Woods M et al (2012) LiDAR sampling density for forest resource inventories in Ontario, Canada. Remote Sens (Basel) 4(4):830–848
Vauhkonen J, Tokola T, Maltamo M et al (2008) Effects of pulse density on predicting characteristics of individual trees of Scandinavian commercial species using alpha shape metrics based on ALS data. Can J Remote Sens 34(Suppl.2):S441–S459
Vauhkonen J, Tokola T, Packalén P et al (2009) Identification of Scandinavian commercial species of individual trees from airborne laser scanning data using alpha shape metrics. For Sci 55(1):37–47
Vauhkonen J, Korpela I, Maltamo M et al (2010) Imputation of single-tree attributes using airborne laser scanning-based height, intensity, and alpha shape metrics. Remote Sens Environ 114(6):1263–1276
Vauhkonen J, Ene L, Gupta S et al (2011) Comparative testing of single-tree detection algorithms under different types of forest. Forestry 85(1):27–40
Vauhkonen J, Maltamo M, McRoberts RE et al (2014) Introduction to forestry applications of airborne laser scanning. In: Maltamo M, Næsset E, Vauhkonen J (eds) Forestry applications of airborne laser scanning: concepts and case studies. Springer, Dordrecht, pp 1–18
Vosselman G (2000) Slope based filtering of laser altimetry data. Int Arch Photogramm Remote Sens 33(B3/2 Part 3):935–942
Wallace L, Lucieer A, Watson C et al (2012) Development of a UAV–LiDAR system with application to forest inventory. Remote Sens (Basel) 4:1519–1543
Wang CK, Tseng YH, Chu HJ (2014) Airborne dual-wavelength LiDAR data for classifying land cover. Remote Sens (Basel) 6(1):700–715
Watt PJ, Donoghue DNM (2005) Measuring forest structure with terrestrial laser scanning. Int J Remote Sens 26:1437–1446
Weinacker H, Koch B, Heyder U et al (2004) Development of filtering, segmentation and modelling modules for lidar and multispectral data as a fundament of an automatic forest inventory system. Int Arch Photogramm Remote Sens Spatial Inf Sci 36(Part 8):50–55
White JC, Wulder MA, Vastaranta M et al (2013) The utility of image-based point clouds for forest inventory: a comparison with airborne laser scanning. Forests 4(3):518–536
Woods M, Pitt D, Penner M et al (2010) Operational implementation of a LiDAR inventory in boreal Ontario. For Chron 87(4):512–528
Wulder MA, Seemann D (2003) Forest inventory height update through the integration of lidar data with segmented Landsat imagery. Can J Remote Sens 29(5):536–543
Yao T, Yang X, Zhao F et al (2011) Measuring forest structure and biomass in New England forest stands using Echidna ground-based lidar. Remote Sens Environ 115(11):2965–2974
Yao W, Krzystek P, Heurich M (2012) Tree species classification and estimation of stem volume and DBH based on single tree extraction by exploiting airborne full-waveform LiDAR data. Remote Sens Environ 123:368–380
Yu X, Hyyppä J, Kaartinen H et al (2004) Automatic detection of harvested trees and determination of forest growth using airborne laser scanning. Remote Sens Environ 90(4):451–462
Yu X, Liang X, Hyyppä J et al (2013) Stem biomass estimation based on stem reconstruction from terrestrial laser scanning point clouds. Remote Sens Lett 4(4):344–353
Zhao K, Popescu S, Nelson R (2009) Lidar remote sensing of forest biomass: a scale-invariant estimation approach using airborne lasers. Remote Sens Environ 113(1):182–196

Regression Tree Modeling of Spatial Pattern and Process Interactions

Trisalyn A. Nelson, Wiebe Nijland, Mathieu L. Bourbonnais, and Michael A. Wulder

Abstract In forestry, many fundamental spatial processes cannot be measured directly and data on spatial patterns are used as a surrogate for studying processes. To characterize the outcomes of a dynamic process in terms of a spatial pattern, we often consider the probability of certain outcomes over a large area rather than on the scale of the particular process. In this chapter we demonstrate data mining approaches that leverage the growing availability of forestry-related spatial data sets for understanding spatial processes. We present classification and regression trees (CART) and associated methods, including boosted regression trees (BRT) and random forests (RT). We demonstrate how data mining or machine learning approaches are useful for relating spatial patterns and processes. Methods are applied to a wildfire data and covariate data are used to contextualize the quantified patterns. Results indicate that fire patterns are mostly related to processes influenced by people. Given the growing number of multi-temporal and large area datasets on forests and ecology machine learning and data mining approaches should be leveraged to quantify dynamic space-time relationships.

Spatial Pattern and Processes

Many scientific disciplines are interested in quantifying the relationships between spatial patterns and spatial processes (Nelson 2012). Intuitively, we understand that geographic patterns present at a given date can tell a story about a prior sequence of

T.A. Nelson (✉)
School of Geographical Sciences and Urban Planning, Arizona State University, Tempe, AZ 85287, USA
e-mail: trisalyn.nelson@asu.edu

W. Nijland • M.L. Bourbonnais
Spatial Pattern Analysis and Research (SPAR) Laboratory, Department of Geography, University of Victoria, PO Box 3060, Victoria, BC, Canada, V8W 3R4

M.A. Wulder
Canadian Forest Service, Natural Resources Canada, Pacific Forestry Centre, 506 West Burnside Road, Victoria, BC, Canada, V8Z 1M5

T.K. Remmel, A.H. Perera (eds.), *Mapping Forest Landscape Patterns*,
DOI 10.1007/978-1-4939-7331-6_5

events, or reveal information on the functioning of a system. Investigation of spatial patterns may lead to a better understanding of processes otherwise obscured from measurement. Popular culture is full of examples of the link between pattern and process. A familiar example of the link between mapped patterns and processes is the use of maps to plot and link events to solve a crime in TV dramas. Perhaps more relevant is the use of pattern-based jargon like hot spots, a spatial pattern of abundance, to indicate locations with important ecological processes (Nelson and Boots 2008).

Owing to the popularity of global positioning systems (GPS), geographic information systems (GIS), remote sensing, and increasingly user-friendly access to these data via Internet applications like Google Maps, there has been an explosion in the availability and interest of mappable data (Nelson 2012). Citizens, managers, and scientists have an increasing interest in extracting the information available from maps. As the amount of spatial data increases concurrently with improved access, the spatial co-location of data from different sources is becoming increasingly powerful for the analysis of spatial patterns and processes. Users from novices to experts, confronted with spatial data, are posing questions and requesting applications that continue to drive a need for new analytical methods. Rather than considering single points or events, increasingly knowledgeable users desire integration and refinement of the data available. We consider a spatial pattern as the expression of one or many spatial processes and a spatial process as a sequence of events carried out in some definite manner (Haining 1990). While spatial patterns are typically considered a snapshot in time, processes are temporally dynamic and expected to change (Getis and Boots 1978).

Describing Spatial Patterns

To describe the outcomes of a dynamic process in terms of a spatial pattern, we often consider the probability of certain outcomes over a large area rather than on the scale of the particular process. Forest fire occurrence is a good example of this. Fires are short-lived, localized events that are highly dependent on weather conditions and the presence of a source of ignition. The specific location of fires is therefore not readily predictable. However, over larger areas patterns in fire occurrence become clear with more frequent fires in areas with high fuel loads, dry climates, and more ignition sources (Parisien and Moritz 2009; Gralewicz et al. 2011; Gralewicz et al. 2012). Measurement of fuel or climate conditions on a landscape may therefore inform us about the probability of wildfire occurrence in a specific place (Chuvieco and Salas 1996; Parisien and Moritz 2009).

A similar ratio is used to map the ecological niche of a plant or tree species. We can create a model of the preferred climate and soil condition of a particular tree showing where it can grow and where it is more likely to thrive (Nijland et al. 2014; Waring et al. 2014), but within that range it still may or may not be present because of past events like the presence of a seed source and the availability of any free space.

Spatial patterns can either be directly related to the spatial processes by occurring in the same place or have a more complex relationship acting as a function of distance to another location. The nature of the relationship influences how we map the spatial patterns involved. The spatial pattern of forest logging operation provides an example of co-located and more complex spatial interactions: Forest logging is done in places with merchantable wood present, a simple co-location. Logging also occurs in places with access to a mill or other processing plant; the spatial relationship between logging and mills is more complex because they interact dependent on distance or connectivity to the mill. Even for simple co-location of spatial variables we still need to consider what "the same place" actually means.

When working with data from different sources, as is often the case, the data needs to be unified to a common spatial unit. The size of the principal spatial unit is dependent on the process we are studying (Nijland et al. 2009). In some cases a predefined or natural spatial unit is available, such as tenure areas in forests or watersheds in hydrology; in other cases, we may need to find our own solution by imposing a grid or other regular pattern, or else segment the area by other boundaries, perhaps roads or administrative areas (Dark and Bram 2007). Naturally, the detail of our environmental data needs to match the principal spatial unit and process. If we use climate patterns only to model the distribution of a plant or tree species, the results will be limited to regional patterns. If more detail is required, additional information with finer detail, for example, soil conditions or forest structure, may be included in the model (Nijland et al. 2014).

With complex spatial relationships we need to define the connection between our location and nearby features. In many cases our best approximation is by simple distance. With other cases the connection may be limited by physical boundaries or operate over a network. For instance, when transporting logs to a mill, the travel time or cost over the road network is more relevant than the simple Euclidean distance to the mill (Anderson et al. 2011).

Process Complexity

In forestry, many fundamental spatial processes cannot be measured directly and data on spatial patterns are used as a surrogate for studying processes (Levin 1992; Sokal et al. 1998; Jacquez 2000). As an example, consider landscape-scale forest insect infestations (e.g., Bone et al. 2013). The spatial processes of large-area insect infestations cannot be measured directly and the pattern of infested trees is the expression of the process of infestation (Robertson et al. 2008). By quantifying the spatial and temporal patterns of insect infestation, we generate new hypotheses or knowledge on the spatial processes of infestation. For instance, knowing the distance at which beetle infestation patterns are aggregated on the landscape provides information on the spatial scale of infestation processes (Powers et al. 1999) and identifying hot spots of infestation in space and time provides evidence of how forest susceptibility changes as an infestation progresses (Nelson et al. 2006).

Relating spatial pattern and process can be complex. Spatial patterns and processes are connected through a positive feedback; patterns are an expression of process, but processes are influenced by pattern (Fortin et al. 2003). For instance, in a forestry context the spatial pattern of the forest age is known to influence the risk of fire, while at the same time fire changes the age distribution of the forest (Gralewicz et al. 2012). The constant interplay between fire and forest age distribution is just one example of the feedback between pattern and process that can complicate interpretation. Interactions between pattern and process can be further concealed by the complicated one-to-many relationship between pattern and process. Many processes will express similar spatial pattern on the landscape and it can be near impossible to assign a pattern to a precise process (Fortin and Dale 2005 pp. 3–4; Langford et al. 2006). In reality most patterns are the result of many processes interacting together and through time.

Data Mining

Growing availability of forestry-related spatial data sets is creating an opportunity to use the spatial patterns in those data sets to explore and quantify spatial processes. Well-known spatial methods like kriging or k-nearest neighbor interpolation do use spatial patterns to generate predictions for unmeasured locations, but do not provide information on the process side (Hastie et al. 2009). Data mining methods are specifically suited to model patterns and processes in large volumes of information (Shekhar et al. 2003). Classification and regression trees (CART) (Breiman et al. 1984) and associated methods, such as boosted regression trees (BRT) (Elith et al. 2008; Hastie et al. 2009) and random forests (RT) (Breiman 2001), represent data mining or machine learning approaches useful for relating spatial patterns and processes. Exploratory in nature, CART and CART-based approaches are increasingly used in forestry and ecology research where there is an interest in quantifying and predicting spatial pattern and process dynamics and identifying influential drivers of these dynamics (De'ath and Fabricius 2000; Hawkins 2012). CART-based approaches have been used to understand natural and human drivers of spatial patterns of forest fire ignition (Gralewicz et al. 2012; Bourbonnais et al. 2013a, b), potential spatial variability in vegetation and forest composition under different climate change scenarios (Holmes et al. 2013), spatial patterns of species distributions based on environmental gradients (Elith et al. 2006; Leathwick et al. 2006), and wildlife health in the context of habitat conditions and human disturbance (Bourbonnais et al. 2013b).

CART-based methods are well suited to the study of spatial pattern and process as they can handle large datasets, complex nonlinear relationships, and missing data that are prevalent in spatial datasets, and can accommodate continuous and categorical variables, as well as variable interactions (De'ath and Fabricius 2000). The ability to handle mixed data types and the relatively straightforward interpretation of the resultant model structure put CART at an advantage over neural nets and support vector machines which are other machine learning methods designed to handle large data volumes and complex relationships (Hastie et al. 2009). Unlike parametric regression methods, such as ordinary least-squares regression and generalized linear models, CART-based methods make no assumptions about the structure of the data and the underlying processes, which in forestry and ecology are complex and may be unknown, and as such represent a flexible data-driven nonparametric regression approach.

In this chapter we provide an overview of CART, and two CART-based methods, BRT and RF. We demonstrate each approach, illustrate how to interpret the results, and comment on the strengths of each method through a case study that aims to quantify how spatial patterns of mountain pine beetle infestation are changing fire processes in British Columbia, Canada. While we highlight the utility of these approaches, we refer the reader who requires additional theoretical background to comprehensive reviews provided by Berk (2008) and Hastie et al. (2009). We introduce the theory of each method and then demonstrate how it applies to the case study of the interaction of mountain pine infestations on large forest fires. We conclude with an interpretation of modeling results and highlight future directions.

Methods

CART Models

CART models, originally implemented by Breiman et al. (1984), recursively partition (i.e., split) the response, in our case the spatial pattern of interest, into increasingly homogeneous subsets based on information provided by the predictor variables considered (Berk 2008). At each binary split, a threshold value (for continuous variables) or group level (for categorical variables) that best reduces the error sum of squares in the case of a continuous response, or the Gini index in the case of a categorical response, is selected to partition the data into two subsets. Data partitioning continues in a stagewise manner, meaning earlier split values are not considered in subsequent partitions, until no further meaningful reductions in the error sum of squares or Gini index can be found based on the data. This exhaustive approach generally leads to a very large tree being grown, or many splits, which is then pruned to remove splits that over-fit the data identified through cross-validation (Hastie et al. 2009).

When displayed graphically, a CART model is an inverted tree with the root node representing the undivided data at the top and branches defined by partition values and leaves, or terminal nodes, representing the response values or groups beneath (De'ath and Fabricius 2000). In our case, the branches and partition values represent the spatial processes considered and the terminal nodes the spatial pattern of interest. The hierarchical structure of the CART model is interpreted based on the partition values and terminal node assignments. At each split, observations that satisfy the decision rule are assigned to the group to the left while those that do not are assigned to the group to the right. The split values of the process variables, and associated terminal node value assignments representing the spatial pattern, allow us to infer the directionality of the spatial pattern-process dynamics. Additionally, the hierarchical structure of the CART model automatically incorporates interaction effects among process variables as terminal node assignments are dependent on all the preceding splits. For continuous response variables, CART model (i.e., regression tree) performance can be assessed based on the total sum of squares variance explained or the deviance explained (De'ath and Fabricius 2000). CART model performance for categorical response variables (i.e., classification tree) can be determined using a variety of classification accuracy assessments including misclassification error rates (De'ath and Fabricius 2000), confusion matrices (Berk 2008), and area under the receiver operating characteristic curve (Hastie et al. 2009). However, while CART models are easy to fit and interpret they do have a number of drawbacks. As mentioned, they are prone to over-fitting and defining stopping criteria and pruning large trees is not trivial (Murthy 1998; Berk 2008). CART models are also overly sensitive to changes in input data that can result in major changes in tree structure and split values (Hastie et al. 2009), making them a temporally static modeling approach. As a result, more robust methods that combine multiple stochastic trees have been developed.

BRT

BRTs use boosting algorithms to improve model accuracy by combining and averaging many CART models, rather than relying on a single tree to explain the association among spatial pattern and process(es) (De'ath 2007; Elith et al. 2008). Similar to CART, BRT is a stagewise procedure. However, unlike CART where binary splits are selected at each stage, BRT iteratively fits a completely new tree at each stage in order to minimize a loss function such as the deviance explained. Beginning with the first tree, a random subset of the data was selected and a tree is built that best minimized the loss in deviance explained. At each subsequent stage, a new tree using randomly selected data was built based on the residuals, or the unexplained variance in the response, from the combination of trees that already exist. Only the fitted values are reestimated at each iteration, while the existing trees and split values are unchanged. The final stochastic BRT model is a linear combination of hundreds or thousands of trees, rather than a single tree, resulting in a more robust model compared to the single tree produced by CART (Elith et al. 2008). However, unlike CART, which has few user-defined parameters, BRT models require the user to define the bag fraction, which specifies the proportion of the data randomly drawn at each iteration; the model learning rate, which determines the contribution of each tree to the model; and the tree complexity, which specifies the complexity of interaction effects included in the model (De'ath 2007; Elith et al. 2008). Combined, the learning rate and tree complexity determine the optimal number of trees required in the BRT model to minimize the loss in deviance explained while avoiding over-fitting the data. We refer the reader to Elith et al. (2008) for an in-depth review of BRT parameters and model fitting procedures.

Similar to CART, the performance of BRT models can be assessed using the deviance explained or a classification accuracy assessment usually based on cross-validation using withheld data (De'ath 2007; Elith et al. 2008). While variable importance and directionality in CART models are easily interpreted using a tree diagram, no such output is produced by BRT as it combines numerous trees. Instead, the influence of each variable is determined based on the number of times each variable is chosen as a split, weighted by its improvement to the model at each split averaged over the total number of trees in the model (Friedman 2001; Friedman and Meulman 2003; Elith et al. 2008). Partial dependence plots, which average out the influence of all other variables besides the variable selected, are used to visualize the associations between influential process variables and the spatial pattern response (Friedman 2001; Friedman and Meulman 2003).

RF Models

RF models (for the statistical background see Breiman 2001) are another machine learning approach that combine and average many CART models. RFs use bootstrap samples of the data to fit numerous (generally 500–2000) individual regression or classification trees. Unlike BRT, a limited number of predictor variables are also drawn at random in each bootstrap sample and used for the recursive partitioning to fit each tree. The number of variables to be randomly selected in each bootstrap sample is the only user-defined parameter in a RF. Also, bootstrap sampling and recursive partitioning of individual trees are usually not done in a stagewise manner meaning influential predictors and thresholds may be selected more than once. However, as observations from all the trees are aggregated through averaging, RFs are quite robust to over-fitting. Observations in the data that do not occur in the bootstrap samples are referred to as the out-of-bag data (Cutler et al. 2007). Each tree is grown to its maximum size and used to predict the out-of-bag data, eliminating the need to retain data for cross-validation (Prasad et al. 2006; Cutler et al. 2007).

The comparison of predicted values or classes from the bootstrap aggregation of trees used to build the RF with those retained in the out-of-bag data provide the mean square error (regression) or misclassification error rate based on the Gini index (classification) of the RF model. Similar to BRT, partial dependence plots are used to visualize associations between process-based variables and the spatial pattern response (Cutler et al. 2007; Hastie et al. 2009), and variable importance is assessed based on the number of times a variable is included as a split in the model and how well it performs. In the case of RF, the accuracy of a variable is determined by randomly permuting values from the out-of-bag data and then comparing predictions made with these new data to those of the model. The difference, divided by the standard error, between the permuted and original out-of-bag data values or misclassification rate represents the importance of the variable (Cutler et al. 2007).

Case Study Context: Influence of Beetle Infestation Spatial Patterns on Fire Spatial Processes

Disturbance plays an important role in defining landscape pattern and can cause substantial change in ecosystem processes (Turner 1989). In Canadian forests, anthropogenic and natural disturbances, such as harvesting (Masek et al. 2011), forest fires (Stocks et al. 2002), and insect infestations and disease (Hall and Moody 1994; Volney and Fleming 2000), are the primary determinants of forest structure.

While mountain pine beetle (*Dendroctonus ponderosae* Hopkins [Coleoptera: Scolytidae]) infestations are endemic in North American lodgepole pine ecosystems (Amman 1977), the infestation that occurred in western Canada during the 1990s and 2000s was the largest on record and affected over 16 million ha (Walton 2010), leading to widespread mortality of adult lodgepole pine trees in the region. Across the large area affected the severity of infestation varied substantially (Robertson et al. 2009a; Wulder et al. 2010). However, the spatial pattern of forests affected by the infestation has been altered considerably, generally resulting in smaller, more complex, more numerous forest patches (Robertson et al. 2009b; Coops et al. 2010).

Impacts of mountain pine beetle infestation on forest fire regimes are of particular concern, as outbreak-related tree mortality is anticipated to increase the frequency and severity of forest fires (Shore et al. 2006; Negrón et al. 2008). The theory that mountain pine beetle-induced tree mortality results in more severe fires has only recently been tested empirically (Jenkins et al. 2012), and initial results of retrospective studies and empirical testing of mountain pine beetle-fire dynamics have been contradictory (e.g., Page and Jenkins 2007; Simard et al. 2011).

The complex spatial pattern-process interaction between mountain pine beetle infestations and fires seems dependent on the severity of mountain pine beetle attack (Hawkes et al. 2004). When trees are killed, foliar moisture content of both needles and fine fuels decreases (Reid 1961; Shore et al. 2006), causing severely affected mountain pine beetle stands to have increased flammability, higher capacity to support sustained crown fires, and high rates of spread (Turner et al. 1999, Page and Jenkins 2007; Jenkins et al. 2008, 2012; Jolly et al. 2012). However, a decreased amount and spatial continuity in crown fuel loading and contiguity, due to forests having a range of attack severity (i.e., light to severe), have also been found to lessen the probability of crown fire ignition (Klutsch et al. 2011; Simard et al. 2011).

Given the extent of mountain pine beetle damage in British Columbia, Canada, and availability of spatial data, the interaction between infestation patterns and forest fire processes is an ideal case study for demonstrating how interactions between spatial pattern and process can be quantified using spatial data and regression trees.

Study Area

The study area includes the spatial extent of the mountain pine beetle infestation in British Columbia from 1999 to 2009, an area of ~16 million ha (Walton 2010). In order to represent the temporal variability observed in the spread of mountain pine

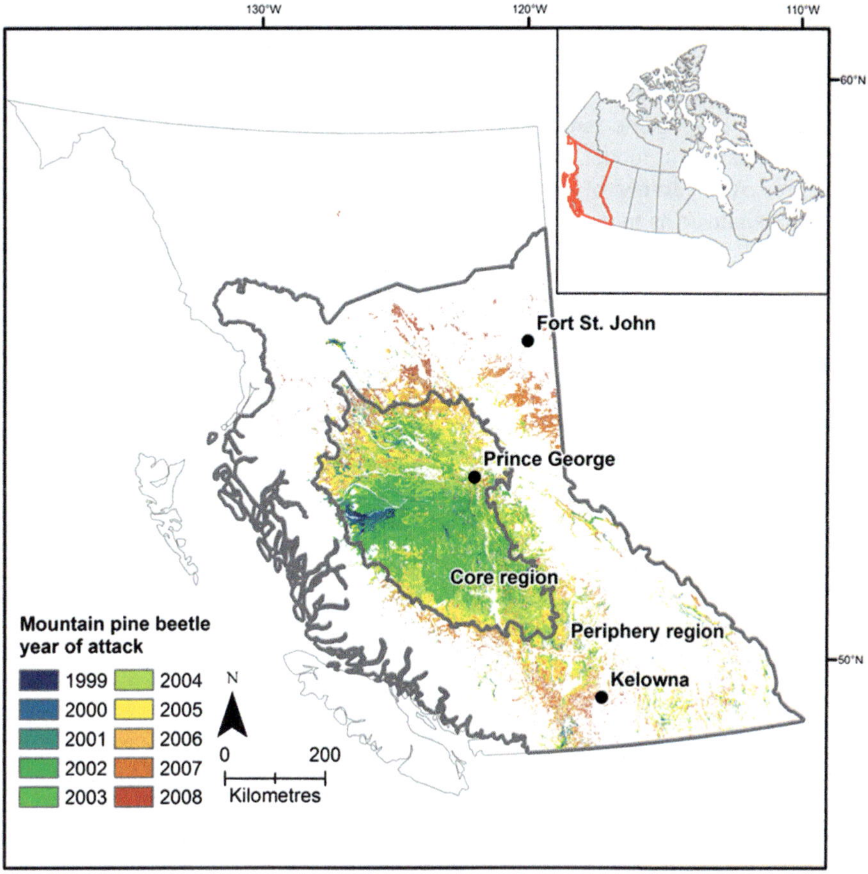

Fig. 1 The cumulative area impacted by mountain pine beetle from 1999 to 2009, divided into three regions (northern Periphery, Core, and southern Periphery) to account for spatial and temporal variability in the spread of the outbreak

beetle across the province, we divided the study area into Core and Periphery regions, based on ecoregions (Fig. 1). Ecoregions partition the province into regions of homogenous vegetation structure (Demarchi 2011). The Core region encompasses the epicenter of the mountain pine beetle epidemic, which transitioned from an incipient mountain pine beetle population in the mid-1990s to an epidemic population in 1999, and has experienced the most severe and widespread mortality (Aukema et al. 2006). While synchronous outbreaks were seen early in the outbreak in the south (Aukema et al. 2006), the epidemic predominantly spread from the study area center towards the south and north (Robertson et al. 2009b; Wulder et al. 2010). The Periphery region has a large abundance of host trees remaining, and mountain pine beetle-induced tree mortality is expected to continue (Walton 2010; Wulder et al. 2010).

Spatial Data

We use a spatial database of past fire activity as well as covariate data sets represented as anthropogenic influences, climate, terrain, elevation, and mountain pine beetle infestation to model the most influential predictors of the spatial pattern of large fires. An overview of data characteristics is provided in Table 1 and all data sets are visualized as maps in Fig. 2.

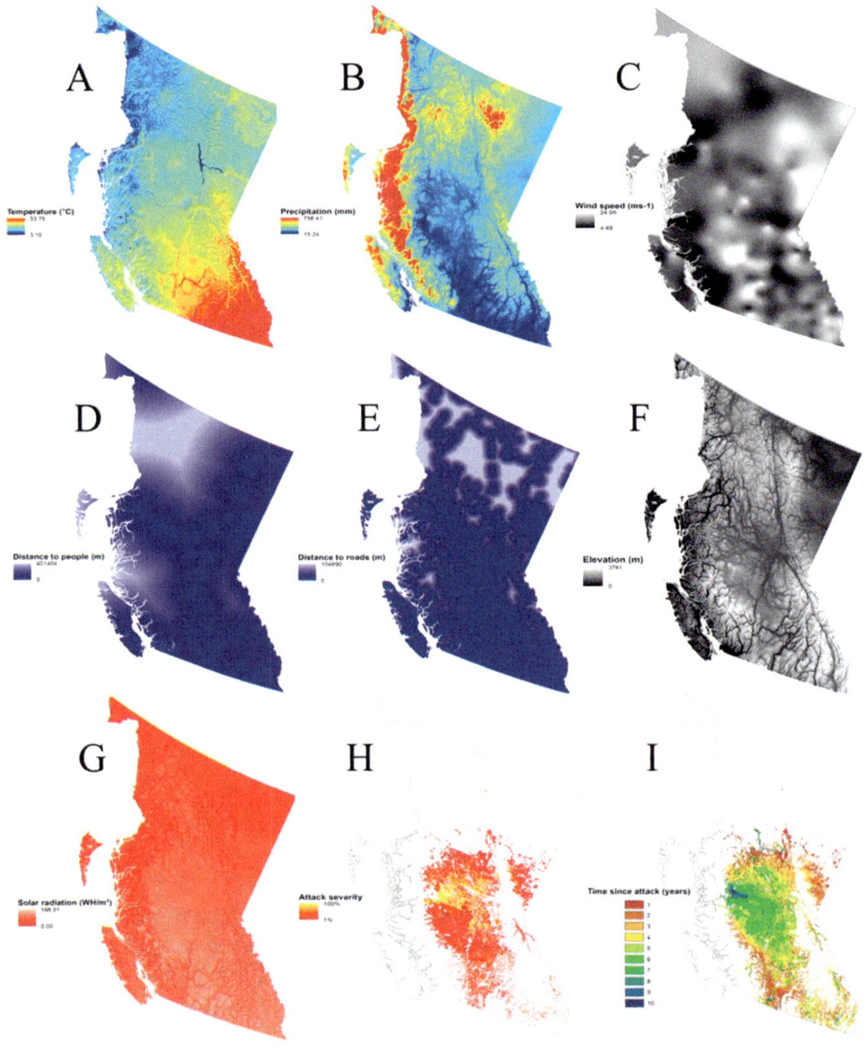

Fig. 2 Covariate layers used in classification tree modeling: (**a**) Temperature. (**b**) Precipitation. (**c**) Maximum average wind speed. (**d**) Distance to people. (**e**) Distance to roads. (**f**) Topography. (**g**) Solar radiation. (**h**) Time since mountain pine beetle attack. (**I**) Severity of mountain pine beetle attack

Wildfire Data

The Canadian National Fire Database (NFDB), a national repository of fire data from provincial, territorial, and Parks Canada fire agencies, provides spatial data for forest fires occurring in Canada (see Stocks et al. 2002). For this study, NFDB polygon data from 1999 to 2009 were used (Canadian Forest Service 2010). To promote reliable fire data only large fires were used. Large fires were defined as ≥32 ha. The 32 ha threshold removed the stochastic influence of small fires from the fire data distribution while providing an adequate number of mapped fires. Large fires were aggregated into the three study regions and stratified based on size into three classes: (a) 32 ha–200 ha; (b) 200 ha–1000 ha; and (c) > 1000 ha. In the Core region there were 82 large fires: 51 in Class A, 18 in Class B, and 13 in Class C. In the Periphery region there were 444 large fires: 256 in Class A, 119 in Class B, and 69 in Class C.

Climate Data

Climate is an important determinant of spatial patterns and processes of forest fire (Table 1) (e.g., van Wagner 1977; Flannigan and Harrington 1989; Flannigan et al. 2005; Parisien et al. 2006). In order to account for topographic variation in climate, ClimateWNA (Hamann and Wang 2005; Wang et al. 2011) was used to create weather variables for only the fire season (e.g., April 1–September 30). For the fire season, averages of maximum temperature and precipitation were calculated from 1999 to 2009 at a 1 ha spatial resolution. Hourly wind speed data were interpolated using spline interpolation from a provincial network of nearly 200 fire weather stations maintained by the BC Wildfire Management Branch. Data were stored in a 1 ha grid cell.

Anthropogenic Covariates

Human proximity and access have been found to be drivers of elevated fire incidence in Canada (Gralewicz et al. 2012). However, areas with high densities of human settlement are also subject to extensive fire suppression efforts, which can be a limiting factor of fire size (Parisien et al. 2006). In order to assess anthropogenic influence on fire size, proximity to the nearest populated place was calculated for each 1 ha cell in British Columbia based on persistent nighttime light derived from the DMSP Operational Linescan System (see Wulder et al. 2011). Similarly, the Euclidean distance to the nearest road of any size was calculated for each 1 ha cell in British Columbia using the 2008 road network file from Statistics Canada (2008).

Table 1 Relevance of covariates as determinants of fire size

Covariate	Driver (units)	Abbreviation	Relevance to fire severity	Reference
Weather	1. Temperature (degrees Celsius) 2. Precipitation (mm) 3. Wind speed (m per second)	1. temp_avg 2. precip_avg 3. Wind	1. Contributes to drying of fuels and increased fire behavior. 2. Influences fuel moisture content. Acts as a moderator of fire severity. 3. Direct determinant of fire intensity, shape, and size.	Parisien et al. (2006) Flannigan et al. (2005) Flannigan and Harrington (1989) van Wagner (1977)
Mountain pine beetle	1. Time since attack (years) 2. Percent pine infested (%)	1. tsa 2. comp_mpb	1 and 2. Stand structure following mountain pine beetle infestation is changed significantly over a temporal scale. Changes in fuel loading, continuity, and moisture caused by mountain pine beetle mortality are believed to be key determinants of fire severity.	Simard et al. (2011) Klutsch et al. (2011) Axelson et al. (2010) Jenkins et al. (2008) Page and Jenkins (2007) Lynch et al. (2006) Shore et al. (2006) Bigler et al. (2005) Turner et al. (1999)
Topography	1. Elevation(m) 2. Solar radiation (WH/m^2)	1. Elev 2. Rad	1. Impacts temperature, precipitation, wind speed, and vegetation type. 2. Impacts air temperature and composition of vegetation.	Parisien et al. (2006) Díaz-Avalos et al. (2001) Miller and Urban (2000) Kumar et al. (1997) Franklin (1995)
Anthropogenic Land use	1. Proximity to roads (md) 2. Proximity to populated places (m)	1. dist2rd 2. dist2lt	1 and 2. Regions with greater anthropogenic influence can both contribute (i.e., ignition source) and restrict (i.e., fire suppression) fire size.	Gralewicz et al. (2012) Parisien et al. (2006)

Topographic Data

Elevation may influence temperature, precipitation, and wind speed, as well as vegetation type and contiguity, which have an impact on fire incidence (Díaz-Avalos et al. 2001; Gralewicz et al. 2012) and fire size (Miller and Urban 2000). We used a digital elevation model (DEM) obtained from the Government of Canada portal Geobase and resampled to 1 ha grid cells. Annual shortwave radiation (Watt hours per m^2—WH/m^2) (Wulder et al. 2010) as derived from the elevation data was also used because solar energy has been found to influence fire size (Kumar et al. 1997).

Mountain Pine Beetle Data

Previous studies of mountain pine beetle and fire dynamics used attack severity (e.g., Turner et al. 1999; Page and Jenkins 2007; Simard et al. 2011) and time since mortality (e.g., Bigler et al. 2005; Lynch et al. 2006) to predict interaction with fire. We employed spatial products generated by Robertson et al. (2009a) at a spatial resolution of 1 ha to represent the spatial pattern of infestation severity, as percent of pixel infested, and time (year) since infestation. Robertson et al. (2009a) integrated aerial overview surveys (AOS) and ground surveys of mountain pine beetle infestation with data on percent pine to map the annual percent pine infested, from 1999 to 2009 within a 1 ha pixel. Time since mortality was also calculated based on when the forest in a pixel reached 50% mortality.

Model Evaluation

In this section we evaluate the three regression tree methods, CART, RF, and BRT, and demonstrate how each can be applied to explore the impact of mountain pine beetle infestation on the spatial patterns of forest fires. Within our modeling framework we assume that covariate data are surrogates for spatial processes. Climate and topography are associated to fire by direct co-location, while anthropogenic influences are modeled as distance to roads and populated areas. By including covariate data sets that represent mountain pine beetle infestation conditions, as well as anthropogenic influences, climate and weather, and elevation, we can determine which processes are the most influential predictors of the spatial pattern of large fires.

We generated separate regression models for the Core and Periphery geographic regions in order to determine how and if various levels of infestation severity and/or duration influenced the spatial pattern of fire. To assess the accuracy of each method we used 70% of data for training and held back a random sample of 30% of data for testing each model. Confusion matrices for classes A, B, and C were provided for each of the models as well as overall accuracy as a percentage.

Cart

The CART model had 82.9% classification accuracy in the Core (Table 2) and 65.5% accuracy in the Periphery (Table 3). In the Core model, the largest fire class (>1000 ha) was most accurately predicted (92.3%). Some of the smallest fires (32–200 ha) were misclassified as midsized fires (11.8%) or as the largest fires (5.9%). The midsize fires (200–1000 ha) were classified 77.8% accurately with the remainder split between the smaller and larger fire classes, 16.7% and 5.6%, respectively. In the Periphery, the midsize fires (200–1000 ha) were most accurately predicted, with 87.5% correct. The smallest fires (32–200 ha) were classified with 66.1% accuracy and most of the misclassified fires were predicted to be midsize (26.5%). The largest fires were predicted with 60.0% accuracy and misclassified as both small (18.6%) and midsize (21.4%) fires.

In Fig. 3 we show CART results for both the Core and Periphery. In the model of the Core, the primary predictor of fire size class was percent pine infested. Values less than 35.0 percent pine infested (comp_mpb) were generally associated with the small and midsize fires (32–200 ha and 200–1000 ha), and the smallest fire class (32–200 ha) was predicted near roads (<3306 m) in locations with higher rainfall (precip_avg $\geq$ 52.4). Variable associations could indicate that wetter conditions and access that enables quick response to fires are helping to limit the size of fires in the Core, when percent pine infested was less than approximately one-third of a forest stand.

Table 2 For the Core, confusion matrices for classification and regression trees (CART), random forests (RF), and boosted regression trees (BRT)

	CART	CART	CART	BRT	BRT	BRT	RF	RF	RF
	A	B	C	A	B	C	A	B	C
A	82.4	11.8	5.9	72.1	19.8	8.1	63.0	23.5	13.6
B	16.7	77.8	5.6	16.1	79.0	4.8	30.0	20.0	50.0
C	0.0	7.7	92.3	12.2	8.2	79.6	25.0	56.3	18.8

Due to small sample sizes the confusion matrices are made from training and test data and are percentages. The overall accuracy of classification is calculated from only test data and for CART = 82.9%, RF = 49.6%, and BRT = 88.0%. A, B, and C are categories of smaller to larger fire size: 32–200 ha, 200–1000 ha, and >1000 ha

Table 3 For the Periphery, confusion matrices for classification and regression trees (CART), random forests (RF), and boosted regression trees (BRT)

	CART	CART	CART	BRT	BRT	BRT	RF	RF	RF
	A	B	C	A	B	C	A	B	C
A	66.1	26.5	7.4	85.7	10.0	4.3	64.7	25.5	9.8
B	12.5	87.5	0.0	3.4	86.2	10.3	39.6	52.1	8.3
C	18.6	21.4	60.0	0.0	0.0	100.0	15.4	7.7	76.9

Due to small sample sizes the confusion matrices are made from training and test data and are percentages. The overall accuracy of classification is calculated from only test data and for CART = 65.5%, RF = 64.4%, and BRT = 73.9%. A, B, and C are categories of smaller to larger fire size: 32–200 ha, 200–1000 ha, and >1000 ha

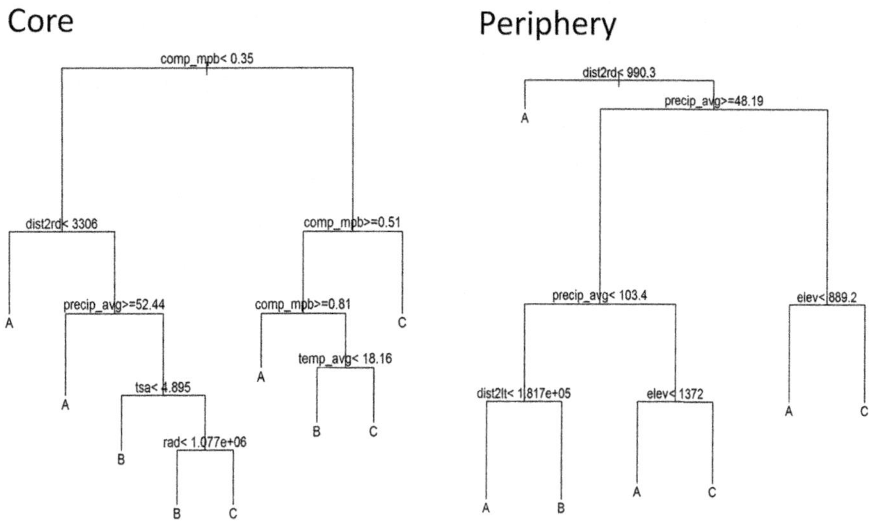

Fig. 3 CART results for Core and Periphery

A benefit of CART is shown by the nuanced predictions associated with the right branches. By allowing multiple splits on a single variable CART can represent complex relationships. When percent pine infested is ≥81.0% the smallest fire class is predicted, which may be explained because very dead stands may limit the fuel load available for fire. However, the largest fires also occur when there is more fuel available, only 35.0% to 51.0% of the stand infested, or at high temperature (<18.2 degrees).

Compared to the CORE, different variables in the Periphery were found to be important predictors of fire size, indicating that the fire processes in the Core and Periphery likely vary. Most notably, no mountain pine beetle infestation variables were important predictors of fire size in the Periphery. The lack of importance of beetle infestation indicates that in the Periphery from 1999 to 2009 the mountain pine beetle infestation processes were not sufficiently severe to be a dominant driver of fire process. Distance to road, average precipitation, and elevation were the only variables used to predict fire size. The smallest fires (32–200 ha), which are more plentiful, occurred near roads. The largest fires occurred far from roads (≥990.3 m) with average precipitation <48.2 mm, and at elevations >889.2 m, or else far from roads (≥990.3 m), with average precipitation between 48.2 and 103.4 mm and elevation >1372 m.

BRTs

The BRT model had the best overall classification accuracy with 88.0% of fires in Core being correctly classified (Table 2) and 73.9% in the Periphery (Table 3). In the Core, the smallest (32–200 ha), midsize (200–1000 ha), and largest (>1000 ha)

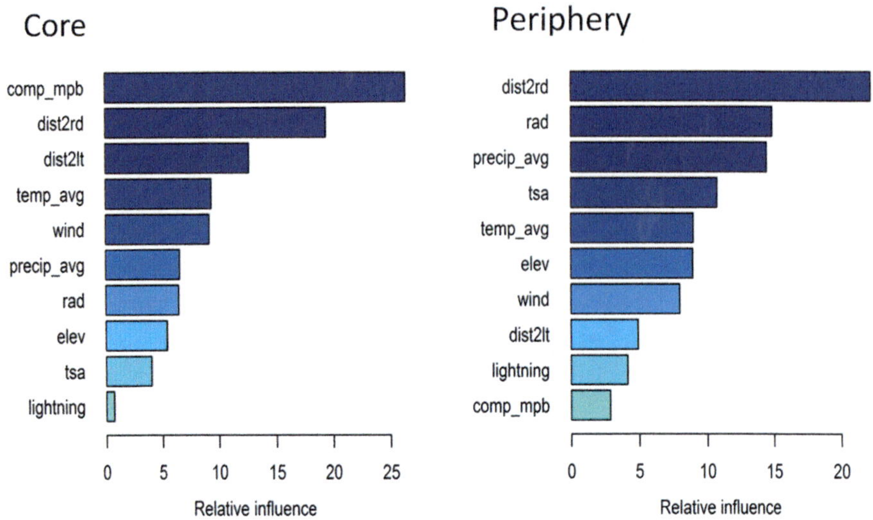

Fig. 4 Boosted regression tree results for Core and Periphery

fire classes had accurate predictions for 72.1%, 79.0%, and 79.6% of fires, respectively. In the Periphery the accuracies were even higher with 85.7%, 86.2%, and 100.0% of the smallest, midsize, and largest fires accurately predicted.

Examples of BRT outputs are shown in Fig. 3. The variable importance plot (Fig. 4) indicates the importance, in terms of rank and strength, of each variable for prediction. In the case of the Core, the percent pine infested was the most important predictor of fire size. Distance to road and populated place were the next strongest predictors, followed by average temperature and wind. In the Periphery, distance to road was the most important predictor. Weather variables were the next most important (solar radiation and average precipitation). Time since attack was the fourth most important variable, while percent pine infested was the least important predictor.

Directionality of associations between fire size and each predictor variable can be explored in the partial dependence plots which are shown for the Core in Fig. 5. Each class has a unique line and color: green, red, and black are the smallest (32–200 ha), midsize (200–1000 ha), and largest (>1000 ha) fire classes, respectively. Partial dependence plots can provide very useful information. For instance, where average temperatures were approximately 19 °C and higher, there was an increased probability of the largest fires. The highest probabilities of midsize fires occurred when temperatures ranged from 16 to 18 °C. Average temperature did not have a large influence on prediction of the smallest fire class. As indicated by the improved accuracy of prediction, there are statistical benefits to using BRT over CART. However, the CART regression trees allowed for intuitive exploration of variable relationships. Though similar information is available from partial dependence plots, the visualization was not as easy to interpret.

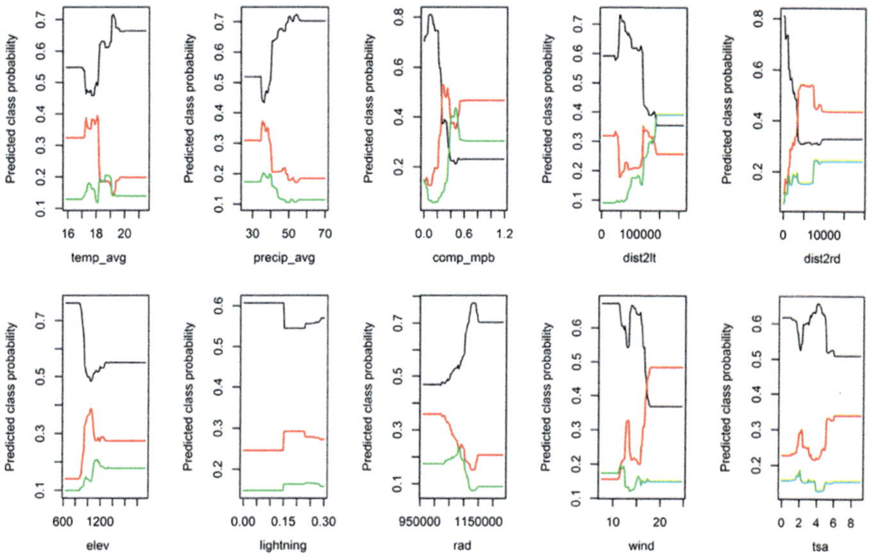

Fig. 5 Boosted regression tree partial dependence plots for Core. The *green*, *red*, and *black* lines are the smallest (32–200 ha), midsize (200–1000 ha), and largest (>1000 ha) fire classes, respectively

RF Models

The RF model had 49.6% classification accuracy in the Core (Table 2) and 64.4% accuracy in the Periphery (Table 3). Compared to both the CART and BRT models, the RF model performed poorly, particularly in the Core. The smallest fires (32–200 ha) were most accurately predicted, 63.0%, and the largest fires (>1000 ha) had the lowest number of correct classifications (18.8%), which may reflect sensitivity to sample size. In the Periphery, where the sample size was much larger, the largest fires (>1000 ha) were most accurately predicted (76.9%).

The RF results are shown in Fig. 6 and included both the mean square error or accuracy plot and the plot of misclassification error rate-based change in the Gini index. We have not included partial dependence plots, though they were available in similar format to the plots shown for the BRT. The accuracy plot was similar to the BRT variable importance plot. In the Core, the percent pine infested is the most important variable for accurately predicting fire size, followed by distance to road and average temperature. Typically, the misclassification plot will rank variables similarly to the variable importance plots. The key difference is how much the prediction is influenced by the removal of a variable. As with all the models, the results for the Periphery are quite different from the Core and indicate different spatial processes operating in each region. In terms of variable importance, distance to road was the most important variable. The second most important variable was time since attack. However, time since attack has a much lower impact on the Gini index, suggesting that it does not impact the misclassification rate. The next highest ranked variables in the accuracy plot were all weather related.

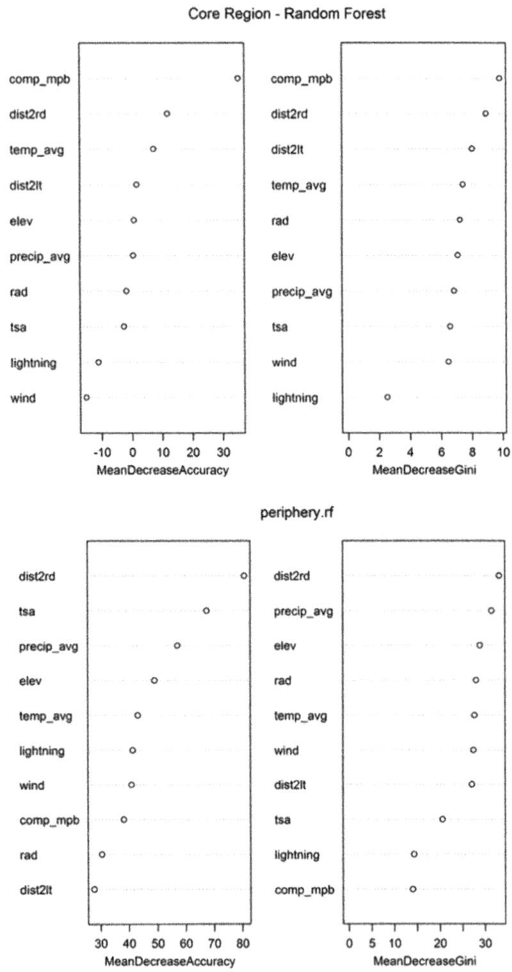

Fig. 6 Random forest results for Core and Peripheral regions showing differences in the predictive ability of each variable

Comparing Modeling Approaches

CART, BRT, and RF each have unique strengths and weaknesses. CART results are easy to interpret. Relationships between variables are intuitively observed making it possible to develop hypotheses about spatial pattern and process relationships. Information on the directionality of variable relationships is available from partial dependence plots, but they are not as intuitive and require a careful eye to examine and summarize. The difficulty with CART is that a different tree may be produced each time the model is run and as such the results may not be robust. The BRT and

RF have statistical benefits. While they are similar in that they fit many forests, the algorithms are sufficiently unique that the accuracy of each was quite different in our case study. The key difference is that BRT fits each tree to a different subsample of unique data. Given that each sample is unique, each new fit is to the residuals of the previous model. In contrast, RFs create a cross validation from the bootstrapped samples and therefore do not require separate testing data to determine how well the model fits. As a general guide, we recommend using CART for exploratory analysis and BRT for prediction. RF may be advantageous when samples are small.

Beyond the statistical fits, it is important to consider the different results obtained from each model that will impact our interpretation of pattern and process. However, it is encouraging that there was consistency in the results. In the Core, consistently, percent pine infested and distance to road were important predictors. Weather variables also emerged consistently, though depending on the model average temperature or average precipitation they may be more important. The variables that were lowest ranked on the variable importance plots were also not included in the pruned trees. In the Periphery, the distance to road variable was consistently the most important predictor. It is interesting that in both the CART and BRT, which performed better that the RF, weather variables were the next most important predictors. The RF includes time since attack as the second most important predictor, which was not included in the pruned CART though it is the fourth most important variable in the BRT. Given the complexities of relating spatial patterns and processes, scientists have advocated for confirmatory analysis whereby several analyses are carried out and the consistent trends become the strongest signal of pattern and process interactions. Ensemble modeling, for example, is a similar idea and allows the strengths of many models to be leveraged together (Grenouillet et al. 2011). In many instances, we strongly advocate for using CART in conjunction with either BRT or RF to provide a more complete understanding of interactions.

Interpreting Regression Tree Results within the Context of Spatial Pattern and Process

In pattern and process studies, often the most difficult part of the research is the interpretation. By including spatial data sets representing different spatial processes in a model, we invite the spatial patterns to indicate which processes are most important. While CART and related methods are very flexible in taking on large volumes of data, it remains of crucial importance to have a theoretical basis for each of the covariate data sets or in the presented study for stratifying the area in a Core and Periphery. One of the most notable results of our models was the difference in the variables that were important for predicting fire in the Core and Periphery. The Core has experienced extensive mountain pine beetle attack (Wulder et al. 2010) and the level of severity impacts the spatial pattern of large fires. From CART, we see evidence that the relationship is not linear. Rather, moderate levels of beetle infestation in locations with high temperature were predicted to support the largest fires. As well, large fires occurred where beetle infestations were relatively low, but forests were far from roads and had dry conditions, and trees had been attacked by beetles about 5 years previously.

In the Periphery, our results indicated that the mountain pine beetle infestation did not impact the spatial pattern of fire size. However, distance to road and weather were both consistently important predictors of fire size.

The relationship between roads and fire patterns we found in our spatial data is consistent with existing knowledge. Gralewicz et al. (2011, 2012) documented that fire patterns in Canada were more related to people than any other variable. Forests closer to roads have more opportunities for a human-induced fire to start. On the other hand, fires may burn longer and larger in areas with little access for fire suppression and a lower economic stake in the standing wood. Given the extensive logging road networks in British Columbia, access to forested areas is greatly influenced by roads. Fuel moisture content (Hayes 1942) and temperature have also been well documented to directly impact surface fire intensity and crown fire initiation (van Wagner 1977; Turner and Romme 1994; Bessie and Johnson 1995; Turner et al. 1999; Hély et al. 2000; Simard et al. 2011). In general, our model supports weather as an important driver of spatial patterns of large fires (Parisien et al. 2006; Parisien and Moritz 2009).

We included topographical variables as generic spatial covariates in the spatial model, but hypothesized that their relation to fire severity is indirect through weather conditions or vegetation composition (Table 1). In the model results they have low relative importance compared to more the more direct measures included. The inclusion of generic variables provides the model with a way to account for spatial patterns unresolved by the selected theory-loaded data sets. Elevation is a commonly used generic variable, but even latitude and longitude or projected coordinates can function as generic covariates (Guisan and Zimmermann 2000; Michaud et al. 2014; Nijland et al. 2014). A high importance of generic spatial patterns in selected models is an indication that spatial processes are present that are not well represented by more specific covariates (Cressie and Chan 1989). In such cases model inputs and underlying hypotheses should be reevaluated. The low importance of generic descriptors in our models strengthens our confidence that all relevant processes are included in the spatial covariates.

In our own work we have found CART, BRT, and RF as useful tools for making linkages between spatial patterns and processes. Data mining methods are usually bound by the assumption of spatial stationarity. Spatial stationarity occurs when the mean of a spatial pattern, the expression of a process, is similar in all parts of the study area (Bailey and Gatrell 1995, pp. 33–35). When data sets are small, spatial stationarity is possible, but as study extents increase this assumption often becomes invalid. In our case study we accounted for spatial non-stationarity by using two study regions, each with unique mountain pine beetle infestation processes and patterns. However, other methods such as geographically weighted regression are gaining momentum due to inherent ability to deal with variation in process interactions that are expected at landscape and regional scales (Brunsdon et al. 1996; Wang et al. 2005).

Future Direction

Geographers have developed a host of statistical techniques for quantifying spatial pattern (Nelson and Boots 2005; Robertson et al. 2007) and ecologists are adept at developing and applying methods that explore how covariate data predict patterns. There are many existing useful statistical methods for analyzing interaction between pattern and process including the tree-based models we focus on in this chapter. Inherent to spatial pattern analysis is the assumption that patterns observed in data represent a process. Historically, we have been limited to a few snapshots of spatial patterns in time. With the growing availability of satellite remotely sensed data, the temporal resolution and extent of data have changed. Archives of Landsat data are freely available from the mid-1980s at a temporal resolution of 16 days and a spatial resolution of 30 m, and back to 1972 with a slightly more coarse spatial resolution (resampled into products at 60 m). The United States Geological Survey Landsat archive has over 500,000 images of Canada (White and Wulder 2014). Landsat data are of special interest due to the capture of relatively large areas over a single imaging footprint at a level of detail that is informative of anthropogenic activities (Wulder et al. 2012). Moderate-resolution imaging spectroradiometer (MODIS) is also freely available but collects data with a more coarse spatial resolution, and images the entire earth every 1– 2 days. Higher spatial resolution images are available from a number of commercial vendors, including RapidEye that can be captured daily. Changing temporal resolution of imagery requires that we consider the temporal resolution maps of a spatial pattern in the measurement of a process. Remote sensing science is moving away from temporally static representations of space to more dynamic representations of pattern (Verbesselt et al. 2010; Gómez et al. 2011). Multi-temporal spatial pattern data sets are better representing spatial processes, and in some cases, the temporal resolution is allowing broad-scale measurement of dynamic spatial process. As has been a constant issue in spatial sciences, development of methods to harness the content of new data sets has not kept pace with data acquisition technology. As data sets continue to grow, it is our view that data mining approaches, such as regression trees and related methods, will become more heavily utilized.

In addition to data mining approaches, it is common to model spatial processes and compare the patterns generated by models to observed data (Nelson and Boots 2005). Research is required to explore the benefits of integrating data-driven and modeling-based approaches for studying pattern and process interactions. Bayesian statistics are gaining momentum and offer a unique mechanism for linking spatial data and process modeling perspectives (Ghazoul and McAllister 2003; van Oijen et al. 2005). Bayesian statistics represent variables using distributions, allowing patterns to be represented by a range of values (Gelman et al. 2009). Given that patterns are only one possible realization of a process, representing patterns as distributions of values is more realistic. From a practical standpoint, Bayesian methods deal well with uncertainty as the distribution can be thought of as a mechanism for providing a confidence interval around observed values (van Oijen and Thomson 2010). Given uncertainty in both spatial data and our understanding of forestry spatial processes, Bayesian approaches offer support for multiple issues. Growing spatial data archives provide data for building informative priors. It is common in Bayesian statistics to use uninformed priors based on uniform distributions, but archives, such as that available with Landsat (Wulder et al. 2012), are a mechanism for informing priors and analysis with multi-temporal representations of spatial patterns.

References

Amman G (1977) Role of the mountain pine beetle in lodgepole pine ecosystems: impact on succession. In: Mattson WJ (ed) The role of arthropods in forest ecosystems. New York, Springer-Verlag, pp 3–18

Anderson NM, Germain H, Bevilacqua E (2011) Geographic information system-based spatial analysis of sawmill wood procurement. J For 109:34–42

Aukema BH, Carroll AL, Zhu J, Raffa KF, Sickley TA, Taylor SW (2006) Landscape level analysis of mountain pine beetle in British Columbia, Canada: spatiotemporal development and spatial synchrony within the present outbreak. Ecography 29:427–441

Axelson JN, Alfaro RI, Hawkes BC (2010) Changes in stand structure in uneven-aged lodgepole pine stands impacted by mountain pine beetle epidemics and fires in central British Columbia. For Chron 86:87–99

Bailey TC, Gatrell AC (1995) Interactive spatial data analysis, vol 413. Longman Scientific & Technical, Essex

Berk RA (2008) Statistical learning from a regression perspective, vol 14. Springer, New York

Bessie A, Johnson E (1995) The relative importance of fuels and weather on fire behaviour in subalpine forests. Ecology 76:747–762

Bigler C, Kulawkowski D, Veblen TT (2005) Multiple disturbance interactions and drought influence fire severity in rocky mountain subalpine forests. Ecology 86:3018–3029

Bone C, Wulder MA, White J, Robertson C, Nelson TA (2013) A GIS-based risk rating of forest insect outbreaks using aerial overview surveys and the local Moran's I statistic. Appl Geogr 40:161–170

Bourbonnais ML, Nelson TA, Cattet MRL, Darimont CT, Stenhouse GB (2013b) Spatial analysis of factors influencing long-term stress in the grizzly bear (*Ursus arctos*) population of Alberta, Canada. PLoS One 8(12):e83768. doi:10.1371/journal.pone.0083768

Bourbonnais ML, Nelson TA, Wulder MA (2013a) Geographic analysis of the impacts of mountain pine beetle infestation on forest fire ignition. Can Geogr/Le Géographe Canadien 58(2):188–202. doi:10.1111/j.1541-0064.2013.12057.x

Breiman L (2001) Random forests. Mach Learn 45:5–32. Retrieved from http://www.springerlink.com/index/U0P06167N6173512.pdf

Breiman L, Friedman J, Olsen R, Stone C (1984) Classification and regression trees. Chapman and Hall/CRC, Boca Raton, p 368

Brunsdon C, Fotheringham AS, Charlton ME (1996) Geographically weighted regression: a method for exploring spatial nonstationarity. Geogr Anal 28(4):281–298

Canadian Forest Service (2010) National Fire Database – agency fire data. Natural Resources Canada, Canadian Forest Service, Northern Forestry Centre (Edmonton, AB). http://cwfis.cfs.nrcan.gc.ca/ha/nfdb. Accessed 28 Apr 2014

Chuvieco E, Salas J (1996) Mapping the spatial distribution of forest fire danger using GIS. Int J Geogr Inf Syst 10:333–345. doi:10.1080/02693799608902082

Coops NC, Gillanders SN, Wulder MA, Gergel SE, Nelson T, Goodwin NR (2010) Assessing changes in forest fragmentation following infestation using time series Landsat imagery. For Ecol Manag 259:2355–2365

Cressie NA, Chan NH (1989) Spatial modeling of regional variables. J Am Stat Assoc 84:393–401. doi:10.2307/2289922

Cutler DR, Edwards TC, Beard KH, Cutler A, Hess KT, Gibson J, Lawler JJ (2007) Random forests for classification in ecology. Ecology 88(11):2783–2792

Dark SJ, Bram D (2007) The modifiable areal unit problem (MAUP) in physical geography. Prog Phys Geogr 31:471–479. doi:10.1177/0309133307083294

De'ath G (2007) Boosted trees for ecological modeling and prediction. Ecology 88(1):243–251. Retrieved from http://www.ncbi.nlm.nih.gov/pubmed/17489472

De'ath G, Fabricius K (2000) Classification and regression trees: a powerful yet simple technique for ecological data analysis. Ecology 81(11):3178–3192. Retrieved from http://www.esajournals.org/doi/pdf/10.1890/0012-9658(2000)081[3178:CARTAP]2.0.CO;2

Demarchi, D. A. (2011) The British Columbia Ecoregion classification, third edition. Ecosystem information section, Ministry of Environment. (Victoria, BC)

Díaz-Avalos C, Peterson DL, Alvarado E, Ferguson SA, Besag JE (2001) Space-time modelling of lightning-caused ignitions in the Blue Mountains, Oregon. Can J For Res 31:1579–1593

Elith J, Graham CH, Anderson RP, Dudık M, Ferrier S, Guisan A et al (2006) Novel methods improve prediction of species ' distributions from occurrence data. Ecography 29(2): 129–151

Elith J, Leathwick J, Hastie T (2008) A working guide to boosted regression trees. J Anim Ecol 77:802–813. doi:10.1111/j.1365-2656.2008.01390.x

Flannigan MD, Harrington JB (1989) A study of the relation of meteorological variables to monthly provincial area burned by wildfire in Canada (1953–80). J Appl Meteorol 27:441–452

Flannigan MD, Logan KA, Amiro BD, Skinner WR, Stocks BJ (2005) Future area burned in Canada. Clim Chang 72:1–16

Fortin M-J, Dale MRT (2005) Spatial analysis: a guide for ecologists. Cambridge University Press, Cambridge

Fortin MJ, Boots B, Csillag F, Remmel TK (2003) On the role of spatial stochastic models in understanding landscape indices in ecology. Oikos 102(1):203–212

Franklin J (1995) Predictive vegetation mapping: geographic modelling of biospatial patterns in relation to environmental gradients. Prog Phys Geogr 19:474–499

Friedman J (2001) Greedy function approximation: a gradient boosting machine, 1999. Ann Statist 29(5):1189–1232. Retrieved from http://scholar.google.com/scholar?hl=en&btnG=Search&q=intitle:Greedy+Function+Approximation:+A+Gradient+Boosting+Machine#3

Friedman JH, Meulman JJ (2003) Multiple additive regression trees with application in epidemiology. Stat Med 22(9):1365–1381. doi:10.1002/sim.1501

Gelman A, Carlin JB, Stern HS, Rubin DB (2009) Bayesian data analysis, 2nd edn. Chapman and Hall/CRC, Boca Raton, FL

Getis A, Boots BN (1978) Models of spatial processes: an approach to the study of point, line, and area patterns. Cambridge University Press, Cambridge

Ghazoul J, McAllister M (2003) Communicating complexity and uncertainty in decision making contexts: Bayesian approaches to forest research. Int For Rev 5(1):9–19

Gómez C, White JC, Wulder MA (2011) Characterizing the state and processes of change in a dynamic forest environment using hierarchical spatio-temporal segmentation. Remote Sens Environ 115(7):1665–1679

Gralewicz NJ, Nelson TA, Wulder MA (2011) Spatial and temporal patterns of wildfire ignitions in Canada from 1980 to 2006. Int J Wildland Fire 21:230–242

Gralewicz NJ, Nelson TA, Wulder MA (2012) Factors influencing national scale wildfire susceptibility in Canada. For Ecol Manag 265:20–29

Grenouillet G, Buisson L, Casajus N, Lek S (2011) Ensemble modelling of species distribution: the effects of geographical and environmental ranges. Ecography 34(1):9–17

Guisan A, Zimmermann NE (2000) Predictive habitat distribution models in ecology. Ecol Model 135:147–186. doi:10.1016/S0304-3800(00)00354-9

Haining R (1990) Spatial data analysis in the social and environmental sciences. Cambridge University Press, Cambridge

Hall JP, Moody B (1994) Forest depletions caused by insects and diseases in Canada 1982–1987. Natural Resources Canada, Canadian Forest Service, Ottawa, ON
Hamann A, Wang TL (2005) Models of climatic normals for genecology and climate change studies in British Columbia. Agric For Meteorol 128:211–221
Hastie T, Tibshirani R, Friedman J (2009) The elements of statistical learning: data mining, inference, and prediction, 2nd edn. Springer, p 745
Hawkes B, Taylor SW, Stockdale C, Shore TL, Alfaro RI, Campbell R, Vera P (2004) Impact of mountain pine beetle on stand dynamics in BC. In: Shore TL, Brooks JE, Stone JE (eds) Mountain pine beetle symposium: challenges and solutions. October 30–31, Kelowna, British Columbia. Natural Resources Canada, Canadian Forest Service, Pacific Forestry Centre, Information Report BC-X-399, Victoria, BC
Hawkins BA (2012) Eight (and a half) deadly sins of spatial analysis. J Biogeogr 39(1):1–9. doi:10.1111/j.1365-2699.2011.02637.x
Hayes GL (1942) Differences in fire danger with altitude, aspect, and time of day. J For 40:318–323
Hély C, Flannigan M, Bergeron Y, McRae D (2000) Role of vegetation and weather on fire behavior in the Canadian mixedwood boreal forest using two fire behavior prediction systems. Can J For Res 31:430–441
Holmes K, Nelson T, Coops N, Wulder M (2013) Biodiversity indicators show climate change will alter vegetation in parks and protected areas. Diversity 5(2):352–373. doi:10.3390/d5020352
Jacquez GM (2000) Spatial analysis in epidemiology: nascent science or a failure of GIS? J Geogr Syst 2:91–97
Jenkins MJ, Hebertson E, Page W, Jorgensen A (2008) Bark beetles, fuels, fires and implications for forest management in the intermountain west. For Ecol Manag 254:16–34
Jenkins MJ, Page WG, Hebertson EG, Alexander ME (2012) Fuels and fire behavior dynamics in bark beetle-attacked forests in western North America and implications for fire management. For Ecol Manag 275:23–34
Jolly WM, Parsons RA, Hadlow AM, Cohn GM, McAllister SS, Popp JB, Hubbard RM, Negron JF (2012) Relationships between moisture, chemistry, and ignition of *Pinus contorta* needles during the early stages of mountain pine beetle attack. For Ecol Manag 269:52–59
Klutsch JG, Battaglia MA, West DR, Costello SL, Negrón JF (2011) Evaluating potential fire behavior in lodgepole pine-dominated forests after a mountain pine beetle epidemic in north-central Colorado. West J Appl For 26:101–109
Kumar L, Skidmore AK, Knowles E (1997) Modelling topographic variation in solar radiation in a GIS environment. Int J Geogr Inf Sci 11:475–497
Langford WT, Gergel SE, Dietterich TG, Cohen W (2006) Map misclassification can cause large errors in landscape pattern indices: examples from habitat fragmentation. Ecosystems 9(3):474–488
Leathwick JR, Elith J, Francis MP, Hastie T, Taylor P (2006) Variation in demersal fish species richness in the oceans surrounding New Zealand: an analysis using boosted regression trees. Ecol Prog Ser 321:267–281.
Levin SA (1992) The problem of pattern and scale in ecology: the Robert H. MacArthur award lecture. Ecology 73(6):1943–1967
Lynch HJ, Renkin RA, Crabtree RL, Moorcroft PR (2006) The influence of previous mountain pine beetle (*Dendroctonus ponderosae*) activity on the 1988 Yellowstone fires. Ecosystems 9:1318–1327
Masek, J. G., Cohen, W. B., Leckie, D., Wulder, M. A., Vargas, R., de Jong, B., Healy, S., et al. (2011) Recent rates of forest harvest and conversion in North America. J Geophys Res 116, G00K03, doi: 10.1029/2010JG001471
Michaud JS, Coops NC, Andrew ME, Wulder MA, Brown GS, Rickbeil GJM (2014) Estimating moose (*Alces alces*) occurrence and abundance from remotely derived environmental indicators. Remote Sens Environ 152:190–201. doi:10.1016/j.rse.2014.06.005
Miller C, Urban DL (2000) Connectivity of forest fuels and surface fire regimes. Landsc Ecol 15:145–154

Murthy S (1998) Automatic construction of decision trees from data: a multi-disciplinary survey. Data Min Knowl Disc 2:345–389. Retrieved from http://link.springer.com/article/10.1023/A:1009744630224

Nelson TA (2012) Trends in spatial analysis. Prof Geogr 64(1):1–12

Nelson TA, Boots B (2008) Detecting spatially explicit hot spots in landscape-scale ecology. Ecography 31(5):556–566

Nelson T, Boots B (2005) Identifying insect infestation hot spots: an approach using conditional spatial randomization. J Geogr Syst 7(3–4):291–311

Nelson T, Boots B, Wulder MA, Shore T, Safranyik L, Ebata T (2006) Rating the susceptibility of forests to mountain pine beetle infestations: the impact of data. Can J Forest Res 36(11):2815–2825.

Negrón JF, Bentz BJ, Fettig CJ, Gillette N, Hansen EM, Hayes JL, Kelsey RG et al (2008) US Forest Service bark beetle research in the western United States: looking toward the future. J For 106:325–331

Nijland W, Addink EA, De Jong SM, Van der Meer FD (2009) Optimizing spatial image support for quantitative mapping of natural vegetation. Remote Sens Environ 113:771–780. doi:10.1016/j.rse.2008.12.002

Nijland W, Nielsen SE, Coops NC, Wulder M a, Stenhouse GB (2014) Fine-spatial scale predictions of understory species using climate- and LiDAR-derived terrain and canopy metrics. J Appl Remote Sens 8:083572. doi:10.1117/1.JRS.8.083572

Page W, Jenkins MJ (2007) Predicted fire behaviour in selected mountain pine beetle-infested lodgepole pine. For Sci 53:662–674

Parisien M-A, Moritz MA (2009) Environmental controls on the distribution of wildfire at multiple spatial scales. Ecol Monogr 79:127–154

Parisien M-A, Peters VS, Wang Y, Little JM, Bosch EM, Stocks BJ (2006) Spatial patterns of forest fires in Canada, 1980–1999. Int J Wildland Fire 15:361–374

Powers JS, Sollins P, Harmon ME, Jones JA (1999) Plant-pest interactions in time and space: Douglas-fir bark beetle outbreak as a case study. Landsc Ecol 14:105–120

Prasad AM, Iverson LR, Liaw A (2006) Newer classification and regression tree techniques: bagging and random forests for ecological prediction. Ecosystems 9(2):181–199. doi:10.1007/s10021-005-0054-1

Reid RW (1961) Moisture changes in lodgepole pine before and after attack by the mountain pine beetle. For Chron 37:368–375

Robertson C, Nelson TA, Boots B (2007) Mountain pine beetle dispersal: the spatial-temporal interactions of infestation. For Sci 53(3):395–405

Robertson C, Wulder MA, Nelson TA, White JC (2008) Risk rating for mountain pine beetle infestation of lodgepole pine forests over large areas with ordinal regression modelling. For Ecol Manag 256:900–912

Robertson C, Farmer CJQ, Nelson TA, Mackenzie IK, Wulder MA, White JC (2009a) Determination of the compositional change (1999–2006) in the pine forests of British Columbia due to mountain pine beetle infestation. Environ Monit Assess 158:593–608

Robertson C, Nelson TA, Jelinski DE, Wulder MA, Boots B (2009b) Spatial–temporal analysis of species range expansion: the case of the mountain pine beetle, *Dendroctonus ponderosae*. J Biogeogr 36:1446–1458

Shekhar S, Zhang P, Huang Y, Vatsavai RR (2003) Trends in spatial data mining. In: Kargupta H, Joshi A, Sivakumar K, Yesha Y (eds) Data mining: next generation challenges and future directions. AAAI/MIT Press, Cambridge, MA, pp 357–380

Shore TL, Safranyik L, Hawkes BC, Taylor SW (2006) Effects of the mountain pine beetle on lodgepole pine stand structure and dynamics. In: Safranyik L, Wilson B (eds) The Mountain pine beetle: a synthesis of biology, management, and impacts on lodgepole pine. Natural Resources Canada, Canadian Forest Service, Pacific forestry Centre, Victoria, BC, pp 95–114

Simard M, Romme WH, Griffin JM, Turner MG (2011) Do mountain pine beetle outbreaks change the probability of active crown fire in lodgepole pine forests? Ecol Monogr 81:3–24

Sokal RR, Oden NL, Thomson BA (1998) Local spatial autocorrelation in biological variables. Biol J Linn Soc 65:41–62

Statistics Canada (2008) .Road Network File, Reference Guide 92-500-GWE Ottawa, On. Available at: www.statcan.ca/bsolc/english/bsolc?catno=92-500-x

Stocks BJ, Mason JA, Todd JB, Bosch EM, Wotton BM, Amiro BD, Flannigan MD, Hirsch KG, Logan KA, Martell DL, Skinner WR (2002) Large forest fires in Canada, 1959–1997. J Geophys Res Atmos 108:5.1–5.12

Turner MG (1989) Landscape ecology: the effect of pattern on process. Annu Rev Ecol Syst 20:171–197

Turner MG, Romme WH (1994) Landscape dynamics in crown fire ecosystems. Landsc Ecol 9:59–77

Turner MG, Romme WH, Gardner RH (1999) Prefire heterogeneity, fire severity, and early post-fire plant reestablishment in subalpine forests of Yellowstone National Park, Wyoming. Int J Wildland Fire 9:21–36

van Oijen M, Thomson A (2010) Toward Bayesian uncertainty quantification for forestry models used in the United Kingdom greenhouse gas inventory for land use, land use change, and forestry. Clim Chang 103(1-2):55–67

van Oijen M, Rougier J, Smith R (2005) Bayesian calibration of process-based forest models: bridging the gap between models and data. Tree Physiol 25(7):915–927

van Wagner CE (1977) Conditions for the start and spread of crown fire. Can J For Res 7:23–33

Verbesselt J, Hyndman R, Newnham G, Culvenor D (2010) Detecting trend and seasonal changes in satellite image time series. Remote Sens Environ 114(1):106–115

Volney WJA, Fleming RA (2000) Climate change and impacts of boreal forest insects. Agric Ecosyst Environ 82:283–294

Walton, A. (2010). Provincial-level projection of the current mountain pine beetle outbreak: update of the infestation projection based on the 2009 provincial aerial overview of forest health and the BCMPB model (year 7). Research branch, BC Forest Service. (Victoria, BC)

Wang T, Hamann A, Spittlehouse DL, Murdock TQ (2011) ClimateWNA – high-resolution spatial climate data for western North America. Am Meteorol Soc 51:16–29. doi:10.1175/JAMC-D-11-043.1

Wang Q, Ni J, Tenhunen J (2005) Application of a geographically-weighted regression analysis to estimate net primary production of Chinese forest ecosystems. Glob Ecol Biogeogr 14(4):379–393

Waring RH, Coops NC, Mathys A, Hilker T, Latta G (2014) Process-based modeling to assess the effects of recent climatic variation on site productivity and forest function across western North America. Forests 5:518–534. doi:10.3390/f5030518

White JC, Wulder MA (2014) The Landsat observation record of Canada: 1972–2012. Can J Remote Sens 39(06):1–13

Wulder MA, Ortlepp SM, White JC, Nelson TA, Coops NC (2010) A provincial and regional assessment of the mountain pine beetle epidemic in British Columbia: 1999–2008. J Environ Inform 15:1–13

Wulder MA, White JC, Coops NC (2011) Fragmentation regimes of Canada's forests. Can Geogr 55:288–230

Wulder MA, Masek JG, Cohen WB, Loveland TR, Woodcock CE (2012) Opening the archive: how free data has enabled the science and monitoring promise of Landsat. Remote Sens Environ 122:2–10

Mapping the Abstractions of Forest Landscape Patterns

Evelyn Uuemaa and Tõnu Oja

Abstract The evaluation of landscape patterns is necessary to explain the relationships between ecological processes and spatial patterns and between the processes and patterns and the factors that control them or that they control. For decades, landscape metrics have been used to measure and abstract landscape patterns. Since the emergence of FRAGSTATS in 1993, the measures and methods incorporated in this software have become widely used and are now a *de facto* standard tool for calculating landscape metrics. However, there are no special metrics unique to forest landscapes. The selection of metrics depends on the purpose of the study rather than on the land use or cover type. However, some metrics are more often used for forested landscapes (e.g., core area metrics). Forest landscape patterns are changing fast due to both natural and human disturbances. Remote sensing offers a rapid method of acquiring up-to-date information over a large geographical area and is therefore widely used as a source of the data needed for pattern assessment and the calculation of landscape metrics. However, to obtain meaningful results, correct preparation of the data is essential. In this chapter, we review the various metrics used to measure forest landscapes for different purposes. We deal with five main issues from the perspective of forest landscape patterns: (1) data preparation before the calculation of metrics (e.g., vector vs. raster data, scale, classification) and the associated uncertainties, (2) measurements of a landscape's configuration and composition using metrics, (3) interpretation of the results, (4) possible uses of the outcomes, and (5) future perspectives (e.g., 3D and 4D landscape metrics).

Introduction

Sustainable forest management practices aim to promote multifunctionality in forest ecosystems (Diaz-Varela et al. 2009) and increasingly integrate aspects of biodiversity such as the abundance and type of dead wood, monitoring of endangered species, use of natural regeneration, and establishment of stands of mixed tree species (Forest Europe 2011). Therefore, it is necessary to examine forest landscape

E. Uuemaa (✉) • T. Oja
Department of Geography, University of Tartu, Vanemuise 46, Tartu 51014, Estonia
e-mail: evelyn.uuemaa@ut.ee

T.K. Remmel, A.H. Perera (eds.), *Mapping Forest Landscape Patterns*,
DOI 10.1007/978-1-4939-7331-6_6

patterns so we can explain the relationships between ecological processes and spatial patterns they control or that control them.

Hundreds of *landscape metrics* have been developed to quantify landscape patterns. The term is generally used to describe all measures that quantify the spatial patterns of landscapes, and range from topographic measures (Vivoni et al. 2005) to the proportions of land use and cover types and a range of shape and area metrics (Palmer 2004). Spatial patterns are represented and quantified in a number of ways, which we discuss in the rest of this chapter. However, most landscape pattern analysis is performed using categorical maps, which tend to ignore the spatial variation within landscape units and trends in system properties across landscapes (Gustafson 1998). A large number of metrics have been developed to quantify spatial heterogeneity in categorical maps.

Landscape-level metrics provide simple measures of a landscape's structure, and their main advantage is that they can be easily calculated with readily available data and free open-source software (Kupfer 2012). These metrics fall into two general categories: those that evaluate the composition of the map without reference to spatial attributes, and those that evaluate the spatial configuration of system properties, and therefore require spatially explicit information to support the calculation (McGarigal et al. 2012; Zaragozi et al. 2012). It is, however, important to note that there are no specific metrics unique to forest landscapes.

Tools for Evaluating Landscape Patterns

Many indices and software packages have been designed to calculate and analyze parameters that describe landscape structure patterns using categorical maps. Spatial metrics may be calculated using either raster or vector data as inputs and the corresponding processing methods. Raster processes are more commonly used because raster data from a range of time periods is more easily available (e.g., satellite imagery), and the variety of spatial metrics developed for raster data is significantly greater than that for vector data (see the more detailed discussion in the section *Raster vs. vector data*). Most of these metrics are included in the FRAGSTATS software (McGarigal et al. 2012). Since the emergence of FRAGSTATS in 1993, it has become the *de facto* standard tool for calculating landscape metrics (Corry 2005). However, there are several other tools for calculating landscape metrics, and most of these are free open-source software. Steiniger and Hay (2009) have prepared an extensive overview of the freely available open-source software for landscape analysis. Table 1 summarizes the available software.

There are two main groups of tools: one based mainly on landscape metrics and another that is more suitable for modeling patterns and the processes that influence or are influenced by these patterns. Most tools require raster data as their input, and this format is also better suited to modeling. Vector-based programs can only compute a limited selection of landscape metrics. The broadest selection of metrics for vector data is available in PolyFrag.

Table 1 A non-comprehensive list of free and open-source GIS tools used for pattern mapping that were released between 1995 and 2014

Program	Description	References	Available since	Format of data
Mainly for calculating landscape metrics				
FRAGSTATS	For computing a wide variety of landscape metrics for categorical map patterns	McGarigal et al. (2012) http://www.umass.edu/landeco/research/fragstats/fragstats.html	1995	Raster
Patch Analyst	Functions for the analysis of patches (in raster and vector formats), such as calculation of landscape metrics. A module for ArcGIS	Elkie et al. (1999) http://www.cnfer.on.ca/SEP/patchanalyst/Patch5_2_Install.htm	1999	Raster/vector
V-LATE	Provides a selected set of the most common metrics. A module for ArcGIS	Lang and Tiede (2003) https://sites.google.com/site/largvlate/gis-tools/v-late	2003	Vector
PolyFrag	A vector-based program for computing landscape metrics. A module for ArcGIS	(MacLean and Congalton 2013) http://www.nhview.unh.edu/polyfrag.html	2013	Vector
LecoS	For calculating the most common landscape metrics. A module for QGis	Jung (2012) http://plugins.qgis.org/plugins/LecoS	2012	Raster
r.le, r.li, r.patch, r.diversity	For multiscale analysis of landscape structure. Modules for the Geographic Resources Analysis Support System (GRASS)	Baker and Cai (1992) http://grass.osgeo.org/grass65/	1992	Raster
Other pattern mapping software				

(continued)

Table 1 (continued)

Program	Description	References	Available since	Format of data
SaTScan	Spatial, temporal, and spatiotemporal scan statistics and tools for cluster detection. Often used in relation to the spatial distribution of diseases in forest landscapes.	Kulldorff (1997) www.satscan.org	1997	Point data
STAMP	For spatial and temporal analysis of polygonal data; can be used to analyze phenomena that change spatially over time, such as the spread of wildfire in a forested area, or the spread of forest insects across a landscape.	Robertson et al. (2007) http://www.geog.uvic.ca/spar/sites/stamp/help/index.html	2007	Vector
SDMTools	A set of tools for post-processing the outcomes of species distribution modeling exercises; also includes a set of landscape metrics. A module for R	http://www.rforge.net/SDMTools/	2009	Raster
Conefor	Allows quantification of the importance of habitat areas and links for the maintenance or improvement of landscape connectivity; based on graph structure theory. Available as standalone software and as command-line code for use in the R software	Saura and Torne (2009) http://conefor.org	2007	Raster

Guidos	For morphological analysis of landscape patterns to classify a landscape at the pixel level into a set of mutually exclusive pattern categories related to fragmentation and connectivity. Morphological spatial pattern analysis (MSPA) is a toolbox designed for Guidos to allow computation of morphological pattern metrics	Vogt et al. (2007) http://forest.jrc.ec.europa.eu/download/software/guidos/	2007	Raster
Dinamica EGO	Dynamic modeling software used for urban growth and tropical deforestation studies. Also enables analysis of landscape structure	Soares et al. (2002) http://www.csr.ufmg.br/dinamica/	2002	Raster
HARVEST	A strategic research and planning tool that makes it possible to assess the consequences of broad timber management strategies for the landscape's spatial pattern	Gustafson and Crow (1996) http://www.nrs.fs.fed.us/tools/harvest/	1996	Raster
LANDIS PRO	Raster-based forest landscape model for simulating landscape changes due to harvesting, windthrow, and fires over large spatial and temporal scales.	Mladenoff et al. (1996) http://landis.missouri.edu/	2005	Raster

(continued)

Table 1 (continued)

Program	Description	References	Available since	Format of data
Qrule	A program for the analysis of landscape patterns, generation of neutral models, and testing of hypotheses related to processes and patterns	Gardner (1999) www.al.umces.edu/Qrule.htm	1986	Raster
SELES	A tool for spatial and temporal landscape simulations that permit simulation of the effects of fire, logging, growth, and succession on the structure of a landscape	Fall and Fall (2001) https://ebmtoolsdatabase.org/tool/seles-spatially-explicit-landscape-event-simulator		Raster
SIMMAP	For simulating landscape spatial patterns through the modified random clusters method	Saura and Martinez-Millan (2001) http://www2.montes.upm.es/personales/saura/software.html	2000	Raster

Based on Steiniger and Hay (2009)

Because of the necessity of handling spatially explicit data, many landscape modules have been integrated into geographical information system (GIS) software. Examples include V-Late and Patch Analyst for ArcGIS, LecoS for QGis, and several packages (r.le, r.li, r.patch, r.diversity) for the Geographic Resources Analysis Support System (GRASS; Zaragozi et al. (2012). Typically, GIS software is needed for data preparation and later for visualization of the results. Often, however, the results of the landscape pattern analysis are simple numerical values rather than visualizations, although most tools make it possible to generate spatially explicit output, for example by using the moving window method in FRAGSTATS or the Patch Analysis layer in Patch Analyst.

Free open-source software is gaining in popularity, and programs such as the free R software environment for statistical computing and graphics (https://www.r-project.org/) are increasingly widely used in Earth sciences and therefore in landscape ecology. The Species Distribution Modelling Tools (SDMTools) library is designed for use in R, and provides a set of tools for post-processing to determine the outcomes of species distribution modeling exercises. The tools include methods for visualizing outcomes, selecting thresholds, calculating measures of accuracy, and calculating landscape fragmentation statistics (VanDerWal et al. 2011). Remmel and Fortin (2013) have produced additional R tools for assessing significant differences between the values of pattern metrics.

In addition to calculating common landscape metrics, several programs focus on modeling changes in spatial patterns (Table 1). SaTScan and STAMP have been used to determine spatial changes in phenomena. SaTScan has been used to identify the spatial and temporal distribution of various diseases (Haque et al. 2009; Naish et al. 2011), and STAMP has been used to support spatial and temporal analysis of mountain pine beetle range expansion (Robertson et al. 2007).

Forests are important habitats, and their connectivity is an important factor because it strongly affects species movement through a landscape. Conefor and Guidos enable analysis of landscape connectivity. Guidos also includes morphological spatial pattern analysis (MSPA), a customized sequence of mathematical morphological operators targeted at describing the geometry and connectivity of the components of an image. Conefor is based on graph theory, which has proven to be a promising approach for analyzing the functional and structural connectivity within landscapes (Decout et al. 2012). For example, it has been used to determine forest habitat connectivity in relation to species movement (Decout et al. 2012; Trainor et al. 2013) and to evaluate the impact of highways on forest connectivity (Gurrutxaga et al. 2011).

Forests change in response to human influences, to processes such as competition and wildfire that are intrinsic to ecosystems, to extrinsic factors such as climate change, as well as to their interactions. Forest landscape models have become increasingly important tools for predicting these changes over large areas (Fraser et al. 2013). These models usually require data in raster format, since the inputs and processes are modeled for each cell of a grid. One of the most widely used tools for simulating forest landscape change is LANDIS PRO (Wang et al. 2014), which makes it possible to simulate processes over a range of scales (from individual trees and stands to the landscape scale). Landscape-scale processes simulated in LANDIS PRO include seed dispersal (e.g., exotic species invasion), fire, windthrow, spread of insects and diseases, and forest harvesting (Wang et al. 2014). LANDIS PRO has been popular for modeling the effects of climate change on forests (Gustafson and Sturtevant 2013; Duveneck et al. 2014). Another model, SELES, has been used to model anthropogenic influences in forest management (Tittler et al. 2012) and to assess forest habitat suitability for a species (Reunanen et al. 2010).

There are also several programs (e.g., SIMMAP, Qrule) for generating neutral landscape models to test the effect of a measured process (e.g., fire, logging) on landscape patterns. Neutral landscape models have mainly been used to test landscape hypotheses, but have also been used to develop and test landscape metrics that can be used to characterize landscape-scale patterns (Gardner and Urban 2007).

Data Preparation and Uncertainties Within Metrics

Multiple factors related to the representation of geographic phenomena have been shown to affect the characterization of landscape patterns, resulting in spatial uncertainty. The uncertainties can be caused during (1) data collection (e.g., errors in fieldwork, problems with satellite sensors), (2) preprocessing (e.g., image classification, pixel size, format conversions), and (3) post-processing (statistical analyses of landscape metrics). In this chapter, we focus only on the problems related to preprocessing of the data. It is crucial to understand the magnitude of the influence of data preprocessing on detection of the real landscape pattern and whether this influence varies among landscapes. If the same processing methods are used, landscapes should retain their ranking when their landscape metrics are compared. For example, highly fragmented landscapes will always have larger values for the landscape metric "number of patches" than less fragmented landscapes. However, the magnitude of the difference between the landscapes may change, and the absolute values of landscape metrics may therefore need to be interpreted with caution (Lechner et al. 2013).

Raster Vs. Vector Data

Different data types can be used to analyze landscape structure. Depending on the data source, one can use either vector or raster data (Zaragozi et al. 2012). The raster format divides data into a grid consisting of individual cells or pixels, each of which is associated with a numeric or descriptive value. Raster data is usually derived from aerial photography and satellite imagery. In contrast, vector data characterize each object explicitly as points, lines, or polygons. Vector data is often obtained from topographic mapping and the vectorizing of historical maps. Both formats have several pros and cons (Table 2; Fig. 1). The problems illustrated in Fig. 1 mostly

Table 2 Pros and cons of using raster and vector data in landscape pattern analysis (Laurent 2006; Wade et al. 2003; Zaragozi et al. 2012)

	Pros	Cons
Raster	Simple data structure, calculations are faster, easy-to-represent continuous data, suitable for modeling	Large files, no topology, objects are generalized based on the cell size, representing points and lines are problematic
Vector	Small files, permits topology, objects are represented explicitly, enables more attribute data than raster	Complex data structure that makes calculations slow

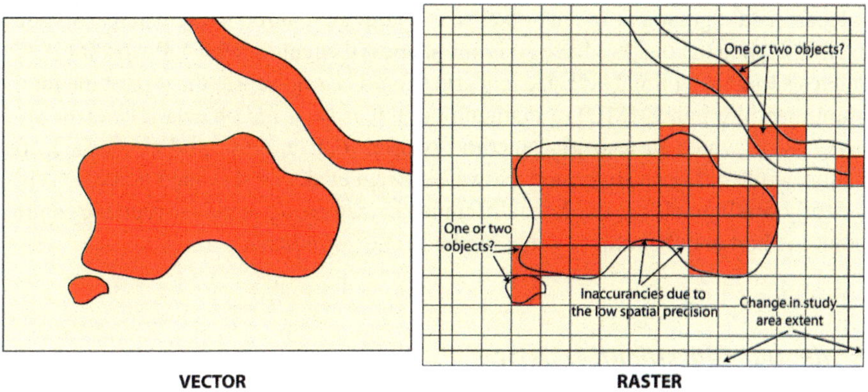

Fig. 1 Possible errors that occur when converting vector data to raster data. Adapted from Laurent (2006)

result from scale issues. Despite its several drawbacks, the raster format is more widely used for landscape analysis because of the ease of conducting complex spatial computations on grids, and because there is a greater variety of landscape metrics designed for use with the raster format (Cushman et al. 2008). Moreover, raster data is more available due to quickly evolving remote sensing technologies. Open data policies are also making more and more remote sensing imagery free, starting with LANDSAT images and ending with the Sentinel program, which enables researchers to download near-real-time high-resolution satellite imagery.

In landscape ecology, edges are among the most important features, and because vector and raster formats represent lines differently, metrics involving edges or perimeters will be affected by the choice of format. Edge lengths will be biased upward in raster data because of the stair-step outline that results from the use of pixels (Cushman et al. 2008). Moreover, the presentation of points and lines in raster format is somewhat problematic. In vector format, points can be represented by explicit x, y coordinates, but in raster format they are represented as single cells, which are the smallest unit of a rasterized image. By definition, points have no area, but with raster data must be converted to cells that have a finite area whose size depends on the pixel size. Lines are represented as spatially connected cells that have been assigned the same attribute value, but a line that is continuous in reality (e.g., a river) may be broken into separate groups of pixels (i.e., into discontinuous lines) when it does not align with the pixels of the grid (as shown for "One or two objects?" at the top of Fig. 1). This can also happen with polygonal objects (Fig. 1). Vector to raster conversion usually also generates a small change in the extent of a study area because the edges of the area may not map precisely to the edges of the pixels, particularly with large pixels (Fig. 1).

It is well known that errors are introduced when converting vector data into raster data or vice versa. However, little work has been done to characterize how raster and vector methods and their associated conversion errors affect landscape metrics, or how those errors affect the final analysis and results (Wade et al. 2003). As a general rule, errors are greater when the pixel size is large in relation to the patch (polygon) size, and when the patches have complex shapes (Congalton 1997). Piwowar (1987) suggests that the optimal cell size in a grid should be one-fourth the size of the minimum mapping unit (MMU) to maintain the integrity of the objects. The error also depends on which rasterization method is used. For land use and cover data, the majority rule has been suggested. However, Bregt et al. (1991) found that when sufficiently small pixels are used, the difference in rasterizing error between the central point method and the majority rule method is not significant.

Scale and Classification Issues

Quantitative geographers long ago recognized that the results of spatial data analysis depend on the data aggregation methods and classification scheme. The general formulation of this issue is known as the *modifiable areal unit problem* (Openshaw and Taylor 1981). The scale dependence of spatial patterns also affects the relationships between ecological processes and the landscape metrics that have been designed to evaluate landscape patterns. Most landscape metrics are scale dependent, and many authors have pointed out that the scale of the data (observations) and the scale of the analysis must be similar in order to calculate and interpret these metrics correctly (Simova and Gdulova 2012).

Two primary scaling factors affect measures of landscape pattern: *grain* refers to the resolution of the data (i.e., the pixel size) and *extent* refers to the size of the area being mapped or studied (Gustafson 1998; Simova and Gdulova 2012). Changes in image scale may affect the landscape metrics in at least three cases: changes in resolution, changes in extent, and changes in both resolution and extent. Wu (2004) divided landscape metrics into three categories based on their responses to changes in scale: simple scaling functions, unpredictable behavior, and staircase patterns.

Saura and Martinez-Millan (2001) noted that the variation in the values of metrics caused by changes of extent depends strongly on the landscape pattern and the magnitude of the change in extent. It is difficult to determine whether the sensitivity to the change in extent comes from the definition of the metric or from the landscape configuration (Baldwin et al. 2004). Kelly et al. (2011) found that several landscape metrics were more sensitive to scale at the level of the land use class, and that this should be especially kept in mind when dealing only with pure forest landscapes. Wu (2004) emphasized that the effect of changes in the extent on the metrics' values is less predictable than the effect of the change in pixel size. Small MMUs lead to larger variation in patch size, whereas large MMUs decrease the classification accuracy (e.g., due to the presence of two or more classes in each MMU) and patches relevant to the study may merge, leading to a loss of important information contained in a mapping unit. For example, an aerial photograph polygon labeled as coniferous may contain many coniferous trees, a few hardwood trees, and a grassy opening. These different stand elements can be resolved if the pixel size is sufficiently small, but if the MMU is large, then they are not classified as separate patches (Fassnacht et al. 2006).

As we noted earlier, raster models are often used for landscape analysis. Therefore, some geometric generalization takes place, and the issue of optimal grain size (i.e., scale) becomes important. It is essential to find the optimal grain size for each study based on the study's scale (e.g., stand vs. landscape). The influence of scale has been widely studied using real and artificial landscapes, but the results often differ among studies (Simova and Gdulova 2012).

Marceau et al. (1994) found that the optimum grain size to obtain the best classification accuracy differed for different image elements. For example, the most accurate classification of maple (*Acer* L.) forest required a grain size of 5 m, whereas poplar (*Populus* L.) and birch (*Betula* L.) had optimal grain sizes of 20 and 30 m, respectively. Based on this result, they suggested that researchers should use a multiscale approach instead of looking for a single overall resolution. Fassnacht et al. (2006) also suggested the use of landscape metric scalograms. Given these sensitivities, they recommended the use of scalograms to describe and compare landscapes rather than seeking a single optimum scale at which to calculate metrics.

Grain size also often determines the thematic resolution, which refers to the number and kind of classes that are defined (i.e., classification). Small or linear objects may not be revealed if the grain size is too big (Lausch and Herzog 2002) and may therefore disappear from a map. Thematic resolution may also be affected by the extent, since a bigger extent provides more opportunities to map rare classes (Fassnacht et al. 2006). Researchers have found that many landscape metrics are more sensitive to classification than to resolution or extent, and that class-level metrics are more sensitive than landscape-level metrics to classification accuracy (Buyantuyev and Wu 2007; Peng et al. 2007). Therefore, the classification scheme chosen by researchers has important effects on the calculation of landscape metrics.

Both composition and configuration metrics are sensitive to the classification scheme. Brown et al. (2004) estimated the errors in an analysis of changes in forest fragmentation, and found that the average patch size and number of patches were more sensitive to characteristics of the satellite image and contained more frequent measurement errors than the percentage forest cover and edge density. Moreover, both the number of classes and their definition are important (Lechner et al. 2012).

The same land cover class can be defined in different ways, and the differences depend on whether a land use or land cover definition is used.

Another problem is that natural phenomena are often not discrete, and present no clear boundaries. Therefore, the classification process becomes even more complicated. Traditional categorical maps require a strong simplification of reality, particularly when they depict a Boolean landscape where any location is represented as having a membership value of 1 in one and only one of the candidate classes and a value of 0 in all other classes. Fuzzy set theory has been proposed as an alternative (Foody 1992). In the application of fuzzy set theory to mapping, all locations have a membership value represented by a real number between 0 and 1 in all classes (see Chapter "Fuzzy Classification of Vegetation for Ecosystem Mapping" for details). If the value is 1, then there is the maximum similarity between the concept of that class and the properties at the real-world location, whereas if the value is 0, then there is no similarity between the two. Arnot et al. (2004) attempted to determine the extent to which a subset of landscape metrics is influenced by the way that the landscape is characterized and found that when landscape metrics are analyzed using a fuzzy approach, their behavior varies; in most cases, the values of the landscape metrics for a Boolean landscape can be considered representative of the landscape, but in some cases they are not. Arnot et al. (2004) suggested that when the ecotone is reliably smaller than the spatial unit of measurement (i.e., the pixel), there may be little point in exploring fuzzy memberships.

Most analyses are planimetric, and assume that landscapes can be adequately described using two-dimensional (2D) metrics. Another approach is to incorporate a third dimension (3D) when determining forest boundaries and calculating landscape metrics. Hou and Walz (2013) used the difference in elevation to identify transitions between forest and grassland. They used a normalized digital surface model to calculate the average heights, and found that forests in their study area were typically taller than 15 m, whereas grassland was shorter than 1 m. Using a moving window analysis, they generated transition zones in which pixels were between 1 and 15 m in height (Fig. 2). This in an interesting approach because it uses simple filtering based on concrete physical values (i.e., height) to classify a generally fuzzy ecotone. It is often difficult to find such clear parameters that can be used for classification. However, increasing availability of high-resolution measurements of the environment provides excellent data that lets us perform such classification. Because Hou and Walz (2013) used a third dimension in landscape classification, they also introduced several landscape metrics suitable for 3D analysis of landscape structures that increased the realism of their representation of landscape diversity (see the section *3D landscape metrics* for details).

When researchers use remote sensing data, the classification method itself can create significant variation in the values of landscape metrics. Langford et al. (2006) demonstrated that misclassification rates typically considered low by remote sensing standards (<15%) led to large errors (over 50%) in landscape pattern analysis, with limited and inconsistent increases in the accuracy of landscape pattern analysis when filtering was applied. Narusk (2014) reached similar results in a comparison of the impact of different classification methods (supervised, unsupervised, and

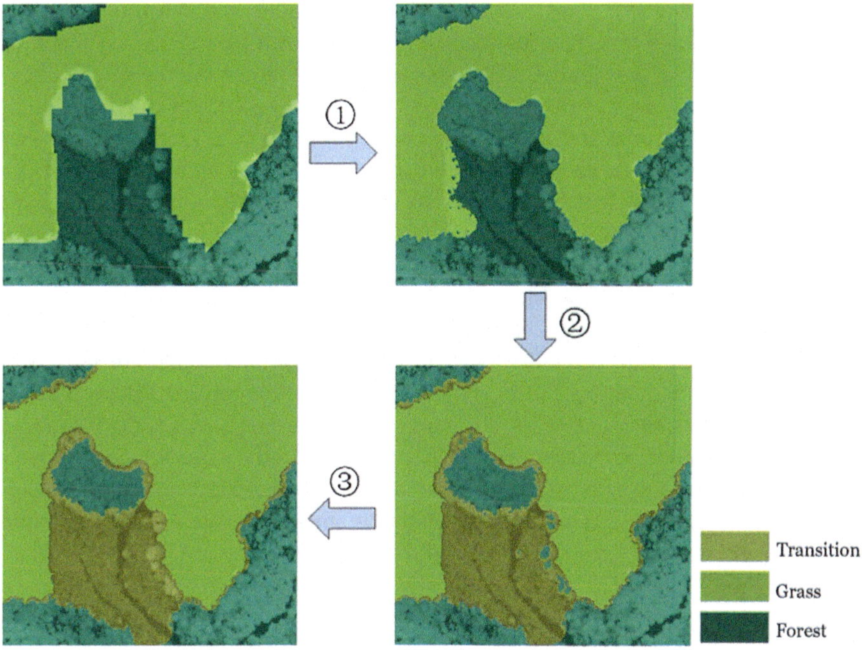

Fig. 2 Detection of transitions between forest and grassland: (*1*) boundary optimization, (*2*) establishment of transition zones, and (*3*) elimination of impurities inside the transition zones. Adapted from Hou and Walz (2013)

object oriented) and different software (Erdas Imagine, ENVI 5.0, IDRISI Selva, and eCognition) on classification and the resulting calculation of landscape indicators. Narusk also found that results of clustering algorithms such as ISODATA and k-means differed in different image processing software and that the classifications produced by these algorithms differed by up to 38%. The main reason for these differences lay in the initialization method used to define the initial clusters, which affected the final results. Narusk found that the most stable group of landscape-level metrics comprised the diversity metrics, whose indices had low coefficients of variation, whereas the core area metrics were most sensitive to differences in the classification methods. Landscape metrics calculated from satellite images classified using an object-oriented method tended to vary more than those calculated by means of pixel-based classification.

Because most studies of landscape pattern analysis use classified thematic maps based on remote sensing data, the accuracy or uncertainty associated with the maps is a critical factor in reliably characterizing spatial patterns, detecting changes, and relating patterns to processes. Without knowing the magnitude of the errors or uncertainties in landscape data, the characterization of landscape patterns is potentially unreliable. Due to the high costs involved in accuracy assessment, the accuracy of many remotely sensed map products has not been assessed; thus, the accuracy of landscape metrics computed based on such remote sensing products is unknown (Shao and Wu 2008). As a rule of thumb, Shao and Wu found that the values of most landscape metrics tend to stabilize when the overall accuracy of image classification approaches 90%. This means that a high degree of classification accuracy is required to ensure the consistency and reliability of calculated landscape metrics. Of course, the degree of the accuracy required for landscape metrics always depends upon the specific objective of the study (Shao and Wu 2004).

Researchers have debated our ability to compare landscape-level metrics (Wu et al. 2002; Remmel and Csillag 2003; Wang and Cumming 2011; Remmel and Fortin 2013). Apart from implementing a suitable method to compare the metrics, it is also beneficial to obtain background information on what the pattern actually represents (Boots and Csillag 2006). Landscape-level metrics are difficult to compare, and because of this difficulty, many researchers consider them to be most useful as diagnostic indicators. For example, Remmel and Fortin (2013) proposed a method for assessing whether an observed pattern formed as a result of random chance or whether some process caused the formation of the pattern. Because relating patterns to processes is a key question in landscape ecology, assessing the significance of pattern metrics is a major step in our ability to relate a spatial structure to the underlying processes that generated the structure.

Mapping Different Aspects of a Landscape Pattern

Landscape composition and configuration can affect ecological processes both independently and interactively. Thus, it is important to understand what aspect of the landscape pattern is being evaluated by each metric (McGarigal et al. 2012).

Composition

Landscape composition describes the amount and type of landscape elements without measuring their spatial arrangement (Farina 2000). Because composition requires integration over all patch types, composition metrics are only applicable at the landscape level (McGarigal et al. 2012). However, when evaluating only forest fragmentation and deforestation, the most common approach in land use classification is to reclassify land cover classes into forest and non-forest types and then focus on the distribution of the forest and configuration of its patterns. Composition is then only measured in terms of the presence or absence of the forest. However, in terms of management of habitat quality, it may also be important to know the structure of the matrix that surrounds the forest (Garmendia et al. 2013), and therefore more land use classes should be used in the study (Fig. 3).

Often, calculating a single numerical value for the whole landscape is insufficiently informative, although it is easy to calculate that number and compare it with other landscapes. To improve visualization of the results and support more

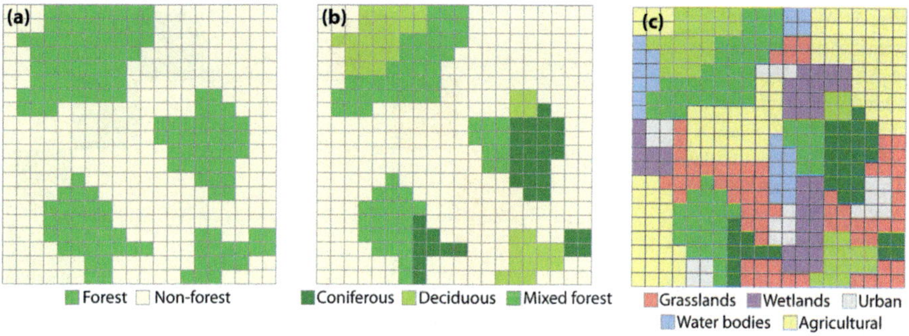

Fig. 3 (**a**) To evaluate deforestation and forest fragmentation, a simple binary forest versus non-forest classification may be sufficient. However, to determine habitat quality, it is often essential to know (**b**) the inner structure of the forest and (**c**) the structure and quality of the matrix that surrounds the forest

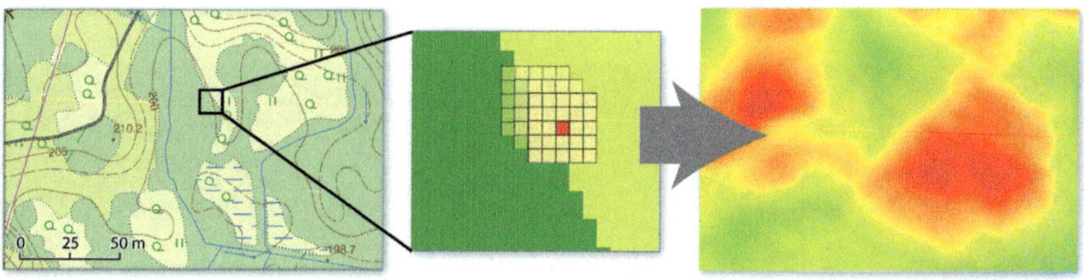

Fig. 4 Using moving window analysis to calculate landscape-level metrics

complicated spatial analysis, the moving window approach is often used, in which a kernel of a specified size and shape is passed over the landscape, and for each focal cell in the window, the software returns a value for each selected metric by means of averaging, kriging, or some other method (McGarigal et al. 2012). Figure 4 illustrates how this can be done. This method can be used to evaluate both the composition and the configuration of the spatial pattern.

This approach is problematic because the window of one focal pixel overlaps that of the next focal pixel (Hagen-Zanker 2016); this makes the data processing time consuming and resource intensive. In moving window analysis, the size of the window determines the scale of the analysis, and constraints on computer speed and working memory often limit the window to a small size, so that only small areas fit in the window. Hagen-Zanker (2016) proposed an algorithm that counts and sums up the pixel values in a window centered on the first pixel. But for the second and subsequent pixels, it only updates the count and summation by adding the pixels that are in the window surrounding the next focal pixel, and not the pixels surrounding the previous focal pixel; it then subtracts the pixels that are in the window surrounding the previous, but not subsequent, focal pixel (Fig. 5). This significantly shortens processing times. However, it does not solve the problem of overlapping values. To overcome this problem, block analysis has been used in several studies

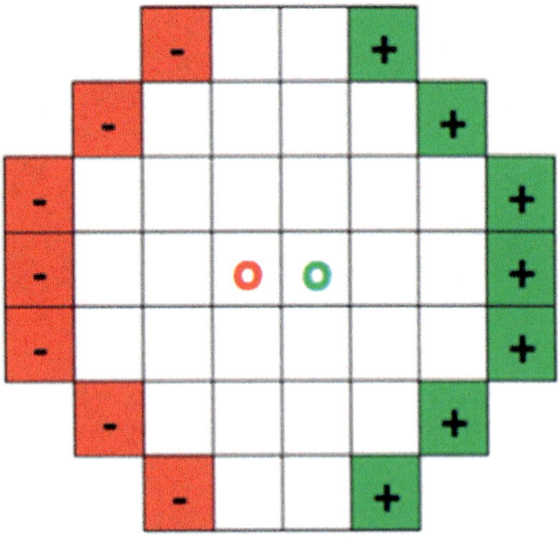

Fig. 5 As the circular window moves from *left* to *right*, *green pixels* are added, *red pixels* are subtracted, and *white pixels* are not recalculated. The red "o" marks the center of the window before the move and the *green* "o" marks its position after the move. Based on Hagen-Zanker (2016)

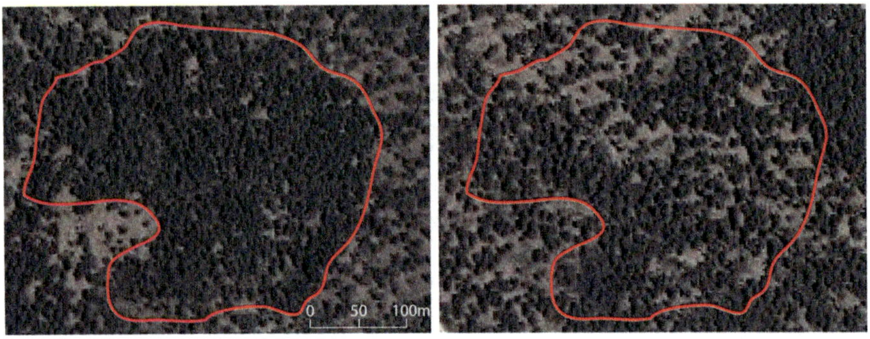

Fig. 6 Canopy closure at different times: >80% on the *left* and < 60% on the *right*. Based on Jomaa et al. (2009)

(Albeke et al. 2010; Mokhtari et al. 2014). In block analysis, a transect is divided into equal-sized but nonoverlapping blocks (e.g., 100 m × 100 m) and landscape metrics are calculated separately for each block.

In addition to fragmentation at the landscape level, it is also essential to investigate changes in the inner structure of patches, such as the species composition inside the patch. The degree of canopy closure is used for this purpose, and this parameter can also describe the degree of degradation of a forest patch (Rikimaru et al. 1999). Jomaa et al. (2009) observed forest loss over time by allocating patch boundaries in a series of years and measuring the changes inside and outside of the boundaries (Fig. 6). They used patch shapes and sizes to determine whether patch configuration is related to forest loss inside the patches, and found that the smaller patches had a higher likelihood of forest loss over time.

Hou and Walz (2013) combined open-source software with RapidEye satellite data to analyze small groves and tree lines. To obtain forest height information, they calculated the normalized digital surface model by subtracting a digital elevation model (DEM) from a digital surface model. They also used the normalized model to analyze transitions between forest and grassland and used the difference in elevation to represent the environmental gradient at the forest boundary and calculate the edge contrast.

Configuration

Landscape *configuration* describes the spatial arrangement of landscape elements (Farina 2000), and Fig. 7 illustrates key components of the configuration. In terms of forest fragmentation, both the size of the patches and their distance from each other are crucial. Configuration can be measured at the patch, land use or cover class, and landscape levels.

Most studies measuring forest fragmentation and deforestation have used a patch-based approach (De Chant et al. 2010). Patch metrics are appropriate at large scales, but many processes are only revealed at smaller scales. The influence of the

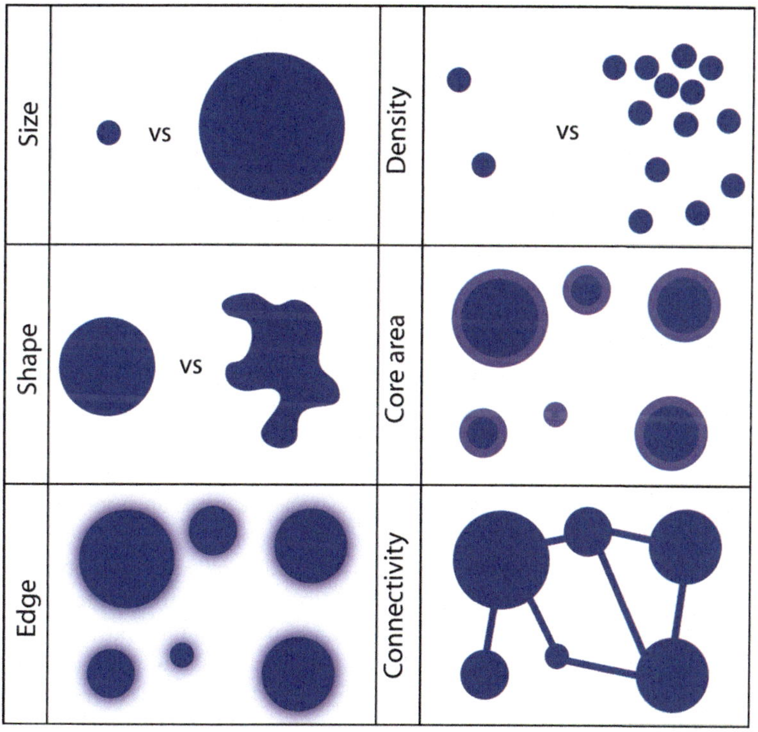

Fig. 7 Examples of patch configurations. The size, density, shape, core area, edge, and connectivity are commonly used to characterize the configuration of patches in a landscape. Adapted from Spearman (2003)

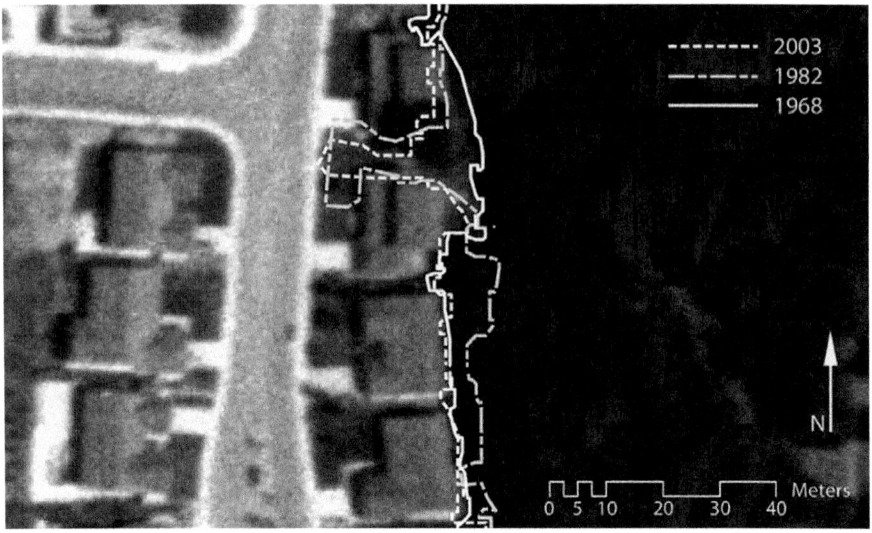

Fig. 8 An example of changing edge sinuosity from De Chant et al. (2010). The lines represent the maximum continuous canopy extent in 1968, 1982, and 2003

adjacent non-forest environment on forest structure and species biodiversity at created edges is also well known (Harper et al. 2005). For example, many species need core areas to survive or reproduce, but if the fragmentation is too extensive, the area influenced by edges may be the dominant characteristic of the landscape (i.e., there may be insufficient core areas). Moreover, forest edges change more frequently and more rapidly than core areas at a landscape scale, and they are also highly dynamic structures at small, physiologically relevant scales (De Chant et al. 2010).

De Chant et al. (2010) used an interesting approach to evaluate the effects of urbanization on forest edges at a local scale. They quantified the complexity of urban forest edges (measured by their sinuosity, calculated as the total path length of each segment divided by its straight-line length between endpoints) in 3 years, and found that edges exhibited low sinuosity immediately after development, but grew significantly more complex over time (Fig. 8). This is an example of a fine-scale study of forest edge structure. Such studies increase our understanding of how urban edges influence forest responses to disturbance and can improve planning strategies.

To promote or protect biodiversity, both the landscape's physical configuration (i.e., structural connectivity) and its functional connectivity are important. Functional connectivity is a complex concept and depends on more than the landscape pattern. It also depends on interactions between the landscape pattern and the biological characteristics of the target species (e.g., their ability to move between patches) and of the corridors, which must have characteristics suitable for the species (Sieving et al. 1996). As the ability to exploit a given type of corridor differs among species, functional connectivity is scale dependent and species specific. For example, several studies have shown that under some conditions, birds can consider disconnected forest patches to be connected homogeneous patches (Uezu et al. 2005; Mueller et al. 2014); in contrast, those patches may not be connected for species that cannot fly between them. This makes functional connectivity harder to measure than other landscape characteristics, and obtaining data on this parameter may require tracking movements of the species.

Criteria for Selecting Metrics

One of the most critical issues with landscape metrics is that there are so many, and it is difficult to interpret them all. It is desirable to use the smallest number of independent metrics that can sufficiently quantify landscape structure and that are also linked with the ecological process under study (Leitao and Ahern 2002; Cushman et al. 2008). Finding a suitable set of metrics is complicated, because the metrics may simultaneously measure multiple aspects of the same structure (Cushman et al. 2008), or may measure the same aspect from a different perspective (e.g., patch density and mean patch size are reciprocals). Many studies have attempted to determine the most parsimonious suite of independent metrics (Riitters et al. 1995; Cain et al. 1997; Lausch and Herzog 2002; Cushman et al. 2008; Schindler et al. 2008). Although these studies agreed about some metrics (e.g., all included patch density and edge density as key metrics), they generally disagreed on the other members of the set. This suggests that there is no universally appropriate set of metrics, and that the optimal metrics are those that best support the goals of a given study.

As we mentioned earlier, there are no specific metrics for evaluating forest patterns, and the selection of metrics depends on the purpose of the study. Abdullah and Nakagoshi (2007) and Sitzia et al. (2010) reported that the mean patch size, connectivity, and edge length are the most commonly used metrics for evaluating changes in forest landscapes. Wulder et al. (2008) and Cardille et al. (2012) used the following landscape metrics to identify representative forest landscapes and represent land cover in other ecozones: proportion of that area covered by forest, proportion of patches that contain forest, number of forest patches, mean forest patch size, standard deviation of the forest patch size, length of forest edge, forest edge density, forest–forest connections, and forest/non-forest connections. They selected these metrics because they depicted fragmentation accurately and were easily interpretable. In our opinion, this selection of metrics is generally good for evaluating fragmentation and deforestation and also for evaluating changes in forest landscapes over time and space. We would, however, omit the length of forest edge, as it duplicates information included in the forest edge density. It is also important to know that the amount of forest edge and the number of forest patches are only comparable across study sites if the study sites have the same area. Otherwise, densities (i.e., values per unit area) are more effective. In addition, Echeverria et al. (2006) and Kouba and Alados (2012) used the mean distance between forest patches, which is an important parameter for measuring habitat structural connectivity.

Bogaert et al. (2000), Butler et al. (2004), and Abdullah and Nakagoshi (2007) combined multiple forest fragmentation metrics into a single index to quantify fragmentation and allow comparisons across landscapes. They based their indices on the same components: the proportions of forested and non-forested area, length of forest edge compared to the length of edge of other land use types, and patch size (measured by the coefficient of variation of the patch size). Fragmentation itself consists of two aspects: the amount of forest in the study area and how it is distributed within that area (i.e., the composition and configuration). There may be much forest in the area, but if it is divided into small and scattered patches, then the forest is still fragmented. Therefore, the proportion of forest alone is not an adequate indicator of fragmentation, and the proportion of the area that contains forest (composition) and patch size (configuration) should both be measured. Although the proportion of forest alone may be sufficient in studies of deforestation, deforestation is rarely observed separately from fragmentation, as these two phenomena are closely related from a forest management perspective.

All patch metrics can be summarized at the land use or cover class level or at the landscape level using various distribution statistics (McGarigal et al. 2012): the median, mean, area-weighted mean, range, standard deviation, and coefficient of variation (Fig. 9). The most commonly used statistic is the frequency distribution for patch size, in which each patch is considered in describing the landscape's structure, regardless of its size. In evaluating habitats, however, larger patches often have greater influence, and area-weighted statistics would be more meaningful for assessing habitats (McGarigal et al. 2012). Moreover, the distribution of metric values is often does not follow a normal distribution, and therefore the mean value gives a somewhat biased evaluation for the whole landscape. In that case, the median will be more meaningful, but we also suggest calculating the coefficient of variation, which describes the variability of the metric for the forest patches. One of the basic symptoms of forest fragmentation is an increase in the number of smaller patches (Echeverria et al. 2006). In these cases, the distribution of patch areas provides important information about the degree of fragmentation (Fig. 10).

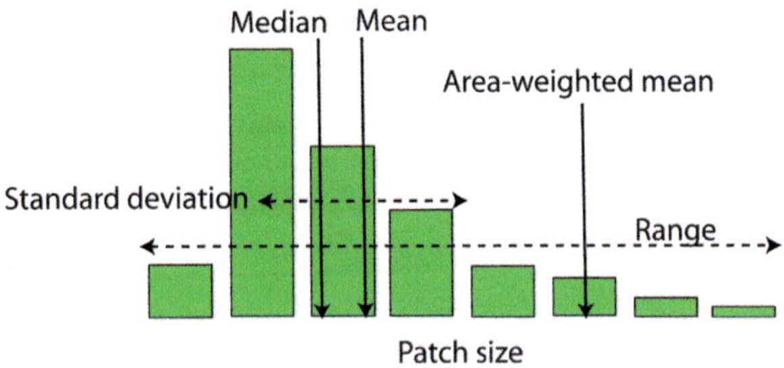

Fig. 9 Distribution of values for a landscape metric (here, patch size) and the associated statistical parameters. Adapted from McGarigal et al. (2012)

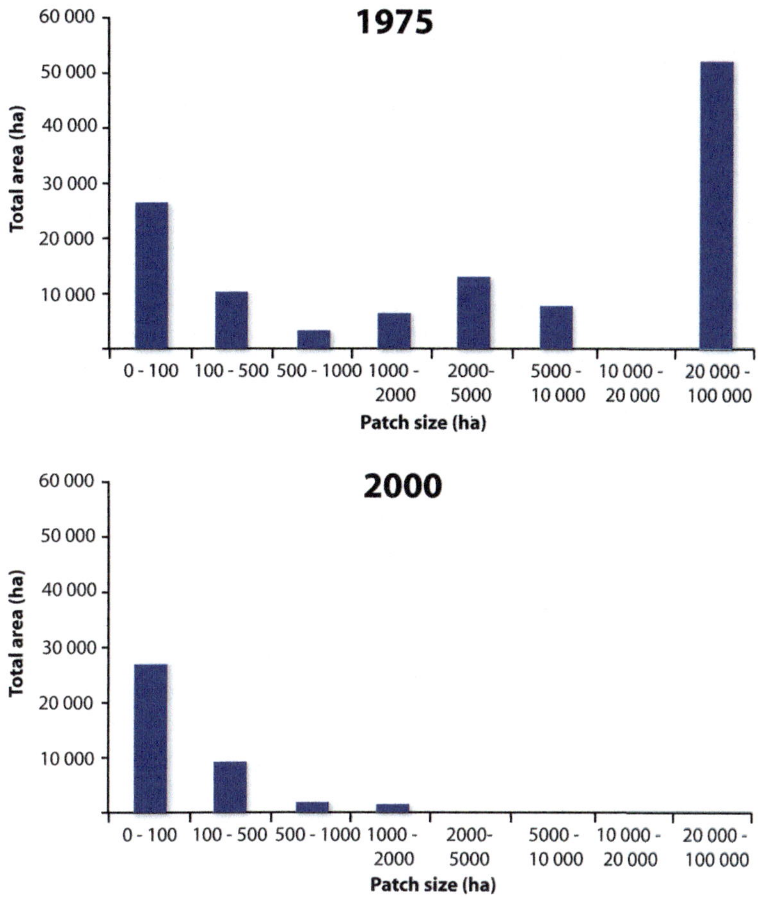

Fig. 10 Temporal changes in the distributions of forest fragment size in a Chilean temperate forest (Echeverria et al. 2006). In 1975, the landscape was dominated by large forest patches, but by 2000, the area of forest had decreased significantly and the large forest patches had disappeared

When selecting metrics for a study, consider the following aspects:

1. **Is it necessary to evaluate composition, configuration, or both**? Previous studies have shown that composition is more important for the relationships between patterns and processes, especially for biodiversity (Uuemaa et al. 2013). However, in assessing fragmentation and deforestation, it is relevant to measure configuration (e.g., patch size, contagion). In many cases, there is little difficulty analyzing both landscape characteristics, and both should be evaluated because of the different insights they provide.
2. **What are the impacts of a metric's range of values?** Metrics differ in their range of values, with some being range limited and others not (Plexida et al. 2014). Comparing metrics with different ranges can be difficult, and it is

challenging to define an ecologically significant change in index values (Remmel and Csillag 2003; Lafortezza et al. 2005). It may be better to use range-limited metrics, as it is easier to understand their expected behavior (Neel et al. 2004).

3. **Should landscape metrics be computed at multiple scales?** To adequately quantify spatial heterogeneity and detect the characteristic scales of landscapes, it can be useful to evaluate metrics at more than one scale (Li and Wu 2004). The behavior of some metrics in response to increasing grain size can be accurately predicted, and this makes them applicable at multiple scales (Li and Wu 2004; Uuemaa et al. 2005).
4. **Should a metric focus on only a single characteristic?** Most indices account for more than one aspect of the spatial pattern, which makes them more difficult to interpret. Therefore, we recommend using simple metrics that are easy to calculate and understand.

Applications of Forest Pattern Mapping

Improving Forest Management

Global deforestation is widely recognized as one of the world's leading environmental problems (Yu et al. 2011). Deforestation not only decreases forest area, but also changes the landscape's configuration (Skole and Tucker 1993), which can exacerbate habitat degradation (Gasparri and Grau 2009). However, several studies have shown that habitat loss is more important than fragmentation itself (Fahrig 2003), which suggests that composition metrics can be more meaningful than configuration metrics for evaluating deforestation. Therefore maps of the proportion of forest are useful in decision making. It is common to analyze forest patterns in fishnet polygons, with the land use map overlaid with the rectangular cells. For each cell, different indicators can be calculated. The simplest indicator (e.g., the

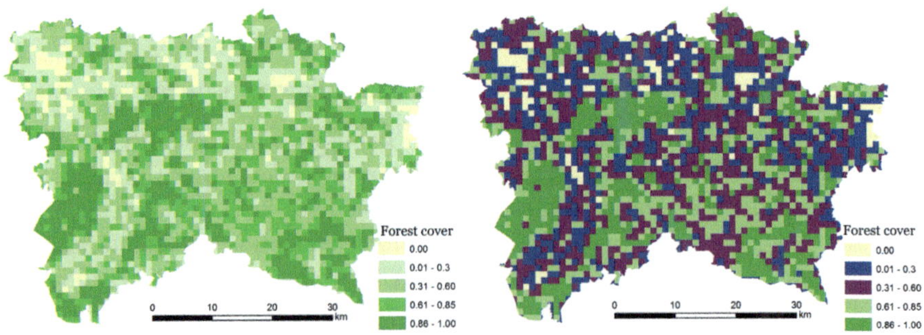

Fig. 11 Spatial distribution of forest cover visualized with a quantitative legend on the *left* and a categorical legend on the *right*. The distribution of forest cover is more easily perceived using the color scheme on the *left*

proportion of forest in each cell) could be useful in determining core areas (Fig. 11). Large areas with a high proportion of forest (>85%) can be considered to be core areas. Although landscape metrics are calculated for qualitative data (mostly land use and cover types), the outcome is always quantitative. Therefore, the results should be presented using visualization methods for quantitative data—a point that is often forgotten. It is advisable to use graduated colors to present proportions of forest cover, as the map is then more easily comprehensible. Colors can be used in many ways to enhance the meaning and clarity of data display, but when used improperly, some important information may be lost (Fig. 11).

Another phenomenon that can be mapped and visualized using a grid is the change over time (Fig. 12). It is possible to show trends for the proportion of forest, its spatial configuration (e.g., edge length, core area), and connectivity in this manner. This method makes it possible to determine the critical areas in which negative changes took place and thus which areas need immediate action or improvement.

Remote sensing data has a relatively long time series, dating back to the 1970s (Landsat), and therefore provides an excellent opportunity to study historical changes in forest cover and relate the spatiotemporal pattern of such changes to other environmental and human factors. Several studies have evaluated the effects of different management strategies on forest pattern (Staus et al. 2002; Miyamoto and Sano 2008). It is also possible to evaluate the spatial pattern of forest fragmentation globally. Li et al. (2010) assessed the differences in forest fragmentation patterns and drivers between China and the conterminous United States using a 300-m-resolution global land cover product (GlobCover) and the moving window method, and found that forests were more fragmented in China than in the United

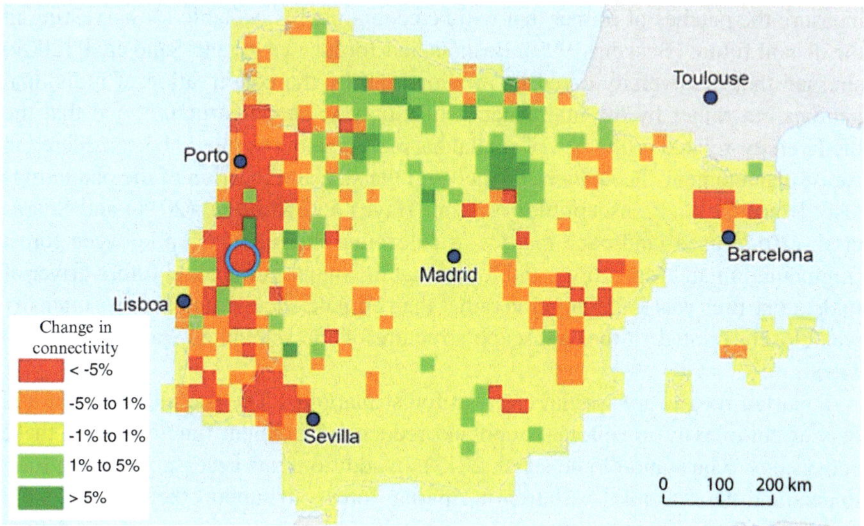

Fig. 12 Trends in forest connectivity in the Iberian Peninsula from 1990 to 2006 using 25 km × 25 km landscape units. Source Estreguil et al. (2012)

States because of a lack of management. Miyamoto and Sano (2008) also found that the size and spacing between patches depend on the nature of the forest management.

Using landscape metrics in pattern evaluation makes it easy to monitor future landscape conditions to detect changes, and their present values can be used as baseline data to create alternative scenarios for future management plans and to compare the resulting changes in spatial processes. Moreover, if these studies are linked with species habitat data, the analysis may help develop more effective management strategies and programs for maintaining biodiversity. Identifying which forested areas in a particular landscape are highly threatened is essential information for decision makers, as is assessing their value for one or more species. The value is either positive, in which case the forest should be conserved, or negative, in which case the forest area may be too small to justify protection at the expense of further development (Abdullah and Nakagoshi 2007).

A particular challenge is how to best achieve landscape-scale restoration and forest defragmentation (Quine and Watts 2009). Plantations make it possible to decrease fragmentation, and with pattern analysis one can determine where the new plantations should be located so that they can be most effective at improving connectivity. However, plantations usually have low conservation value in terms of biodiversity. Sano et al. (2009) provided detailed suggestions on how to improve management in study areas with different spatial characteristics. For example, the size of the core area in their study area indicated that the areas of most natural forest cover types were too small to support selective cutting, which requires larger areas. Baskent (1999) indicated that harvesting patterns could be used to design management regimes for the creation of alternative forest landscapes with significantly different spatial structures. The sizes of modern clear-cut areas determine in large measure the patches of timber that will be economically available for harvesting in the distant future (Baskent 1999). Baskent and Jordan (1996) and Sano et al. (2009) stressed that biodiversity could be provided not by the conservation of individual habitats but rather by the maintenance of a diverse spatial structure, and that the biodiversity-related properties of spatial harvesting patterns should be considered in forest management. The pattern created and the size and location of the plantations also determine their susceptibility to fire. Hayes and Robeson (2011) and Soares et al. (2012) used landscape metrics to determine the relationship between forest fragmentation and forest fires, and found that the major current and future driver of understory fires was fragmentation rather than climate change, and that fire intensity was closely related to the landscape structure of the remaining (post-harvesting) forest.

Riparian forests are special cases in forest mapping. The fragmentation of and loss of complexity in riparian woodland reduce the efficient functioning of these ecosystems (Garofano-Gomez et al. 2013). In addition, it is necessary to more fully understand the structural variation in riparian forests to support their management and restoration (Fernandes et al. 2011, 2013). Field surveys are the usual data source in such cases, but they may take an infeasible amount of time, and because of their expense, they are spatially limited (Baker et al. 2007; Fernandes et al. 2013). A rela-

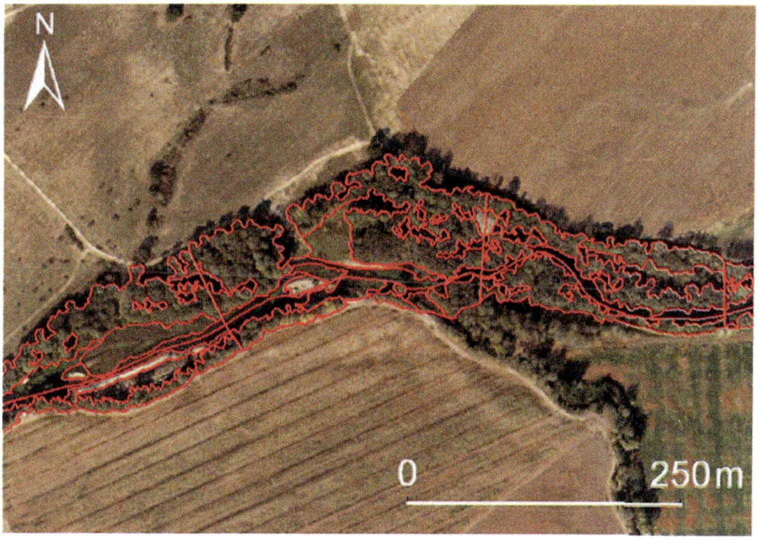

Fig. 13 Illustration of riparian patches, with perpendicular lines used to divide contiguous sampling units. Source: Fernandes et al. (2011)

tively common approach is to use fixed-distance riparian metrics (e.g., buffers), but these do not correctly depict the variation in the spatial configuration of riparian patterns within watersheds (Baker et al. 2007), and fixed-distance metrics are also strongly correlated with whole-watershed land cover proportions (Baker et al. 2006). The limited width of the riparian zone and the complexity of its internal structure mean that imagery with high spatial resolution (<5 m) is needed to permit spectral analysis of riparian vegetation (Aguiar et al. 2011; Fernandes et al. 2011). Fernandes et al. (2011) demonstrated that the spatial patterns of riparian vegetation are heterogeneous, but can nonetheless be identified from digital images with resolutions of 5 m or better, and landscape metrics can then be used to describe the spatial patterns of riparian vegetation (Fig. 13).

In terms of conservation efforts, landscape metrics enable managers to quickly evaluate the influence of interventions such as the creation of natural parks on forest fragmentation (Southworth et al. 2004). Soverel et al. (2010) found that all national parks in Canada have at least one significantly different aspect of spatial configuration compared with other parks in the national park system. These metrics also provide a baseline for evaluating the effectiveness of management inside national parks. However, Remmel and Csillag (2003) found that comparisons of landscape metrics and testing for significant differences among various landscapes or studies produced uncertain results when the distributions of the landscape metrics were not known. Fortin et al. (2003), Remmel and Csillag (2003), and Remmel and Fortin (2013) provided insights into how statistical differences between landscape metrics can be determined. They stressed the importance of knowing the expected range of variation about landscape metrics' values so that statistical comparisons can be made. This issue is related to the wider topic of comparing categorical maps. Boots and Csillag (2006) comprehensively studied map comparisons and concluded that although pattern characterization and comparisons benefit from the choice of appropriate methods, meaningful hypotheses can only be tested when we remember that patterns are the results of spatial processes.

Assessment of Forest Habitats

The conservation of fragmented landscapes requires an understanding of the value of the remaining forest patches for the organisms that inhabit the forest (Holland and Bennett 2009). The quality of forest habitats in landscape mosaics can be assessed based on the forest's patch size, quality, connectivity, and boundary configuration (Hardt et al. 2013). In evaluating habitat quality, scale plays an essential role; for example, the landscape elements that are important for a mouse are not relevant for a deer (McGarigal et al. 2012). The general rule is that the smaller the organism's body, the more detailed the spatial resolution that is needed to map the corresponding habitats. However, as home ranges interact with the size and configuration of landscape features to determine the actual use of the available habitats (Macdonald and Rushton 2003; Zapponi et al. 2013), the scale of the analysis must be species specific. Moreover, it is necessary to assess existing natural areas and prioritize conservation actions across multiple spatial scales. For habitat analyses, knowledge of the forest's vertical structure and its compositional structure (i.e., plant species richness) is often essential. Light detection and ranging (LiDAR) and synthetic aperture radar (SAR) are emerging as important tools for vegetation with a pronounced vertical structure, but for assessments of a forest's compositional structure, vegetation sampling is often needed (Corona et al. 2011; Chirici et al. 2012). Plant species richness is a useful measure of biodiversity, but because of the cost and the time requirements for field sampling can typically only be evaluated in small areas of a given landscape (Schetter et al. 2013). It is often useful to combine field data with remotely sensed data and landscape pattern metrics to evaluate habitat quality or predict species richness (Schetter et al. 2013).

Many studies have been performed to study how fragmentation affects the diversity and abundance of mammals (Corona et al. 2011; Garmendia et al. 2013). However, in most research, only the land use or cover type was used as the basis for landscape analysis. However, Li et al. (2000) found that simple land use and cover boundaries often lack ecological relevance and are therefore unsuitable for evaluating habitat quality, because no single map can represent the diverse habitat requirements of many species. Kintz et al. (2006) have also pointed out that land use maps derived from satellite imagery may not contain all the necessary information to support a biodiversity assessment. For example, it is not always possible to distinguish between natural forest and plantations, and this distinction is important for habitat quality. Plantations often have a lower habitat value than natural forest, and their spatial configuration is often much more regular than that of natural forests. To overcome this drawback, Li and Wu (2004) suggested creating habitat rank maps for each target species by aggregating adjacent patches with the same suitability rank for a given species.

Previous studies of mammals have shown that forest composition is the most significant factor that influences species diversity (Uuemaa et al. 2013). However, many authors stress that increasing fragmentation caused by anthropogenic land-use change is one of the most common threats to species and genetic diversity worldwide (McAlpine and Eyre 2002; Martinez et al. 2010). Therefore, the effect of fragmentation on animal diversity has also been widely analyzed. Different species have different correlations with landscape metrics that depend on their landscape preferences. For example, large, compact (simple) patches are preferred by wild hogs (*Sus scrofa*) (Gaines et al. 2005), moose (*Alces alces*) (Maier et al. 2005), and deer (*Odocoileus virginianus*) (Plante et al. 2004), whereas ocelots (*Leopardus pardalis*) (Jackson et al. 2005) and gliders (*Petaurus* spp.) (McAlpine and Eyre 2002; Masse and Cote 2012) preferred areas that had a greater degree of fragmentation (i.e., a larger number of smaller patches with more edge).

In addition to the shape and size of the habitats, connectivity is an essential indicator for biodiversity. Forest connectivity combines the availability of forest with the distance between patches and characteristics of the corridors between patches; it refers to the degree to which the landscape facilitates or impedes the movement of species with specific dispersal capabilities and requirements (Estreguil et al. 2012). Several landscape metrics have been designed to measure connectivity (e.g., the connectance index, cohesion), but special software has also been developed to analyze connectivity, including Conefor and Guidos (see the section *Tools for evaluating landscape patterns* for details).

Many studies on mammals are performed with low-resolution data (Corona et al. 2011), but more detailed data is often needed, as low-resolution satellite imagery (e.g., Landsat data) often has low overall accuracy for identifying habitat types and corridors. However, maps derived from high-resolution imagery are expensive (Lu and Weng 2007). Ramezani et al. (2010) proposed using point sampling as an alternative to wall-to-wall mapping to estimate landscape metrics on a more detailed scale. They found that the sample-based estimates are competitive in terms of time consumption (thus, cost), as they required less time than wall-to-wall mapping.

To evaluate habitat quality for birds, more detailed information about canopy closure and the forest's vertical structure is needed (Kirk et al. 2012). Moreover, Saveraid et al. (2001) argued that satellite data are useful for identifying areas where certain species may be located, but more detailed vegetation and habitat data collected in the field are necessary to accurately determine nesting and breeding habitats. The combination of remotely sensed and ground-based data provides a researcher with more complete information and the ability to determine species occurrences. For example, canopy gaps caused by natural disturbances such as treefall are a significant source of heterogeneity in intact forests, and avian species richness and abundance are influenced by these gaps (Gharehaghaji et al. 2012). Bird censuses are usually undertaken using point counts, and buffer zones are generated around each counting point based on home ranges; landscape metrics are then calculated only for these buffer zones and not for the entire landscape (Gillespie and Walter 2001).

Landscape metrics have been used most widely to evaluate the influence of environmental changes on bird species richness (Thompson et al. 2008). Similarly to the case of mammals, studies have shown that bird species generally also respond more strongly to the composition of the land use and cover classes than to the landscape's configuration (Uuemaa et al. 2013). Among the configuration metrics, edge density has given good results in predicting bird abundance (Fauth et al. 2000). The shape of the patches appears to play less of a role for bird diversity (Uuemaa et al. 2013).

Patch area and isolation alone are often weak determinants of habitat quality, since the distance to the feeding or breeding locations and the distance to major disturbances have a greater effect on species abundance and richness (McGarigal et al. 2012). There must be enough bodies of water near the home forest patch, and it must be sufficiently far from sources of disturbance such as highways. Moreover, the forest's inner structure (e.g., stand age, vertical structure) is often more important than the size of the forest patch or the isolation between patches (Uuemaa et al. 2013). Obtaining information about a forest's inner structure assumes that the forests are mapped at a detailed scale and that fieldwork is also conducted. In addition, using the moving window approach helps researchers to map the inner structure of the forest patches. Many forest species are edge specific or core specific, so mapping core areas and edges would be of special interest. However, most previous studies have only measured the structural edges of the habitats and did not account for the habitat's functional characteristics (Uuemaa et al. 2013).

Researchers should focus more on incorporating the magnitude of the change that occurs at an edge (edge contrast) into habitat analyses (Uuemaa et al. 2013). For many species, there is a significant difference between whether the edge separates forest from artificial or natural land use types (McGarigal et al. 2012). There are multiple ways of determining the contrast between edges, such as using light conditions in the weighting system. However, although there have been many studies about the habitat preferences of different species in relation to forest fragmentation, there has been little analysis of conservation effectiveness based on this context (Uuemaa et al. 2013).

It is therefore necessary to evaluate whether protected areas actually support biodiversity conservation. More and more focus has recently been placed outside of protected areas, since reserves alone often cannot ensure the necessary habitat quality required to maintain biodiversity and the matrix that contains a reserve can have an important effect. For example, even a relatively severely logged forest outside a reserve may represent a significant resource for biodiversity conservation, and secondary forests are an often-overlooked resource that could be managed to help reduce pressures elsewhere (Gillespie et al. 2008). Off-reserve areas complement existing reserves by minimizing edge effects, reducing fragmentation, and increasing habitat quality and extent (Lethbridge et al. 2010). There has been some progress in developing probabilistic habitat quality models that use parameters based on species responses to the landscape configuration (i.e., landscape metrics) to determine habitat restoration priorities (Westphal et al. 2007). For example, Watts et al. (2009) used the Marxan algorithm to prioritize habitat restoration using landscape metrics as parameters that described the landscape preferences of different species. Westphal et al. (2007) and Lethbridge et al. (2010) also developed the conservation decision support tool "Optimal Restoration of Altered Habitats" (OPRAH), which combines species preferences for certain landscape configurations with habitat quality information to select the optimal habitat restoration priorities for single or multiple species.

Mapping Landscape Metrics by Using GIS

There are many simple ways to evaluate forest patterns using ordinary GIS software without requiring the installation of specific landscape analysis extensions. In all GIS software, it is possible to calculate the area and perimeter of polygons. From these parameters, several distribution statistics are easily calculated (e.g., patch and edge density; the mean, median, standard deviation, and range of patch areas and shapes; the nearest-neighbor distance).

There are also simple ways to generate different types of buffers inside forest habitat patches, thereby supporting an analysis of core areas (Fig. 14). In most cases, a single buffer zone is enough, but the width of the zone depends on the species under study. The advantage of using GIS software instead of classical programs for calculating landscape metrics is the ability to create multiple buffers. Sometimes the edges have a complex structure (Fig. 15), especially in riparian forests. All the same simple statistics can be calculated, in addition to statistics on the core areas and the edges (i.e., buffers).

To evaluate a forest's habitat quality, additional habitat parameters such as the distance to bodies of water and roads may be essential, as we noted earlier. GIS software can easily perform proximity analysis and find the distances to the closest defined objects (e.g., the nearest lake to each forest patch). Raster data enables the creation of a proximity map (Fig. 16), which identifies the distances from all loca-

Fig. 14 Buffers created for forest patches at (*left*) a specified distance, (*middle*) at multiple distances, and (*right*) at different distances based on an attribute of the forest

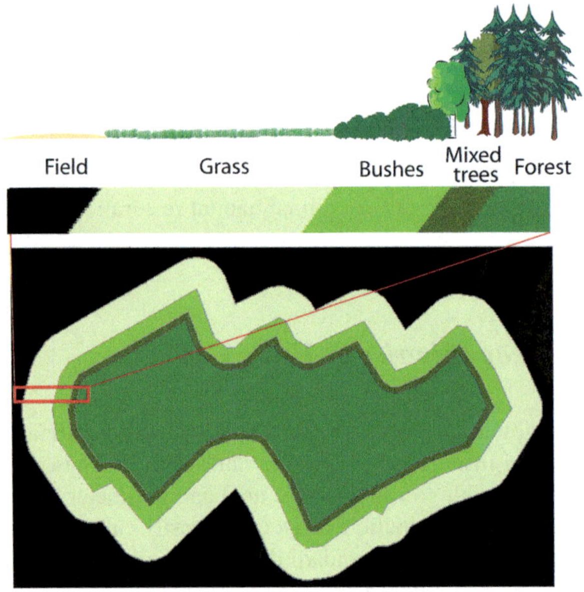

Fig. 15 Multiple-layered edge structures can be presented as multiple buffers

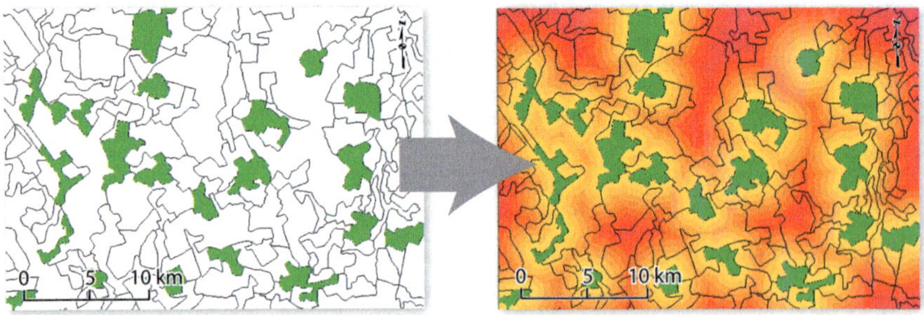

Fig. 16 Creating a proximity map from a land use and cover map by first identifying forests and then calculating Euclidean distances from all locations to the forest patches

tions in the study area to the nearest polygon containing a given feature (e.g., a river). Proximity maps are useful for identifying equidistant "ridges" that bisect a set of forest parcels (Berry 2007). Proximity maps also better handle changes in landscape structure. If one of the forest patches is removed (e.g., by timber harvesting or wildfire), the proximity map shows the influence on the surrounding areas (Berry 2007).

The moving window approach is also easily applicable in GIS software for simple focal statistics. It is also possible to use different kernel sizes and shapes to calculate a variety of statistics.

Using Landscape Metrics in Modeling

There are many ways to use landscape metrics in modeling. The most common approach is to use landscape metrics to capture the patterns of real and modeled landscape attributes. This approach is mostly used to evaluate forest landscapes (Fig. 17), especially in terms of deforestation (McConnell et al. 2004). Another group of studies have used landscape metrics as indicators of species richness, with the landscape metrics used as input parameters for the models (Gillespie et al. 2008).

Echeverria et al. (2008) used landscape metrics in forest fragmentation modeling and predicted the future trends in deforestation patterns. Analysis of landscape patterns revealed that the loss of forests was concentrated around the edges of forest fragments located in slightly undulating terrain because local people use trees near the borders of forest patches to produce fuelwood and then clear these areas for crops and pasture. McConnell et al. (2004) used proximity maps created based on the distance from the forest edge and villages as one of the input factors to model deforestation in Madagascar, and found that the prior land use was the best predictor of deforestation (Fig. 18). Their results were similar to those of Echeverria et al. (2008), who found that the clearing of the land took place closer to settlements and roads, and was most likely to start at forest edges. These results suggest that forest edges are important predictors of future deforestation. Moreover, Echeverria et al.

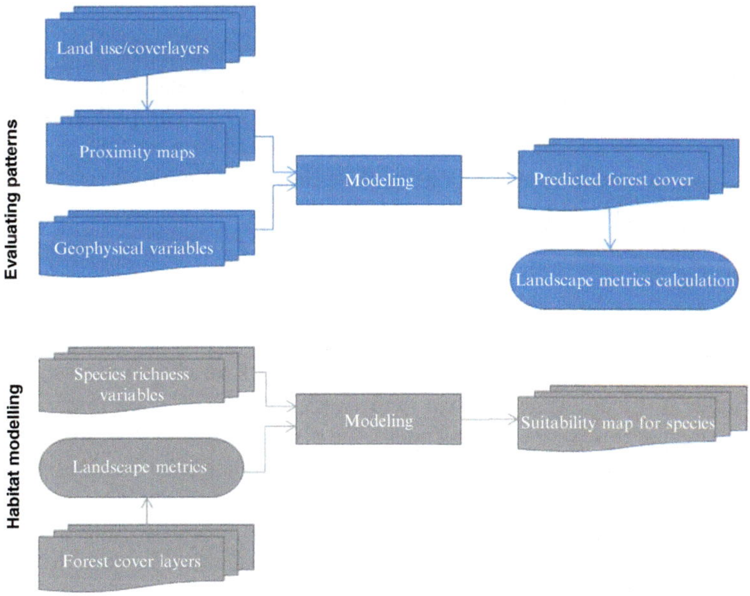

Fig. 17 Illustrations of two approaches to using landscape metrics in modeling: (*top*) as model outputs and (*bottom*) as model inputs

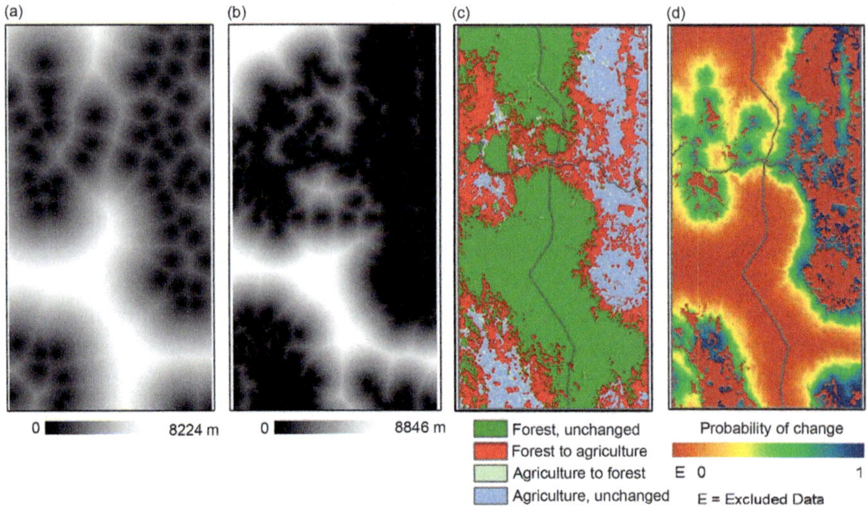

Fig. 18 Modeling deforestation probabilities (McConnell et al. 2004). The explanatory factors were (**a**) distance from the nearest village (centroids) and (**b**) the distance from the forest's edge. The resulting images show the predicted changes (**c**) in land use and cover type based on cross-tabulation of land cover maps from 1957 to 2000 and (**d**) in the probability of change based on a logistic regression model, in which the probability is expressed ranging from 0 (*low*) to 1 (*high*). Cells that were not candidates for conversion were excluded from the regression

(2008) also found that patch size was an important factor, since small patches appeared to be more vulnerable to deforestation.

Proximity maps are widely used as suitability maps for modeling the spread of forest fires (Clarke et al. 1994) and forest succession (Favier and Dubois 2004), with cellular automata and landscape metrics together enabling a comprehensive evaluation of the modeling results. However, landscape metrics can also be used to calibrate cellular automata by incorporating them during multiple runs (trial solutions) of genetic algorithms (Li et al. 2013). Moreover, maps of the values of landscape metrics created by a moving window approach can also be used as suitability maps for modeling inputs.

As we mentioned earlier, landscape metrics are often used as predictors in habitat modeling. Schindler et al. (2013) found that landscape metrics were good indicators for overall species richness and for the richness of woody plants, orthopterans, and reptiles in Mediterranean forest landscapes. They also noted that the performance of the metrics was scale dependent; the diversity of woody plants, orthopterans, and small terrestrial birds was usually better predicted by using larger buffer zones, whereas the diversity of reptiles was frequently predicted better using larger buffer zones. Monitoring landscape metrics may help to identify critical changes in forest patterns that might contribute to a loss of forest biodiversity. It has been argued that linking sample data on biodiversity indicators to ecologically meaningful forest type units has substantial advantages for forest biodiversity assessment (Corona et al. 2011).

Choosing input parameters for modeling can be improved by classifying forest areas into different forest types (e.g., evergreen coniferous vs. deciduous broad-leaved) to reframe biodiversity indicators collected over wide areas into smaller, more homogeneous units characterized by similar key determinants of biodiversity (Larsson 2001). Moreover, this classification also helps to improve our understanding, interpretation, and communication of data on biodiversity variables by enabling comparison of ecologically similar forests (Rego et al. 2004). This can be done by integrating forest inventories into biodiversity assessment studies.

Future Perspectives on Mapping Patterns

3D Landscape Metrics

In the past several years, one of the most important developments has been integration of the third dimension into the analysis of landscape metrics. There are two aspects that should be considered when dealing with 3D in mapping forest landscapes: topography as the third dimension and forest's vertical structure. Topography plays an important role in ecosystem functions and structure, even though traditional pattern analysis only considers a planimetric surface, which can produce misleading results in mountainous areas. The technological progress in the field of remote sensing has led to a rapid improvement in the quality of DEMs created from LiDAR measurements, which provide elevation data good enough for use in landscape analysis. A variety of methods exist for measuring terrain irregularity, ranging from the concept of a "fractal dimension" (Mandelbrot 1983) to the widely used terrain ruggedness index (Riley et al. 1999), which expresses the elevation difference between adjacent cells in a grid. Although the fractal dimension is quite widely used in landscape analysis, it has only been implemented in a planar system and does not account for 3D. There are few approaches that have been used to integrate topography in the calculation of landscape metrics. Jenness (2005) proposed a method for calculating true surfaces that is based on a moving window algorithm and that estimates the true surface area for each grid cell using a triangulation method based on the use of triangular polygons to cover a surface (Fig. 19).

Hoechstetter et al. (2008) and Batista et al. (2012) showed that there is a significant difference between the values estimated using the 2D and 3D forms of most landscape metrics, although shape metrics appear to be not influenced by surface roughness. Therefore, using 3D landscape metrics to evaluate forest patterns in mountainous areas should be seriously considered because these metrics provide more realistic results. For example, Hou and Walz (2013) introduced several metrics suitable for analysis of the 3D landscape structure and found that the basic effect of switching from 2D to 3D metrics was to increase the patch area and perimeter. For diversity metrics, the 3D metrics produced a lower evenness index, leading to a correspondingly lower diversity index. The values of an edge contrast index (ECON) also decreased because the transitions caused by the use of 3D data acted as buffer

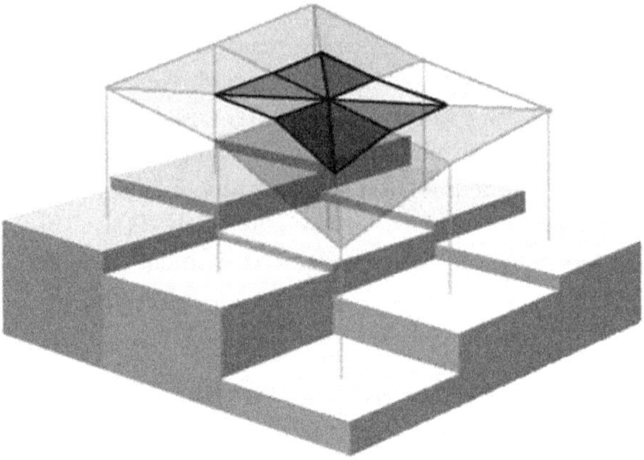

Fig. 19 Illustration of the Jenness (2005) method to determine the true surface area and true surface perimeter of patches in a 3D landscape. The true surface area of each cell in the raster grid is obtained by adding the areas of the eight shaded triangles, and the true surface perimeter is obtained by summation of the lengths of the eight bold line segments

regions that could be used to smooth the "terrain barrier" between vegetation patches, thereby smoothing "edge effects" between patches and leading to more realistic quantification of fragmentation. However, the effective mesh size (MESH), which serves as an example of the group of fragmentation metrics, proved to be sensitive to transitions, and the values of MESH for each land use or cover class or the whole landscape increased when the 3D transition zones were included in the calculation.

McGarigal et al. (2009) proposed using various surface metrics, such as the topographic wetness index of Moore et al. (1993) and the topographic position index of Jenness (2005), because many of these indices had no analogues among patch metrics and had the potential to offer new insights into landscape patterns. In addition, we propose combining surface metrics with patch metrics, which can be easily done using map algebra, to take advantage of the strengths of both types of metric (Fig. 20). The terrain ruggedness index is widely used and easily calculable using free software (e.g., SAGA GIS; http://www.saga-gis.org/). Another way to evaluate topographic complexity is to use patch metrics that can be applied to quantitative data (e.g., contrast metrics). Contrast metrics can be computed based on elevation values.

The forest's vertical structure is another important aspect of habitat analyses, and is also important for assessing fire susceptibility. LiDAR datasets have become widely used to assess variations in leaf area index as a function of height above the ground (Solberg et al. 2009), which can be used to estimate forest canopy fuel loads (Erdody and Moskal 2010). This is important because the continuity of the fuel between the ground surface and the canopy strongly influences the risk of fire reaching the tree crowns, leading to the development of severe fires. Fire also behaves differently in forests with dense patches separated by open ground than in areas

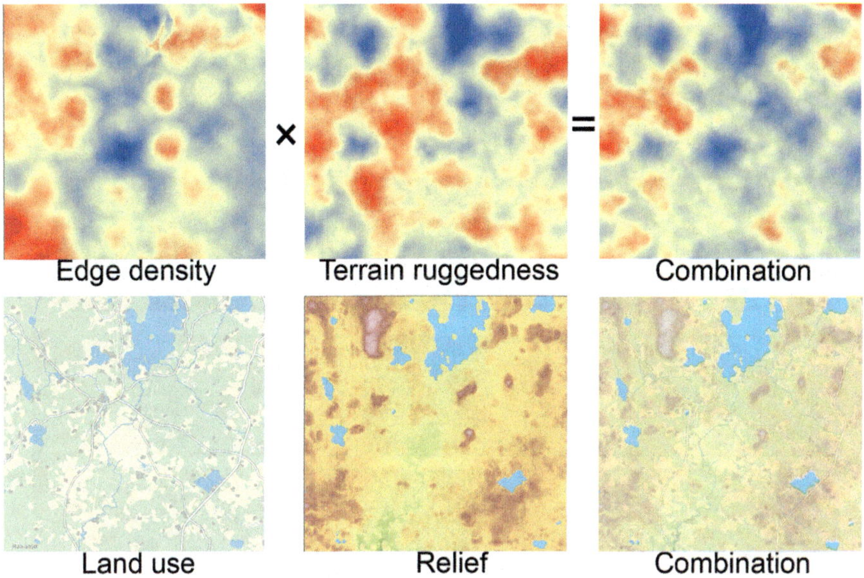

Fig. 20 Integrating topographic characteristics into landscape analysis. The values of the landscape metrics can be combined with terrain ruggedness indices by using simple techniques from map algebra

with equal distances between all trees, such as plantations (Richardson and Moskal 2011). Therefore, both the stand's leaf area distribution and stem density are valuable information for forest managers. The individual-tree-crown approach, in which LiDAR data is first segmented into individual tree crowns, enables researchers to delineate individual trees within the forest and predict tree density. However, there are also limitations because LiDAR tends to underestimate the density of smaller trees. Richardson and Moskal (2011) developed an approach to assess forest structure by estimating the density and spatial configuration of trees in four height classes, and evaluated the results by using landscape metrics (Fig. 21). Landscape metrics let researchers measure spatial patterns of tree densities at different spatial scales. LiDAR-based 3D metrics offer insights into the "within-patch diversity," which improves the assessment of vegetation patterns in terms of potential forest vulnerability to fire or to insect pests (Blaschke et al. 2004) and will help managers to develop management plans and researchers to assess habitat quality.

4D Landscape Metrics

In addition to accounting for topography, more and more time series data (4D) is needed. Location-based environmental information is required in near real time during crisis situations to permit timely responses (Klug and Kmoch 2015). Remote

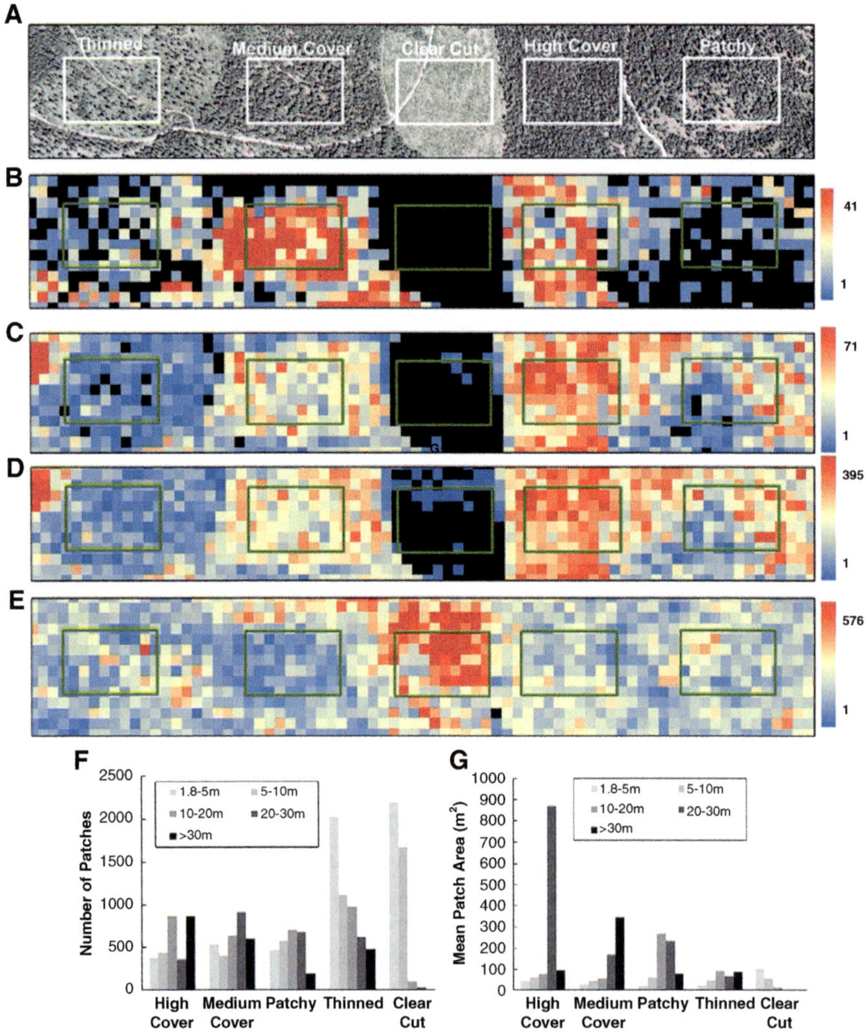

Fig. 21 Tree density photograph and raster images of the data used to calculate landscape metrics. (**a**) Orthophoto of forest with a range of tree densities. (**b**–**e**) Maps of predicted tree density for (**b**) trees taller than 30 m, (**c**) trees 20–30 m in height, (**d**) trees 10–20 m in height, and (**e**) trees 5–10 m in height. Legends to the *right* of (**b**–**e**) describe the number of trees predicted in each 30.5 m × 30.5 m cell of the grid; the green boxes in (**b**–**e**) correspond to the white boxes in (**a**). (**f**) Number of patches, and (**g**) mean patch area for the cells in the grid. Adapted from Richardson and Moskal (2011)

sensing and Web-based technologies are developing fast and will enable sharing and presenting of data as they become more accessible (Granell et al. 2016). This will let researchers perform interactive analysis and modeling using up-to-date datasets. Geosensor networks are specialized applications of wireless sensor network technology in a geographic space that can detect, monitor, and report data on environmental phenomena and processes (Nittel 2009). Sensor networks add the possibility of regional- and even local-scale observations at many points by implementing a higher density of sensor nodes in a given area; thus, they can deliver a more accurate description of the temporal and spatial variations that are occurring.

A particularly important aspect of this approach is the possibility of real-time data delivery. Both the real-time aspect and the increased spatial and temporal resolutions have brought new research challenges. For example, the availability of sensor platforms with different sizes provides huge collections of concurrent streams of georeferenced sensor data in real time (Nittel 2009). Leveraging this technology will let us observe phenomena that were impossible or prohibitively difficult to measure before.

In addition, unmanned aerial vehicles (UAVs) offer new opportunities to obtain near-real-time data from forests. UAVs have already been used to map insect damage. Näsi et al. (2015) used a UAV-mounted hyperspectral imaging sensor to identify mature Norway spruce (*Picea abies*) trees suffering from infestation by the invasive bark beetle *Ips typographus*. They developed a processing approach that let them analyze the spectral characteristics of the sensor images with high spatial and spectral resolution in a forested environment, and were able to identify damaged trees.

Moreover, new space programs such as the Sentinel satellite program (https://sentinel.esa.int/web/sentinel/home) will provide open access to near-real-time high-resolution SAR images. Majasalmi and Rautiainen (2016) demonstrated the potential of Sentinel-2 bands in estimating canopy biophysical properties for boreal forests in Finland. They found that the Sentinel-2 data could be used to estimate the effective leaf area index and the inverted red-edge chlorophyll index.

There is no longer a problem getting access to up-to-date high-resolution data. We are now able to detect and map forest fires and illegal logging in real time. The problem has become one of being able to process the data fast enough. In terms of quickly quantifying changes and effects, landscape metrics offer the advantage of being easily and quickly calculated. The challenge will be to build appropriate data management technologies that can query, process, mine, and analyze the data streams in *real time* to find trends and identify events (Li et al. 2016). One increasingly acknowledged need is for open and standard-compliant distribution and access to data to ensure interoperability among and harmonization of geospatial datasets and time series for use in environmental analysis (Klug and Kmoch 2015).

Conclusions

For decades, landscape metrics have been used to measure and abstract landscape patterns. Because there are hundreds of different metrics available, many programs have been developed to compute these metrics. At present, there are no special-purpose metrics that are uniquely suited to forest landscapes. However, because forests have unique characteristics that differ from those of other ecosystems, and particularly from artificial anthropogenic ecosystems, future research should look for new metrics that capture these unique characteristics. For example, vertical structure is an important characteristic of forests, but has not yet been adequately described. Many researchers have attempted to define an optimal set of metrics, but thus far, it appears that the selection of metrics depends more strongly on the purpose of the study than on the land use or cover type. However, some metrics are used more often for forest habitats than for other ecosystems, such as the edge and core area metrics.

Forest landscape patterns are changing fast due to natural factors (e.g., climate change) and human disturbances (e.g., land use or management changes). Remote sensing offers a rapid, cost-effective method for acquiring up-to-date information over a large geographical area and is therefore widely used as a source of the data needed for pattern assessment. However, to obtain meaningful results from the calculation of landscape metrics, correct preparation of the data is essential. Variations in the value of landscape metrics can result from different pixel sizes, classification methods, or filtering applied to the images.

When researchers map forests, they must always consider whether to map only the forests or to include the matrix in which the forests are embedded, which raises the issue of how to consider the boundaries and internal heterogeneity. The subsequent pattern analysis will depend greatly on how forest is defined (e.g., how it differs from transitional vegetation types). Modern technology lets us map everything in high detail, but we might not always need so much detail; sometimes, too much detail becomes noise. Similarly, it is possible to obtain too much information. In pattern analysis, it is often better to choose fewer and more meaningful indicators than a larger number of metrics that are harder to interpret.

Landscape metrics have a wide array of uses, ranging from forest management and habitat monitoring to modeling of deforestation or forest fire susceptibility. Their advantage is the simplicity of calculation and relatively easy comparison of the resulting numbers. However, in any kind of spatial analysis, visualization of the results becomes important. We believe that this aspect of landscape research has received too little attention in the research literature. Choosing the wrong visualization method can lead to misinterpretation of the results, or a failure to see important patterns.

Although many successes have been achieved using simple planimetric analyses, there is considerable room for improvement. The future of the landscape metrics will lie in finding ways to integrate topographic information (i.e., to move from 2D to 3D analysis) and temporal information (i.e., to move to 3D or 4D analysis). Because of the huge quantities of data that are becoming available, finding ways to perform near-real-time calculations and modeling will be an important challenge, but if that challenge can be met will permit the development of Web-based services that make data more widely available.

Acknowledgments This study was supported by grant no. IUT2-16 of the Ministry of Education and Science of Estonia and by the Marie Skłodowska-Curie Actions individual fellowships offered by the Horizon 2020 Programme under REA grant agreement number 660391. We also thank the reviewers and language editor for constructive and helpful comments and suggestions that helped to significantly improve this manuscript.

Appendix: Technical Terms

Categorical maps—Maps that define each unit of a landscape in terms of a descriptive category (e.g., forest vs. grassland) rather than quantitatively.

Central point method—A method used to convert vector data to raster data by assigning a value to a cell in a grid based on the value for the polygon that overlaps the center of the cell. See also *majority rule method.*

Fractal dimension—A ratio that provides a statistical index of the degree of complexity of a pattern by examining how the level of detail in a pattern changes in response to changes in the scale at which it is measured.

Free open-source software (FOSS)—Software whose source code is made available to anyone under a license in which the copyright holder provides the rights to study, change, and distribute the software to anyone and for any purpose.

Geographical information system (GIS)—A system designed to capture, store, manipulate, analyze, manage, and present all types of spatial or geographical data.

Graph theory—A mathematical description of the properties of graphs and, thus, of the pairwise relationships between variables or objects.

Image rectification—A transformation process used to project an image from a sensor's coordinate system into a geographical coordinate system.

Landscape metrics—Algorithms that quantify specific spatial characteristics of patches, classes of patches, or entire landscape mosaics.

Leaf area index—A dimensionless quantity that characterizes plant canopies by dividing the total one-sided area of leaf tissue by the ground surface area covered by the canopy that contains those leaves.

LiDAR (light detection and ranging)—A remote sensing method that uses pulsed laser light to measure the distance between the sensor and a surface.

Majority rule method—A method used to convert vector data to raster data by using the feature that accounts for the largest area within a cell of a grid to define the attribute value assigned to the cell. See also *central point method.*

Map algebra—A set of primitive operations in a geographic information system (GIS) that allow two or more raster layers ("maps") of similar dimensions to produce a new raster layer (map) using algebraic operations such as addition or subtraction.

Marxan algorithm—An algorithm used in conservation planning that aims to minimize the sum of the site-specific costs and connectivity costs for selected planning units, subject to the constraint that the conservation features in a reserve system must achieve predetermined targets.

Minimum mapping unit (MMU)—The size of the smallest feature that can be delineated within the boundaries of a map.

Modifiable areal unit problem—A challenge that occurs during the spatial analysis of aggregated data, in which the results differ when the same analysis is applied to the same data under different aggregation schemes.

Neutral landscape models—The minimum set of rules required to generate a pattern in the absence of a particular process; neutral models provide a means of testing the effect of the measured process on the patterns that are actually observed.

Open data—Data that can be freely used, reused, and redistributed by anyone, without restrictions from copyright, patents, or other mechanisms of control.

Red-edge chlorophyll index—A method used to estimate canopy chlorophyll and nitrogen contents based on remote sensing data.

SAR (synthetic aperture radar)—A form of radar that can be used to create images of objects, such as landscapes; the images can be two- or three-dimensional representations of the object.

Landscape metric scalograms—The response curves of landscape metrics to changing grain size that allow the detection of the most representative scales.

References

Abdullah SA, Nakagoshi N (2007) Forest fragmentation and its correlation to human land use change in the state of Selangor, peninsular Malaysia. For Ecol Manag 241(1–3):39–48. doi:10.1016/j.foreco.2006.12.016

Aguiar FC, Fernandes MR, Ferreira MT (2011) Riparian vegetation metrics as tools for guiding ecological restoration in riverscapes. Knowl Manag Aquat Ecosyst 402:21. doi:10.1051/Kmae/2011074

Albeke SE, Nibbelink NP, Mu L, Ellsworth DJ (2010) Measuring boundary convexity at multiple spatial scales using a linear "moving window" analysis: an application to coastal river otter habitat selection. Landsc Ecol 25(10):1575–1587. doi:10.1007/s10980-010-9528-4

Arnot C, Fisher PF, Wadsworth R, Wellens J (2004) Landscape metrics with ecotones: pattern under uncertainty. Landsc Ecol 19(2):181–195

Baker WL, Cai YM (1992) The R le-programs for multiscale analysis of landscape structure using the grass geographical information-system. Landsc Ecol 7(4):291–302. doi:10.1007/Bf00131258

Baker ME, Weller DE, Jordan TE (2006) Improved methods for quantifying potential nutrient interception by riparian buffers. Landsc Ecol 21(8):1327–1345. doi:10.1007/s10980-006-0020-0

Baker ME, Weller DE, Jordan TE (2007) Effects of stream map resolution on measures of riparian buffer distribution and nutrient retention potential. Landsc Ecol 22(7):973–992. doi:10.1007/s10980-007-9080-z

Baldwin DJB, Weaver K, Schnekenburger F, Perera AH (2004) Sensitivity of landscape pattern indices to input data characteristics on real landscapes: implications for their use in natural disturbance emulation. Landsc Ecol 19(3):255–271

Baskent EZ (1999) Controlling spatial structure of forested landscapes: a case study towards landscape management. Landsc Ecol 14(1):83–97. doi:10.1023/A:1008071307848

Baskent EZ, Jordan GA (1996) Designing forest management to control spatial structure of landscapes. Landsc Urban Plan 34(1):55–74. doi:10.1016/0169-2046(95)00200-6

Batista T, Mendes P, Carvalho L, Vila-Vicosa C, Gomes CP (2012) Suitable methods for landscape evaluation and valorization: the third dimension in landscape metrics. Acta Bot Gallica 159(2):161–168. doi:10.1080/12538078.2012.696930

Berry JK (2007) Map analysis: understanding spatial patterns and relationships. GeoTec Media
Blaschke T, Tiede D, Heurich M (2004) 3D landscape metrics to modelling forest structure and diversity based on laser scanning data. Remot Sens Spat Inform Sci 8:129–132
Bogaert J, Van Hecke P, Salvador-Van Eysenrode D, Impens I (2000) Landscape fragmentation assessment using a single measure. Wildl Soc Bull 28(4):875–881
Boots B, Csillag F (2006) Categorical maps, comparisons, and confidence. J Geogr Syst 8(2):109–118
Bregt AK, Denneboom J, Gesink HJ, Vanranden Y (1991) Determination of rasterizing error—a case-study with the soil map of the Netherlands. Int J Geogr Inf Syst 5(3):361–367. doi:10.1080/02693799108927861
Brown DG, Addink EA, Duh JD, Bowersox MA (2004) Assessing uncertainty in spatial landscape metrics derived from remote sensing data. Remot Sens Gis Accur Assessm:221–232
Butler BJ, Swenson JJ, Alig RJ (2004) Forest fragmentation in the Pacific Northwest: quantification and correlations. For Ecol Manag 189(1–3):363–373. doi:10.1016/j.foreco.2003.09.013
Buyantuyev A, Wu JG (2007) Effects of thematic resolution on landscape pattern analysis. Landsc Ecol 22(1):7–13. doi:10.1007/s10980-006-9010-5
Cain DH, Riitters K, Orvis K (1997) A multi-scale analysis of landscape statistics. Landsc Ecol 12(4):199–212. doi:10.1023/A:1007938619068
Cardille JA, White JC, Wulder MA, Holland T (2012) Representative landscapes in the forested area of Canada. Environ Manag 49(1):163–173. doi:10.1007/s00267-011-9785-2
Chirici G, McRoberts RE, Winter S, Bertini R, Brandli UB, Asensio IA, Bastrup-Birk A, Rondeux J, Barsoum N, Marchetti M (2012) National forest inventory contributions to forest biodiversity monitoring. For Sci 58(3):257–268. doi:10.5849/forsci.12-003
Clarke KC, Brass JA, Riggan PJ (1994) A cellular-automaton model of wildfire propagation and extinction. Photogramm Eng Remote Sens 60(11):1355–1367
Congalton RG (1997) Exploring and evaluating the consequences of vector-to-raster and raster-to-vector conversion. Photogramm Eng Remote Sens 63(4):425–434
Corona P, Chirici G, McRoberts RE, Winter S, Barbati A (2011) Contribution of large-scale forest inventories to biodiversity assessment and monitoring. For Ecol Manag 262(11):2061–2069. doi:10.1016/j.foreco.2011.08.044
Corry RC (2005) Characterizing fine-scale patterns of alternative agricultural landscapes with landscape pattern indices. Landsc Ecol 20(5):591–608. doi:10.1007/s10980-004-5036-8
Cushman SA, McGariyal K, Neel MC (2008) Parsimony in landscape metrics: strength, universality, and consistency. Ecol Indic 8(5):691–703. doi:10.1016/j.ecolind.2007.12.002
De Chant T, Gallego AH, Saornil JV, Kelly M (2010) Urban influence on changes in linear forest edge structure. Landsc Urban Plan 96(1):12–18. doi:10.1016/j.landurbplan.2010.01.006
Decout S, Manel S, Miaud C, Luque S (2012) Integrative approach for landscape-based graph connectivity analysis: a case study with the common frog (Rana temporaria) in human-dominated landscapes. Landsc Ecol 27(2):267–279. doi:10.1007/s10980-011-9694-z
Diaz-Varela ER, Marey-Perez MF, Rigueiro-Rodriguez A, Alvarez-Alvarez P (2009) Landscape metrics for characterization of forest landscapes in a sustainable management framework: potential application and prevention of misuse. Ann For Sci 66(3). doi:10.1051/Forest/2009004
Duveneck MJ, Scheller RM, White MA, Handler SD, Ravenscroft C (2014) Climate change effects on northern Great Lake (USA) forests: a case for preserving diversity. Ecosphere 5(2):23. doi:10.1890/Es13-00370.1
Echeverria C, Coomes D, Salas J, Rey-Benayas JM, Lara A, Newton A (2006) Rapid deforestation and fragmentation of Chilean Temperate Forests. Biol Conserv 130(4):481–494. doi:10.1016/j.biocon.2006.01.017
Echeverria C, Coomes DA, Hall M, Newton AC (2008) Spatially explicit models to analyze forest loss and fragmentation between 1976 and 2020 in southern Chile. Ecol Model 212(3–4):439–449. doi:10.1016/j.ecolmodel.2007.10.045

Elkie PC, Rempel RS, Carr AP (1999) Patch analyst user's manual—a tool for quantifying landscape structure. NWST Tech Manual TM-002

Erdody TL, Moskal LM (2010) Fusion of LiDAR and imagery for estimating forest canopy fuels. Remote Sens Environ 114(4):725–737. doi:10.1016/j.rse.2009.11.002

Estreguil C, Caudullo G, de Rigo D, San Miguel J (2012) Forest landscape in Europe: pattern. Fragmentation and Connectivity, European Commission

Fahrig L (2003) Effects of habitat fragmentation on biodiversity. Annu Rev Ecol Evol Syst 34:487–515. doi:10.1146/annurev.ecolsys.34.011802.132419

Fall A, Fall J (2001) A domain-specific language for models of landscape dynamics. Ecol Model 141(1–3):1–18. doi:10.1016/S0304-3800(01)00334-9

Farina A (2000) Landscapes ecology in action. Kluwer Academis Publishers, Dordrecht

Fassnacht KS, Cohen WB, Spies TA (2006) Key issues in making and using satellite-based maps in ecology: a primer. For Ecol Manag 222(1–3):167–181. doi:10.1016/j.foreco.2005.09.026

Fauth PT, Gustafson EJ, Rabenold KN (2000) Using landscape metrics to model source habitat for Neotropical migrants in the midwestern US. Landsc Ecol 15(7):621–631

Favier C, Dubois MA (2004) Reconstructing forest savanna dynamics in Africa using a cellular automata model, FORSAT. Lect Notes Comput Sci 3305:484–491

Fernandes MR, Aguiar FC, Ferreira MT (2011) Assessing riparian vegetation structure and the influence of land use using landscape metrics and geostatistical tools. Landsc Urban Plan 99(2):166–177. doi:10.1016/j.landurbplan.2010.11.001

Fernandes MR, Aguiar FC, Ferreira MT, Pereira JMC (2013) Spectral separability of riparian forests from small and medium-sized rivers across a latitudinal gradient using multispectral imagery. Int J Remote Sens 34(7):2375–2401. doi:10.1080/01431161.2012.744491

Foody GM (1992) A fuzzy-sets approach to the representation of vegetation continua from remotely sensed data—an example from lowland heath. Photogramm Eng Remote Sens 58(2):221–225

Forest Europe (2011) State of Europe's Forests 2011. Status and Trends in Sustainable Forest Management in Europe FOREST EUROPE. UNECE and FAO

Fortin MJ, Boots B, Csillag F, Remmel TK (2003) On the role of spatial stochastic models in understanding landscape indices in ecology. Oikos 102(1):203–212

Fraser JS, He HS, Shifley SR, Wang WJ, Thompson FR (2013) Simulating stand-level harvest prescriptions across landscapes: LANDIS PRO harvest module design. Can J For Res 43(10):972–978. doi:10.1139/cjfr-2013-0190

Gaines KF, Porter DE, Punshon T, Brisbin IL (2005) A spatially explicit model of the wild hog for ecological risk assessment activities at the department of energy's Savannah river site. Hum Ecol Risk Assess 11(3):567–589. doi:10.1080/10807030590949654

Gardner RH (1999) RULE: map generation and spatial analysis program. In: Klopatek JM, Gardner RH (eds) Landscape ecological analysis: issues and applications. Springer, New York, NY, pp 280–303

Gardner RH, Urban DL (2007) Neutral models for testing landscape hypotheses. Landsc Ecol 22(1):15–29. doi:10.1007/s10980-006-9011-4

Garmendia A, Arroyo-Rodriguez V, Estrada A, Naranjo E, Stoner K (2013) Landscape and patch attributes impacting medium- and large-sized terrestrial mammals in a fragmented rain forest. J Trop Ecol 29:331–344. doi:10.1017/S0266467413000370

Garofano-Gomez V, Martinez-Capel F, Bertoldi W, Gurnell A, Estornell J, Segura-Beltran F (2013) Six decades of changes in the riparian corridor of a Mediterranean river: a synthetic analysis based on historical data sources. Ecohydrology 6(4):536–553. doi:10.1002/eco.1330

Gasparri NI, Grau HR (2009) Deforestation and fragmentation of Chaco dry forest in NW Argentina (1972-2007). For Ecol Manag 258(6):913–921. doi:10.1016/j.foreco.2009.02.024

Gharehaghaji M, Shabani AA, Feghhi J, Danehkar A, Kaboli M, Ashrafi S (2012) Effects of landscape context on bird species abundance of tree fall gaps in a temperate deciduous forest of Northern Iran. For Ecol Manag 267:182–189. doi:10.1016/j.foreco.2011.12.001
Gillespie TW, Walter H (2001) Distribution of bird species richness at a regional scale in tropical dry forest of Central America. J Biogeogr 28(5):651–662
Gillespie TW, Foody GM, Rocchini D, Giorgi AP, Saatchi S (2008) Measuring and modelling biodiversity from space. Prog Phys Geogr 32(2):203–221. doi:10.1177/0309133308093606
Granell C, Havlik D, Schade S, Sabeur Z, Delaney C, Pielorz J, Uslander T, Mazzetti P, Schleidt K, Kobernus M, Havlik F, Bodsberg NR, Berre A, Mon JL (2016) Future Internet technologies for environmental applications. Environ Model Softw 78:1–15. doi:10.1016/j.envsoft.2015.12.015
Gurrutxaga M, Rubio L, Saura S (2011) Key connectors in protected forest area networks and the impact of highways: a transnational case study from the Cantabrian Range to the Western Alps (SW Europe). Landsc Urban Plan 101(4):310–320. doi:10.1016/j.landurbplan.2011.02.036
Gustafson EJ (1998) Quantifying landscape spatial pattern: what is the state of the art? Ecosystems 1(2):143–156. doi:10.1007/s100219900011
Gustafson EJ, Crow TR (1996) Simulating the effects of alternative forest management strategies on landscape structure. J Environ Manag 46(1):77–94. doi:10.1006/jema.1996.0008
Gustafson EJ, Sturtevant BR (2013) Modeling forest mortality caused by drought stress: implications for climate change. Ecosystems 16(1):60–74. doi:10.1007/s10021-012-9596-1
Hagen-Zanker A (2016) A computational framework for generalized moving windows and its application to landscape pattern analysis. Int J Appl Earth Obs 44:205–216. doi:10.1016/j.jag.2015.09.010
Haque U, Huda M, Hossain A, Ahmed SM, Moniruzzaman M, Haque R (2009) Spatial malaria epidemiology in Bangladeshi highlands. Malar J 8:185. doi:10.1186/1475-2875-8-185
Hardt E, dos Santos RF, de Pablo CL, de Agar PM, Pereira-Silva EFL (2013) Utility of landscape mosaics and boundaries in forest conservation decision making in the Atlantic Forest of Brazil. Landsc Ecol 28(3):385–399. doi:10.1007/s10980-013-9845-5
Harper KA, Macdonald SE, Burton PJ, Chen JQ, Brosofske KD, Saunders SC, Euskirchen ES, Roberts D, Jaiteh MS, Esseen PA (2005) Edge influence on forest structure and composition in fragmented landscapes. Conserv Biol 19(3):768–782. doi:10.1111/j.1523-1739.2005.00045.x
Hayes JJ, Robeson SM (2011) Relationships between fire severity and post-fire landscape pattern following a large mixed-severity fire in the Valle Vidal, New Mexico, USA. For Ecol Manag 261(8):1392–1400. doi:10.1016/j.foreco.2011.01.023
Hoechstetter S, Walz U, Dang LH, Thinh NX (2008) Effects of topography and surface roughness in analyses of landscape structure—a proposal to modify the existing set of landscape metrics. Landsc Online 3:1–14
Holland GJ, Bennett AF (2009) Differing responses to landscape change: implications for small mammal assemblages in forest fragments. Biodivers Conserv 18(11):2997–3016. doi:10.1007/s10531-009-9621-7
Hou W, Walz U (2013) Enhanced analysis of landscape structure: inclusion of transition zones and small-scale landscape elements. Ecol Indic 31:15–24. doi:10.1016/j.ecolind.2012.11.014
Jackson VL, Laack LL, Zimmerman EG (2005) Landscape metrics associated with habitat use by ocelots in south Texas. J Wildl Manag 69(2):733–738
Jenness J (2005) Topographic position index (tip_jen.avx) extension for ArcView 3.x., Jenness Enterprises. http://www.jennessent.com/arview/tpi.htm
Jomaa I, Auda Y, Hamze M, Abi saleh B, Safi S (2009) Analysis of eastern mediterranean oak forests over the period 1965–2003 using landscape indices on a patch basis. Landsc Res 34(1):105–124. doi:10.1080/01426390802591569

Jung M (2012) LecoS-A QGIS plugin to conduct landscape ecology statistics, http://plugins.qgis.org/plugins/LecoS
Kelly M, Tuxen KA, Stralberg D (2011) Mapping changes to vegetation pattern in a restoring wetland: finding pattern metrics that are consistent across spatial scale and time. Ecol Indic 11(2):263–273. doi:10.1016/j.ecolind.2010.05.003
Kintz DB, Young KR, Crews-Meyer KA (2006) Implications of land use/land cover change in the buffer zone of a National Park in the Tropical Andes. Environ Manag 38(2):238–252. doi:10.1007/s00267-005-0147-9
Kirk DA, Welsh DA, Baker JA, Thompson ID, Csizy M (2012) Avian assemblages differ between old-growth and mature white pine forests of Ontario, Canada: a role for supercanopy trees? Avian Conserv Ecol 7(1):4. doi:10.5751/Ace-00503-070104
Klug H, Kmoch A (2015) Operationalizing environmental indicators for real time multi-purpose decision making and action support. Ecol Model 295:66–74
Kouba Y, Alados CL (2012) Spatio-temporal dynamics of Quercus faginea forests in the Spanish Central Pre-Pyrenees. Eur J For Res 131(2):369–379. doi:10.1007/s10342-011-0509-1
Kulldorff M (1997) A spatial scan statistic. Commun Stat Theor Methods 26(6):1481–1496. doi:10.1080/03610929708831995
Kupfer JA (2012) Landscape ecology and biogeography: rethinking landscape metrics in a post-FRAGSTATS landscape. Prog Phys Geogr 36(3):400–420. doi:10.1177/0309133312439594
Lafortezza R, Corry RC, Sanesi G, Brown RD (2005) Quantitative approaches to landscape spatial planning: clues from landscape ecology. Sust Developm Plan 84:239–250
Lang S, Tiede D (2003) vLATE Extension für ArcGIS—vektorbasiertes Tool zur quantitativen Landschaftsstrukturanalyse. Presented at German ESRI Anwenderkonferenz, Innsbruck. https://sites.google.com/site/largvlate/gis-tools/v-late
Langford WT, Gergel SE, Dietterich TG, Cohen W (2006) Map misclassification can cause large errors in landscape pattern indices: examples from habitat fragmentation. Ecosystems 9(3):474–488. doi:10.1007/s10021-005-0119-1
Larsson TB (2001) Biodiversity evaluation tools for European forests. Eur For Inst Proc 38:75–81
Laurent E (2006) Methods for generating patch and landscape metrics. Paper presented at the Conservation Design Workshop, St. Louis, MO, April 11, 2006
Lausch A, Herzog F (2002) Applicability of landscape metrics for the monitoring of landscape change: issues of scale, resolution and interpretability. Ecol Indic 2(1–2):3–15
Lechner AM, Langford WT, Bekessy SA, Jones SD (2012) Are landscape ecologists addressing uncertainty in their remote sensing data? Landsc Ecol 27(9):1249–1261. doi:10.1007/s10980-012-9791-7
Lechner AM, Reinke KJ, Wang Y, Bastin L (2013) Interactions between landcover pattern and geospatial processing methods: effects on landscape metrics and classification accuracy. Ecol Complex 15:71–82. doi:10.1016/j.ecocom.2013.03.003
Leitao AB, Ahern J (2002) Applying landscape ecological concepts and metrics in sustainable landscape planning. Landsc Urban Plan 59(2):65–93
Lethbridge MR, Westphal MI, Possingham HP, Harper ML, Souter NJ, Anderson N (2010) Optimal restoration of altered habitats. Environ Model Softw 25(6):737–746. doi:10.1016/j.envsoft.2009.11.013
Li HB, Wu JG (2004) Use and misuse of landscape indices. Landsc Ecol 19(4):389–399
Li HB, Gartner DI, Mou P, Trettin CC (2000) A landscape model (LEEMATH) to evaluate effects of management impacts on timber and wildlife habitat. Comput Electron Agric 27(1–3):263–292. doi:10.1016/S0168-1699(00)00088-0
Li MS, Mao LJ, Zhou CG, Vogelmann JE, Zhu ZL (2010) Comparing forest fragmentation and its drivers in China and the USA with Globcover v2.2. J Environ Manag 91(12):2572–2580. doi:10.1016/j.jenvman.2010.07.010
Li X, Lin JY, Chen YM, Liu XP, Ai B (2013) Calibrating cellular automata based on landscape metrics by using genetic algorithms. Int J Geogr Inf Sci 27(3):594–613. doi:10.1080/13658816.2012.698391

Li A, Dragicevicb S, Castroc FA, Sesterd M, Wintere S, Coltekinf A, Pettitg C, Jiangh B, Haworthi J, Steinj A, Chengi T (2016) Geospatial big data handling theory and methods: a review and research challenges. ISPRS J Photogramm Remote Sens 115:119–133

Lu D, Weng Q (2007) A survey of image classification methods and techniques for improving classification performance. Int J Remote Sens 28(5):823–870. doi:10.1080/01431160600746456

Macdonald DW, Rushton S (2003) Modelling space use and dispersal of mammals in real landscapes: a tool for conservation. J Biogeogr 30(4):607–620

MacLean MG, Congalton RG (2013) PolyFrag: a vector-based program for computing landscape metrics. GISci Remot Sens 50(6):591–603. doi:10.1080/15481603.2013.856537

Maier JAK, Hoef JMV, McGuire AD, Bowyer RT, Saperstein L, Maier HA (2005) Distribution and density of moose in relation to landscape characteristics: effects of scale. Can J For Res 35(9):2233–2243. doi:10.1139/X05-123

Majasalmi T, Rautiainen M (2016) The potential of Sentinel-2 data for estimating biophysical variables in a boreal forest: a simulation study. Remot Sens Lett 7(5):427–436. doi:10.1080/2150704X.2016.1149251

Mandelbrot BB (1983) The fractal geometry of nature. W. H. Freeman, New York

Marceau DJ, Howarth PJ, Gratton DJ (1994) Remote-sensing and the measurement of geographical entities in a forested environment. 1. The scale and spatial aggregation problem. Remote Sens Environ 49(2):93–104. doi:10.1016/0034-4257(94)90046-9

Martinez S, Ramil P, Chuvieco E (2010) Monitoring loss of biodiversity in cultural landscapes. New methodology based on satellite data. Landsc Urban Plan 94(2):127–140. doi:10.1016/j.landurbplan.2009.08.006

Masse A, Cote SD (2012) Linking habitat heterogeneity to space use by large herbivores at multiple scales: from habitat mosaics to forest canopy openings. For Ecol Manag 285:67–76. doi:10.1016/j.foreco.2012.07.039

McAlpine CA, Eyre TJ (2002) Testing landscape metrics as indicators of habitat loss and fragmentation in continuous eucalypt forests (Queensland, Australia). Landsc Ecol 17(8):711–728

McConnell WJ, Sweeney SP, Mulley B (2004) Physical and social access to land: spatio-temporal patterns of agricultural expansion in Madagascar. Agric Ecosyst Environ 101(2–3):171–184. doi:10.1016/j.agee.2003.09.012

McGarigal K, Tagil S, Cushman SA (2009) Surface metrics: an alternative to patch metrics for the quantification of landscape structure. Landsc Ecol 24(3):433–450. doi:10.1007/s10980-009-9327-y

McGarigal K, Cushman SA, Ene E (2012) "FRAGSTATS v4: Spatial Pattern Analysis Program for Categorical and Continuous Maps." Computer software program produced by the authors at the University of Massachusetts, Amherst. http://www.umass.edu/landeco/research/fragstats/fragstats.html

Miyamoto A, Sano M (2008) The influence of forest management on landscape structure in the cool-temperate forest region of central Japan. Landsc Urban Plan 86(3–4):248–256. doi:10.1016/j.landurbplan.2008.03.002

Mladenoff DJ, Host GE, Boeder J, Crow TR (1996) LANDIS: a spatial model of forest landscape disturbance, succession, and management. In: Goodchild MF, Steyaert LT, Parks BO et al (eds) GIS and environmental modeling: progress and research issues. Wiley, New York, pp 175–180

Mokhtari Z, Sofyanian A, Kajeddin SJ, Ziaee H (2014) Gradient analysis of urban landscape pattern (Case Study: Isfahan, Iran). J Biodivers Environm Sci 4(6):322–333

Moore ID, Gessler PE, Nielsen GA, Petersen GA (1993) Terrain attributes: estimation methods and scale effects. In: Jakeman AJBM, McAleer M (eds) Modeling change in environmental systems. Wiley, London, pp 189–214

Mueller T, Lenz J, Caprano T, Fiedler W, Bohning-Gaese K (2014) Large frugivorous birds facilitate functional connectivity of fragmented landscapes. J Appl Ecol 51(3):684–692. doi:10.1111/1365-2664.12247

Naish S, Hu W, Mengersen K, Tong S (2011) Spatial and temporal clusters of Barmah Forest virus disease in Queensland, Australia. Tropical Med Int Health 16(7):884–893. doi:10.1111/j.1365-3156.2011.02775.x

Narusk K (2014) Sensitivity of landscape pattern metrics to processing methods of satellite images (Maastikumustri indikaatorite tundlikkus satelliidipiltide rasterandmestiku töötlustehnikate suhtes). University of Tartu, Tartu

Näsi R, Honkavaara E, Lyytikäinen-Saarenmaa P, Blomqvist M, Litkey P, Hakala T, Viljanen N, Kantola T, Tanhuanpää T, Holopainen M (2015) Using UAV-based photogrammetry and hyperspectral imaging for mapping bark beetle damage at tree-level. Remote Sens 7(11):15467–15493

Neel MC, McGarigal K, Cushman SA (2004) Behavior of class-level landscape metrics across gradients of class aggregation and area. Landsc Ecol 19(4):435–455

Nittel S (2009) A survey of geosensor networks: advances in dynamic environmental monitoring. Sensors 9(7):5664–5678

Openshaw S, Taylor PJ (1981) The modifiable areal unit problem. In: Wrigley N, Bennett RJ (eds) Quantitative geography: a british view. Routledge and Kegan, London, pp 60–70

Palmer JF (2004) Using spatial metrics to predict scenic perception in a changing landscape: Dennis, Massachusetts. Landsc Urban Plan 69(2–3):201–218. doi:10.1016/j.landurbplan.2003.08.010

Peng J, Wang Y, Ye M, Wu J, Zhang Y (2007) Effects of land-use categorization on landscape metrics: a case study in urban landscape of Shenzhen, China. Int J Remote Sens 28(21):4877–4895. doi:10.1080/01431160601075590

Piwowar JM (1987) Conversion between vector and raster format data for geographic information systems applications. Waterloo, ON, Canada

Plante M, Lowell K, Potvin F, Boots B, Fortin MJ (2004) Studying deer habitat on Anticosti Island, Quebec: relating animal occurrences and forest map information. Ecol Model 174(4):387–399. doi:10.1016/j.ecolmodel.2003.09.035

Plexida SG, Sfougaris AI, Ispikoudis IP, Papanastasis VP (2014) Selecting landscape metrics as indicators of spatial heterogeneity-A comparison among Greek landscapes. Int J Appl Earth Obs 26:26–35. doi:10.1016/j.jag.2013.05.001

Quine CP, Watts K (2009) Successful de-fragmentation of woodland by planting in an agricultural landscape? An assessment based on landscape indicators. J Environ Manag 90(1):251–259. doi:10.1016/j.jenvman.2007.09.002

Ramezani H, Holm S, Allard A, Stahl G (2010) Monitoring landscape metrics by point sampling: accuracy in estimating Shannon's diversity and edge density. Environ Monit Assess 164(1–4):403–421. doi:10.1007/s10661-009-0902-0

Rego F, Godinho-Ferreira P, Uva JS, Cunha J (2004) Combination of structural and compositional factors for describing forest types using NFI data, vol 51. Monitoring and Indicators for Forest Biodiversity in Europe—From Ideas to Operationality. EFI Proceedings

Remmel TK, Csillag F (2003) When are two landscape pattern indices significantly different? J Geogr Syst 5:331–351

Remmel TK, Fortin MJ (2013) Categorical, class-focused map patterns: characterization and comparison. Landsc Ecol 28(8):1587–1599. doi:10.1007/s10980-013-9905-x

Reunanen P, Fall A, Nikula A (2010) Biodiversity and ecological forest-cover domains in boreal landscapes. Biodivers Conserv 19(3):665–678. doi:10.1007/s10531-009-9726-z

Richardson JJ, Moskal LM (2011) Strengths and limitations of assessing forest density and spatial configuration with aerial LiDAR. Remote Sens Environ 115(10):2640–2651. doi:10.1016/j.rse.2011.05.020

Riitters KH, Oneill RV, Hunsaker CT, Wickham JD, Yankee DH, Timmins SP, Jones KB, Jackson BL (1995) A factor-analysis of landscape pattern and structure metrics. Landsc Ecol 10(1):23–39

Rikimaru A, Miyatake S, Dugan P (1999) Sky is the limit for forest management tool. ITTO Trop For Update 9(3):6–9

Riley SJ, DeGloria SD, Elliot R (1999) A terrain ruggedness index that quantifies topographic heterogeneity. Int J Therm Sci 5(1–4):23–27

Robertson C, Nelson TA, Boots B, Wulder MA (2007) STAMP: spatial-temporal analysis of moving polygons. J Geogr Syst 9(3):207–227. doi:10.1007/s10109-007-0044-2

Sano M, Miyamoto A, Furuya N, Kogi K (2009) Using landscape metrics and topographic analysis to examine forest management in a mixed forest, Hokkaido, Japan: guidelines for management interventions and evaluation of cover changes. For Ecol Manag 257(4):1208–1218. doi:10.1016/j.foreco.2008.10.005

Saura S, Martinez-Millan J (2001) Sensitivity of landscape pattern metrics to map spatial extent. Photogramm Eng Remote Sens 67(9):1027–1036

Saura S, Torne J (2009) Conefor Sensinode 2.2: a software package for quantifying the importance of habitat patches for landscape connectivity. Environ Model Softw 24(1):135–139. doi:10.1016/j.envsoft.2008.05.005

Saveraid EH, Debinski DM, Kindscher K, Jakubauskas ME (2001) A comparison of satellite data and landscape variables in predicting bird species occurrences in the Greater Yellowstone Ecosystem, USA. Landsc Ecol 16(1):71–83

Schetter TA, Walters TL, Root KV (2013) A multi-scale spatial analysis of native and exotic plant species richness within a mixed-disturbance oak Savanna landscape. Environ Manag 52(3):581–594. doi:10.1007/s00267-013-0120-y

Schindler S, Poirazidis K, Wrbka T (2008) Towards a core set of landscape metrics for biodiversity assessments: a case study from Dadia National Park, Greece. Ecol Indic 8(5):502–514. doi:10.1016/j.ecolind.2007.06.001

Schindler S, von Wehrden H, Poirazidis K, Wrbka T, Kati V (2013) Multiscale performance of landscape metrics as indicators of species richness of plants, insects and vertebrates. Ecol Indic 31:41–48. doi:10.1016/j.ecolind.2012.04.012

Shao GF, Wu WC (2004) The effects of classification accuracy on landscape indices. Remot Sens Gis Accur Assessm:209–220

Shao GF, Wu JG (2008) On the accuracy of landscape pattern analysis using remote sensing data. Landsc Ecol 23(5):505–511. doi:10.1007/s10980-008-9215-x

Sieving KE, Willson MF, DeSanto TL (1996) Habitat barriers to movement of understory birds in fragmented south-temperate rainforest. Auk 113(4):944–949

Simova P, Gdulova K (2012) Landscape indices behavior: a review of scale effects. Appl Geogr 34:385–394. doi:10.1016/j.apgeog.2012.01.003

Sitzia T, Semenzato P, Trentanovi G (2010) Natural reforestation is changing spatial patterns of rural mountain and hill landscapes: a global overview. For Ecol Manag 259(8):1354–1362. doi:10.1016/j.foreco.2010.01.048

Skole D, Tucker C (1993) Tropical Deforestation and Habitat Fragmentation in the Amazon—Satellite Data from 1978 to 1988 (Vol 260, Pg 1909, 1993). Science 261(5125): 1104–1104

Soares BS, Cerqueira GC, Pennachin CL (2002) DINAMICA—a stochastic cellular automata model designed to simulate the landscape dynamics in an Amazonian colonization frontier. Ecol Model 154(3):217–235. doi:10.1016/S0304-3800(02)00059-5

Soares B, Silvestrini R, Nepstad D, Brando P, Rodrigues H, Alencar A, Coe M, Locks C, Lima L, Hissa L, Stickler C (2012) Forest fragmentation, climate change and understory fire regimes on the Amazonian landscapes of the Xingu headwaters. Landsc Ecol 27(4):585–598. doi:10.1007/s10980-012-9723-6

Solberg S, Brunner A, Hanssen KH, Lange H, Naesset E, Rautiainen M, Stenberg P (2009) Mapping LAI in a Norway spruce forest using airborne laser scanning. Remote Sens Environ 113(11):2317–2327. doi:10.1016/j.rse.2009.06.010

Southworth J, Nagendra H, Carlson LA, Tucker C (2004) Assessing the impact of Celaque National Park on forest fragmentation in western Honduras. Appl Geogr 24(4):303–322. doi:10.1016/j.apgeog.2004.07.003

Soverel NO, Coops NC, White JC, Wulder MA (2010) Characterizing the forest fragmentation of Canada's national parks. Environ Monit Assess 164(1–4):481–499. doi:10.1007/s10661-009-0908-7

Spearman L (2003) The challenges of defining patches & patch dynamics in insect conservation biology, 2015. http://crssa.rutgers.edu/courses/lse/Web_Patch/final/Lauren/spearman_insectpatch_v2.html

Staus NL, Strittholt JR, DellaSala DA, Robinson R (2002) Rate and pattern of forest disturbance in the Klamath-Siskiyou ecoregion, USA between 1972 and 1992. Landsc Ecol 17(5):455–470. doi:10.1023/A:1021274701133

Steiniger S, Hay GJ (2009) Free and open source geographic information tools for landscape ecology. Ecol Inform 4(4):183–195. doi:10.1016/j.ecoinf.2009.07.004

Thompson RG, Warkentin IG, Flemming SP (2008) Response to logging by a limited but variable nest predator guild in the boreal forest. Can J For Res 38(7):1974–1982. doi:10.1139/X08-049

Tittler R, Messier C, Fall A (2012) Concentrating anthropogenic disturbance to balance ecological and economic values: applications to forest management. Ecol Appl 22(4):1268–1277

Trainor AM, Walters JR, Urban DL, Moody A (2013) Evaluating the effectiveness of a Safe Harbor Program for connecting wildlife populations. Anim Conserv 16(6):610–620. doi:10.1111/acv.12035

Uezu A, Metzger JP, Vielliard JME (2005) Effects of structural and functional connectivity and patch size on the abundance of seven Atlantic Forest bird species. Biol Conserv 123(4):507–519. doi:10.1016/j.biocon.2005.01.001

Uuemaa E, Roosaare J, Mander U (2005) Scale dependence of landscape metrics and their indicatory value for nutrient and organic matter losses from catchments. Ecol Indic 5(4):350–369. doi:10.1016/j.ecolind.2005.03.009

Uuemaa E, Mander U, Marja R (2013) Trends in the use of landscape spatial metrics as landscape indicators: a review. Ecol Indic 28:100–106. doi:10.1016/j.ecolind.2012.07.018

VanDerWal J, Falconi L, Januchowski S, Shoo L, Storlie C (2011) SDMTools: species distribution modelling tools: tools for processing data associated with species distribution modelling exercises. 1.1-5. http://cran.r-project.org/

Vivoni ER, Teles V, Ivanov VY, Bras RL, Entekhabi D (2005) Embedding landscape processes into triangulated terrain models. Int J Geogr Inf Sci 19(4):429–457. doi:10.1080/13658810512331325111

Vogt P, Riitters KH, Iwanowski M, Estreguil C, Kozak J, Soille P (2007) Mapping landscape corridors. Ecol Indic 7(2):481–488. doi:10.1016/j.ecolind.2006.11.001

Wade TG, Wickham JD, Nash MS, Neale AC, Riitters KH, Jones KB (2003) A comparison of vector and raster GIS methods for calculating landscape metrics used in environmental assessments. Photogramm Eng Remote Sens 69(12):1399–1405

Wang XL, Cumming SG (2011) Measuring landscape configuration with normalized metrics. Landsc Ecol 26(5):723–736. doi:10.1007/s10980-011-9601-7

Wang WJ, He HS, Fraser JS, Thompson FR, Shifley SR, Spetich MA (2014) LANDIS PRO: a landscape model that predicts forest composition and structure changes at regional scales. Ecography 37(3):225–229. doi:10.1111/j.1600-0587.2013.00495.x

Watts ME, Ball IR, Stewart RS, Klein CJ, Wilson K, Steinback C, Lourival R, Kircher L, Possingham HP (2009) Marxan with zones: software for optimal conservation based land- and sea-use zoning. Environ Model Softw 24(12):1513–1521. doi:10.1016/j.envsoft.2009.06.005

Westphal MI, Field SA, Possingham HP (2007) Optimizing landscape configuration: a case study of woodland birds in the Mount Lofty Ranges, South Australia. Landsc Urban Plan 81(1–2):56–66. doi:10.1016/j.landurbplan.2006.10.015

Wu JG (2004) Effects of changing scale on landscape pattern analysis: scaling relations. Landsc Ecol 19(2):125–138

Wu JG, Shen WJ, Sun WZ, Tueller PT (2002) Empirical patterns of the effects of changing scale on landscape metrics. Landsc Ecol 17(8):761–782

Wulder MA, White JC, Cranny M, Hall RJ, Luther JE, Beaudoin A, Goodenough DG, Dechka JA (2008) Monitoring Canada's forests. Part 1: completion of the EOSD land cover project. Can J Remote Sens 34(6):549–562

Yu DY, Shi PJ, Han GY, Zhu WQ, SQ D, Xun B (2011) Forest ecosystem restoration due to a national conservation plan in China. Ecol Eng 37(9):1387–1397. doi:10.1016/j.ecoleng.2011.03.011

Zapponi L, Del Bianco M, Luiselli L, Catorci A, Bologna MA (2013) Assessing environmental requirements effects on forest fragmentation sensitivity in two arboreal rodents. Mamm Biol 78(3):157–163. doi:10.1016/j.mambio.2012.08.005

Zaragozi B, Belda A, Linares J, Martinez-Perez JE, Navarro JT, Esparza J (2012) A free and open source programming library for landscape metrics calculations. Environ Model Softw 31:131–140. doi:10.1016/j.envsoft.2011.10.009

Towards Automated Forest Mapping

Lars T. Waser, Ruedi Boesch, Zuyuan Wang, and Christian Ginzler

Abstract The need for up-to-date and accurate information on forest resources has rapidly increased in recent years. Forest mapping is an important source of information for the assessment of woodland resources and a key issue for any National Forest Inventory (NFI). Nowadays, new perspectives for automated forest mapping are emerging through the latest developments in remote sensing data and techniques. In this chapter, an overview of current remote sensing data and techniques for mapping woodland and forests, the challenges and requirements for optimization and automation, and the need for validating final products are presented. Special attention is paid to land use—a crucial criterion for forest mapping which, in contrast to land cover, cannot be easily derived from remotely sensed data. Three different approaches for extracting woodland areas (i.e., patches of trees and shrubs) are presented, all of which involve a high degree of automation. Two additional approaches, which are based on NFI forest definitions, are presented. These require the subdivision of woodlands into the classes "used for forestry" and "other use" and implement the criteria "height," "minimum crown coverage," "minimum area," "minimum width," and "land use". Special attention is paid to connecting patches using distance criteria from national forest definitions. The main points of this chapter are as follows: (1) *forest* needs an exact definition which may differ depending on the country, (2) mapping woodland can be highly automated and is indispensable prior to mapping forests, and (3) forest mapping is now feasible using remote sensing data and techniques; however, it is less automated due to the implementation of a forest definition.

Abbreviations

ALS	Airborne laser scanning
CHM	Canopy height model
CIR	Color-infrared

L.T. Waser (✉) • R. Boesch • Z. Wang • C. Ginzler
Swiss Federal Institute for Forest, Snow and Landscape Research,
Zuercherstrasse 111, Birmensdorf CH-8903, Switzerland
e-mail: waser@wsl.ch

T.K. Remmel, A.H. Perera (eds.), *Mapping Forest Landscape Patterns*,
DOI 10.1007/978-1-4939-7331-6_7

DSM	Digital surface model
DTM	Digital terrain model
GLM	Generalized linear model
GSD	Ground sample distance
IGBP	International geosphere biosphere program
IP	Interpretation plot
ITC	Individual tree crowns
k-NN	K-nearest neighbor
LiDAR	Light detection and ranging
NDVI	Normalized difference vegetation index
NFI	National forest inventory
REDD	Reducing emissions from deforestation and degradation
SAR	Synthetic aperture radar
TOF	Trees outside forest
VHM	Vegetation height model
VHR	Very high resolution

Introduction

This chapter is organized into four main sections: (1) the requirements for automated forest mapping regarding forest definitions, remote sensing datasets and techniques, and existing maps; (2) commonly used data sources and processing techniques; (3) three highly automated approaches for woodland mapping; and (4) two highly automated approaches for forest mapping. The content of each section is briefly summarized.

Forest mapping is a critical task because the resulting datasets are fundamental input for a broad range of users and applications, ranging from global environmental change assessment to local forest management planning (see, for example, the "Modelling Wildfire Regimes" and "Mapping Insect Defoliation" Chapters in this volume). Precise, up-to-date, and regularly gathered information on the area covered by trees and shrubs is an important basis for assessing woodland resources and understanding the functionality of forests.

The need for consistency and the reduction of manual workload are key reasons for automation in forest mapping. Thus, a high degree of automation in the process of assessing woodland is essential for governmental authorities, international reporting (including the Kyoto protocol), activities within the Reducing Emissions from Deforestation and Degradation (REDD) framework, forest disturbance assessments, and biodiversity and restoration programs. Based on mapped woodland, other forest-related parameters (e.g., forest structure, biomass, and carbon storage) are estimated, which are, in turn, needed for resource management by public and private authorities.

One particular challenge to forest mapping is the fact that the term *forest* is often defined similarly to *woodland* by the remote sensing community. In most studies, a simplified and generalized definition for *forest* is applied, which is mainly based on minimal tree height, minimal tree area, and width and crown coverage, and less on land use according to National Forest Inventories (NFI)s. Thus, besides the availability, processing, and quality of remotely sensed data, in order to map *forest*, an appropriate definition of forest must be applied. The need to clarify the difference between *woodland* and the varying definitions of the term *forest* is highlighted below.

Definitions

The goal of this section is to emphasize the difference between woodland, trees, and forests since a variety of forest-related terms and a wide range of forest definitions exist. Prior to defining the term *forest*, however, a short explanation of the difference between *land cover* and *land use* should be given, since the criteria used to define forests are usually based on their definitions or sometimes a combination of them. Problems arise regarding land use, which, while a key parameter in the NFI definition of forest, is not easily assessable when using remotely sensed data—in contrast to land cover which is assessable.

Usually, *land cover* is distinct from *land use*, despite the two terms often being used interchangeably. In short, land cover indicates the physical land type, whereas land use indicates how people are using the land. For example, a temporarily unstocked area (e.g., after impacts such as fires, storms, or harvesting) will be identified as a non-forest land cover type when using remote sensing data and techniques, but will in fact maintain its status as a forest land use type within the NFI. According to the Food and Agriculture Organization (FAO) (FAO 1997) *land cover* refers to the physical forms of *land cover* observable from airborne and spaceborne remote sensing data, and to their structure. It includes vegetation (e.g., grassland and trees) which may be natural or planted, and non-vegetated areas (e.g., bare ground, asphalt, and water). One of the major land cover issues is that similarly named categories are often defined in different ways (Alford 1993). For example, areas without trees may be classified as forest if the intention is to replant (such as in the UK and Ireland), and areas with many trees may not be labeled as forests if the trees are not growing fast enough (such as in Norway and Finland).

Land use refers to the function of land and how it is used by people, that is, the activities undertaken on it to produce goods and services. *Land use* is characterized by the arrangements, activities, and inputs people undertake within a certain *land cover* type to produce, change, or maintain the land. Many *land use* classification systems and programs have been developed worldwide. The most commonly used terms in use include *urban*, *agricultural*, *forest*, *water*, and *wetlands* (FAO/UNEP 1999). A good overview and more details may be found in Fisher et al. (2005).

Forest

Even though a precise and unambiguous definition of *forest* is indispensable when mapping and comparing forest areas, this is frequently neglected by using a simplified term *woodland* (Fig. 1). This term can be used to include both trees and shrubs and is one of the most important parameters in forestry. In most NFI, the estimation of forest cover is a crucial parameter and should therefore be easily understood and reproducible. NFI programs are required to produce timely and accurate estimates for a wide range of forest resource variables for a variety of users and applications. They aim to periodically report the current state of forests and changes in forests by providing quantitative statistical information on area, species composition, volume, and growing stock. Clearly, the definition of *forest* is also crucial to assess areas of deforestation which refer to areas where a forest has formerly been. Using exactly the same unambiguous definitions of forest over time is essential—otherwise, the stated deforestation cannot be properly understood.

The term *forest* always implies a definition (usually the percentage of area covered by trees, with a minimum area and tree height)—although the commonly used terms *stocked area*, *tree area*, or *area covered by trees* are often handled in a similar manner. To classify an area as *forest* or as *non-forest*, different forest definitions are available which should be defined based on exact geometric terms. Unfortunately, current definitions of *forest* are imprecise in most cases. Differing from country to country, a standardized and generally valid definition is nonexistent, since the term *forest* can be defined according to different criteria. Moreover, the definition of

Fig. 1 Woodland in the temporal zone of Switzerland consisting of individual trees and shrubs of different ages may belong to a forest or not—depending on the definition of forest used

forest varies depending on the context. For example, there may be differences from a silvicultural, ecological, or legal point of view. According to Lund (2016), over 960 forest definitions exist worldwide, while the minimum thresholds for degree of tree cover in these definitions vary between 5% and 80%.

The FAO of the United Nations has defined *forest* as lands which are larger than 0.5 hectares in area, with a tree cover of more than 10% (FAO 2000, 2001, 2010). In most national and global forest definitions, the key parameters defining *forest* are height, width or area, and crown coverage. According to Tomppo et al. (2010), tree cover thresholds for forest definition in NFIs range from 10% to 50% among different national definitions. For example, this threshold is 10% for France and Scandinavian countries; 20% for Great Britain, Spain, and Switzerland; 25% for the United States, rising to 30% for Austria and New Zealand; and 50% in Germany and Hungary. Even on a national level, *forest* is often defined differently by various authoritative bodies within individual countries. Despite the fact that these measurable items describe land cover, land use cannot be easily calculated out of measurable data. Even in the field, the main use of trees is difficult to determine. Deciding whether extensively used fruit trees, olive trees, or Christmas tree plantations close to a forest's edges belong to forestry land or not is highly subjective.

Remote Sensing for Automated Mapping of Woodland and Forest

Remote sensing used in the forestry sector covers a wide variety of techniques and applications for extracting woodland, and while some have been operational for decades, others have only appeared recently and are undergoing fast development (Koch et al. 2008). Providing consistent, reproducible, and up-to-date information on various forest parameters proves to be the main advantage of using remote sensing as a tool for monitoring, for example, in the framework of NFIs, and for mapping purposes, for instance, in national map products.

Traditionally, woodland mapping approaches have been the product of visual image interpretations and delineation of aerial imagery in combination with field visits. Thus, their development is time consuming, and restricted to relatively small areas. In addition, shadow effects limit the exact detection of forest borders or small gaps.

In the two last decades, NFI data has been combined with remote sensing data and techniques (Tomppo et al. 2008; McRoberts et al. 2014) to map *forest* precisely. This has been achieved mostly by extrapolating estimates from field plot samples using the k-nearest neighbor (k-NN) algorithm (Tomppo and Halme 2004). A good overview of remote sensing support for NFIs can be found in McRoberts and Tomppo (2007) and Barrett et al. (2016). While visual image interpretation is quite time consuming and more subjective—which, depending on the experience of the interpreter, may be appropriate for small-area applications—mapping *woodland* based on entire satellite images is feasible with high accuracy within short timeframes thanks to highly automated approaches. Existing remote sensing-based forest maps vary in scale (global, continental, pan-European, and national levels), and also in the level of detail (sources of information, forest definition, and target interest groups). While forest stand maps provide very detailed information, forest layers of map products are much more generalized, with scales ranging from a few meters up to 1 km. How accurate and up-to-date the maps are may vary in both products due to coarse image resolution and a certain time gap between image acquisition and production. Consequently, forest borders, gaps, and areas with close dense forest may not be sufficiently represented to a large degree.

Over the last decade, many at least partly automated forest mapping approaches have been implemented based on remote sensing techniques and data in the framework of research case studies, national forest inventories, and mapping or resource assessment programs. Highly automated mapping approaches that are mostly restricted to relatively small areas have been used in the framework of change detection studies in general (Waser et al. 2008a; Wang et al. 2015), FAO forest definition applications (Magdon et al. 2014), and the assessment of deforestation such as REDD (Gebhardt et al. 2014). Recently, Waser et al. (2015) presented a wall-to-wall forest mapping approach for all of Switzerland incorporating the NFI forest definition.

At the global scale, particularly worth mentioning are the Global Forest Resources Assessment Remote Sensing Survey (FAO 2012) initiated by the Food and Agriculture Organization (FAO), the Global Forest Watch (GFW) (GFW 2016) initiative of the World Resources Institute, the Global Land Cover mapping approach (GLC) (Bartholomé and Belward 2005), the global forest/non-forest mapping initiated by the Japan Aerospace Exploration Agency (JAXA 2016), and the European 2006 Forest cover map (JRC 2016) from the Joint Research Center (JRC). Recently, Hansen et al. (2013) generated a spatially and temporally detailed global forest and forest change map, which provides valuable information for many land use-related applications at the regional level. More recently, Schepaschenko et al. (2015) presented a global hybrid forest map approach based on remote sensing data, maps, and FAO statistics, and new global forest/non-forest maps based on ALOS PALSAR data (2007–2010) were developed by Shimada et al. (2014).

Data and Preprocessing

In this section, we review the current state of research and technology required for automated mapping of woodland and forest.

Reference Data

Satellite imagery, large mosaics of aerial imagery, or a combination of both have become well-accepted sources of landscape information that contribute to the construction of land cover maps. While forest maps are constructed using such images, forest inventory data or derivatives often come from other existing maps. However, these maps are not always sufficiently validated for various reasons. Often, no adequate reference data is available to evaluate maps. The need for reference data and accurate assessment tools for maps that distinguish between the three land cover classes of non-forest, coniferous forest, and deciduous forest is raised in McRoberts (2012). The minimum requirements of reference datasets commonly used for evaluating woodland and forest maps are given in this section. An adequate reference dataset can be characterized by the following traits:

- Minimal time span between acquisition of reference data and data the mapping is based on
- Minimal differences between the reference data and mapping scales
- Reproducibility (by different users)
- Reliable data source (e.g., from inventory data or field measurements—and not digitized from existing map products)
- Representative of the area under investigation (size, dispersion)

The collection of reference data remains time consuming and is the least automated step in the implementation of any mapping approach. In the following sections, examples of practicable and reliable (best practice) reference datasets with pros and cons are presented:

Digitized Polygons

In these datasets, acquisition is carried out at the individual tree level and suitable for fine-scale applications and small areas (a few to a few dozen square kilometers), for instance, at stand, local, or community levels. In order to be representative of trees within an area of interest, different types of randomly sampled digitized polygons such as tree crowns or crown clusters that belong to entire tree groups anywhere in the woodland—preferably at borders, and in open land as single trees—should be considered. The delineation can be carried out either by digitizing polygons of tree crowns on orthorectified aerial images or by stereo-interpretation of aerial images. While both methods of collecting reference data can be regarded as very effective—depending on the pre-knowledge and expertise of the interpreter and time availability—the latter depends on the extent of the area to be investigated. Figure 2 shows examples of these different types of tree and non-tree polygons, which were digitized on true-color orthoimages in a study area in Switzerland. While a certain degree of experience is needed on the part of the interpreter, little technical or forest-related knowledge is required. The collected reference data is reliable and reproducible. Depending on the area under investigation, the manual workload can increase rapidly.

Regular Point Raster

Tree/non-tree decisions at each grid point in a regular point raster are another very effective method for validating mapped woodland or forest. The regular point raster can either cover the entire area of interest or consist of a limited number of regular

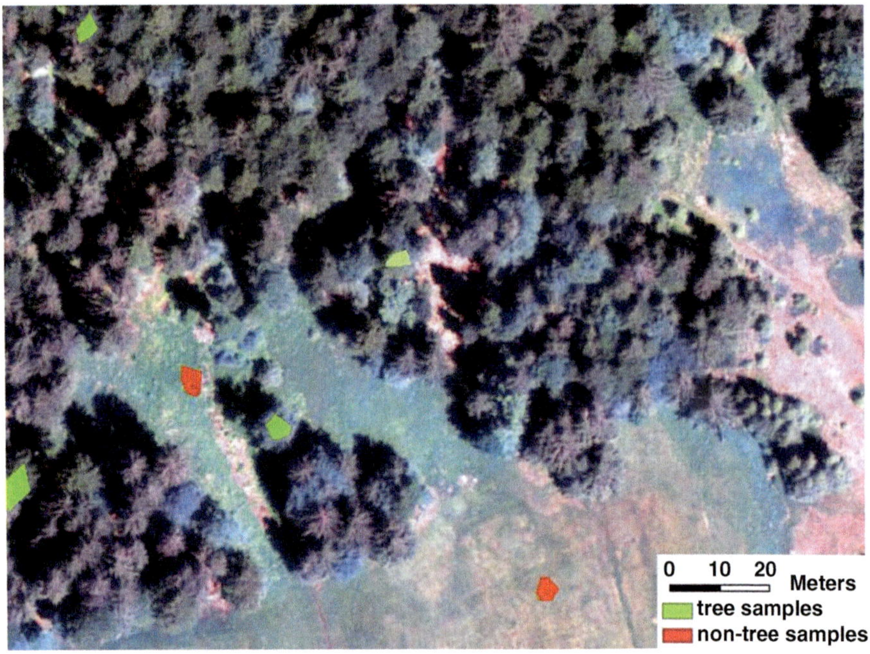

Fig. 2 Examples of digitized tree and non-tree polygons based on a true-color orthoimage in a mixed temperate forest in Switzerland

points at the edges of a regular grid with a larger mesh size. The optimal mesh size depends on the required level of detail, area of interest, and time resources available. If the interpretation is based on stereo-images, the regular grid must be extracted from the digital surface model (DSM)—the same one used for the orthoimage generation. A raster point is assigned as non-tree if the cursor is on the ground or within a shadow on the ground. Problems may occur with areas that have shadows if it is not clear where the exact position of the cursor is set. In these cases, each point must be handled consistently for the entire area of investigation, regardless of the assignment decision. In practice, a 10–50 m mesh size is generally most appropriate for small areas of few square kilometers, and up to 500 m for large areas of several thousand square kilometers. Figure 3 shows an example of aerial image-interpreted tree/non-tree decisions based on a 10 m regular point grid.

The interpreter's requirements for this approach are similar to those needed for the digitized polygons for which little technical and forest-related knowledge is required. Again, the collected reference data is a reliable and reproducible data source. Depending on the area under investigation, the manual workload can increase rapidly.

Another example of the regular point raster approach is the stereo-image interpretation of NFI plots, which usually use a much larger mesh size (see Fig. 4).

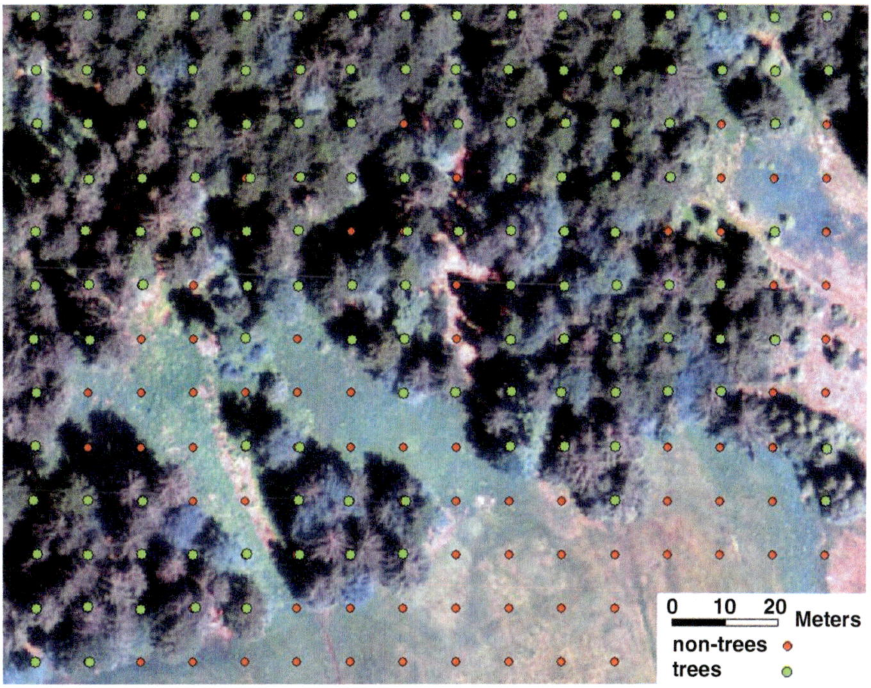

Fig. 3 True-color aerial image-interpreted point raster for the same area with tree/non-tree decisions. The grid has a spacing of 10 m and shadows on the ground are assigned as non-trees

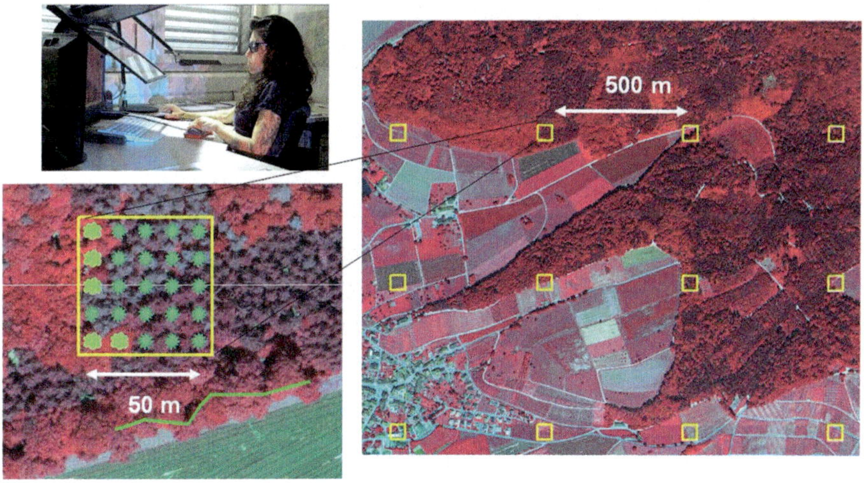

Fig. 4 Stereo-interpretation of 25 points of a regular raster for tree/non-tree decisions and the distinction between deciduous (*flowers*) and coniferous trees (*stars*) on a false-color-infrared aerial image (*left*). The grid has a spacing of 5 m and is part of the 500 m regular raster for stereo-interpretation in the framework of the Swiss NFI (*right*)

External Datasets

Another method is to use external datasets to validate mapped woodlands or forest, either from existing forest-related datasets such as stand maps and tree cover maps or from side products such as layers from any kind of available map (thematic and topographic). The latter are often produced in a context different from forestry. Examples include forest border delineation or the identification of forest areas for layers in topographic maps or for layers in thematic maps, such as land cover/land use vegetation maps or maps from CORINE provided by the European Environmental Agency's International Geosphere Biosphere Program (IGBP—an international research program that studies the phenomenon of global change). The disadvantages of using such datasets are manifold. Issues of concern include dataset up-to-dateness (they are often old and rarely updated regularly), and the fact that, as side products, they tend to be less accurate and less detailed than digitized polygons or raster points. An example of delineated forest borders identified as a side product within the framework of a project—which aimed at updating the measurement of agricultural area to match registered cadastral surveying in Switzerland—is illustrated in Fig. 5.

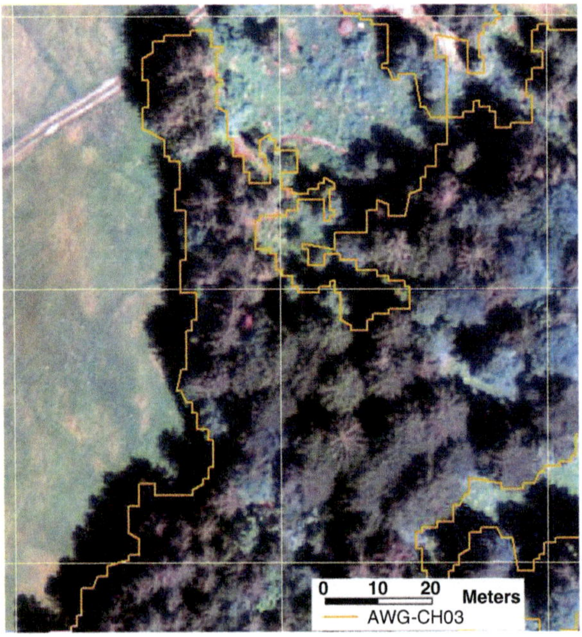

Fig. 5 Forest border delineation (*yellow line*) based on a LiDAR vegetation height model. Large forest gaps are well represented in the true-color aerial image, whereas smaller strips of trees or gaps may not be entirely detected

Remote Sensing Systems

Before giving an overview of existing remote sensing systems, we first provide a short explanation of passive/active systems and their history in the context of mapping woodland areas. Passive sensor systems measure energy wavelengths of the electromagnetic spectrum that are reflected or emitted by objects on the earth's surface. Active sensor systems send out energy, in waves or pulses, in the direction of interest and record the energy coming back from the surface. One advantage in favor of active sensors is the ability to obtain measurcments at any time, regardless of the season or cloud cover. Passive systems are still more common than active systems.

Historically, the first images taken from airplanes date from the beginning of the twentieth century, followed by the initial development of photogrammetry and applications of aerial photography in the 1920s. In the 1950s, as aerial photography became widely used, tree mapping was done by simple delineation or interpretation based on aerial stereo-images, mostly at regional and national levels (Spurr 1960; Gillis and Leckie 1996). Soon after, spaceborne optical systems proved to be particularly useful for mapping and monitoring forests at the national or even global level. Nowadays, the pace of technical advancement of sensors and platforms is more rapid than ever before. New developments in sensor technology for airborne (digital aerial cameras, hyperspectral sensors, airborne laser scanning) and spaceborne (multispectral sensors, radar, airborne laser scanning) systems, such as greatly increased geometric, radiometric, and spectral resolution, has led to new perspectives for mapping woodland. In the last decade, great progress has been made in 3D remote sensing, including digital aerial stereo-imagery, light detecting and ranging (LiDAR), and synthetic aperture radar (SAR) interferometry. The potential and use of currently implemented sensor systems for mapping woodland in general and individual tree species in particular, both passive and active, are described in more detail below. An overview of commercially available and frequently used sensors is given in Table 1. Principally, current digital data from any sensor system enables a high degree of automation in the forest mapping approach. The choice of the most adequate dataset depends on several considerations such as temporal availability, costs, continuity, and technical know-how of the operators. For large-area woodland mapping purposes, 30 m Landsat data might be a good alternative regarding costs and availability, since a huge worldwide archive exists. Archived images can be searched and obtained using the LandsatLook Viewer provided by the U.S. Geological Survey (USGS).

To guarantee a homogenous and integral woodland or forest cover map several factors need to be eliminated or at least minimized. First, the time span between the acquisition of data the mapping is based on and that for validation purposes should be minimized. This is an essential point because forest cover usually changes within a few years. Second, external factors (e.g., atmospheric influence, clouds, shadows, dust, and illumination factors such as sun angle and topography) should be reduced.

Table 1 Overview of studies utilizing airborne and spaceborne remote sensing data for mapping forest and woodland

Study	Country	Data source	Method[a]
Brandtberg (2002)	Sweden	Aerial images	FA
Leckie et al. (2003)	Canada	Aerial images	LM/ML
Laliberte et al. (2004)	USA-New Mexico	Aerial images	NN
Leckie et al. (2005)	Canada	Aerial images	LM/ML
Næsset and Gobakken (2005)	Norway	LiDAR	REG
Reitberger et al. (2006)	Germany	LiDAR	KM
Straub et al. (2008)	Germany	LiDAR	REG
Hirschmugl et al. (2007)	Austria	Aerial images	RM
Wang et al. (2007)	Switzerland	LiDAR, aerial images	RM
Straub et al. (2008)	Germany	LiDAR	REG
Waser et al. (2008a, b)	Switzerland	Aerial images	REG
Koch et al. (2009)	Germany	LiDAR	LM
Waser et al. (2010)	Germany	Aerial images	REG
Waser et al. (2011)	Switzerland	Aerial images	REG
Eysn et al. (2012)	Austria	LiDAR	LM
Kaartinen et al. (2012)	Finland	LiDAR	LM, RM, KM
Waser (2012)	Switzerland	Aerial images	REG
Dalponte et al. (2014)	Norway	LiDAR, aerial images	LM
Eysn et al. (2015)	Central Europe	LIDAR	LM
Waser et al. (2015)	Switzerland	Aerial images	TH
Dees et al. (1998)	Germany	Landsat	REG
Kennedy and Bertolo (2002)	Europe	AVHRR	REG
Stibig et al. (2004)	SE Asia	SPOT-4	KM
Laliberte et al. (2004)	USA-New Mexico	Quickbird	NN
Förster et al. (2005)	Germany	ASTER, SPOT-5	FA
Förster and Kleinschmit (2008)	Germany	Quickbird-2	FA
Pekkarinen et al. (2009)	Europe	Landsat ETM	NN
Hansen et al. (2013)	Global	Landsat TM, ETM	REG

The table provides author references, location, data source, and method used

[a]Methods: *FA* Fuzzy algorithms, *KM* k-means clustering, *kNN* k-nearest neighbor, *LDA* linear discriminant analysis, *LM* local minima/maxima, *ML* maximum likelihood, *NN* nearest-neighbor, *REG* regression techniques, *RM* region growing or merging, *QDA* quadratic discriminant analysis, *US* unsupervised, *TH* thresholds

While clouds and shadows might cover the area of interest, dust or illumination effects affect the reflectance of the forest canopy. Both result in a decrease in the accuracy of the mapped areas. Third, the optimal time of data acquisition is during the vegetation period. Thus, phenology is crucial, as certain differences in woodland can be distinguished by remote sensing techniques only within a restricted period, for example, under leaves-on conditions.

Passive Systems

Over the past decade, airborne photogrammetric film cameras used to extract individual tree crowns (e.g., Brandtberg 2002) or to extract shrub and tree cover (e.g., Waser et al. 2008b) have been replaced by high-resolution digital airborne sensors (Petrie and Walker 2007), which provide higher spectral and radiometric resolution, and are regularly updated and available in almost all countries. Different studies have successfully used digital aerial imagery to extract individual tree crowns (Hirschmugl et al. 2007; Chubey et al. 2009), trees, and shrubs (Waser et al. 2010), and to generate a countrywide wall-to-wall forest cover map (Waser et al. 2015). Other studies (e.g., Darvishsefat et al. 2002; APEX 2011) underscore the advantages of using hyperspectral imagery with a spectrum range from 400 to 2500 nm and between 100 and 300 image bands for tree species classifications and fewer for mapping forest.

Over the past 30 years, spaceborne systems have also been successfully used for mapping woodland or forest, resulting in many mapping products (see Mapping Woodland and Forest). There is a broad range of spaceborne systems, including low-spatial-resolution (GSD 0.5–1 km) satellite systems such as NOAA AVHRR (e.g., Kennedy and Bertolo 2002) and SPOT4-VEGETATION (e.g., Stibig et al. 2004) for global applications, medium-resolution satellites (GSD 10–30 m) such as Landsat TM, ETM+ for national and continental applications (e.g., Keil et al. 1990; Pekkarinen et al. (2009), and ASTER (e.g., Stoffels et al. 2012).

Since the launch of IKONOS at the end of 1999, different series of very-high-resolution (VHR) satellite images exist (e.g., KOMPSAT-2, ORBVIEW-3, QUICKBIRD-2, RAPIDEYE, and WORLDVIEW-2) for mapping forest and forest type (e.g., Förster and Kleinschmit 2008; Immitzer et al. 2012, and Waser et al. 2014). With the exception of RAPIDEYE (only multispectral bands with GSD 6.5 m), they provide all spatial resolutions for panchromatic images between 0.5 and 1 m, and for multispectral images between 1.8 and 6.5 m. New perspectives for woodland and forest mapping regarding temporal and spatial resolution are provided by a group of upcoming new satellite sensors (e.g., WORLDVIEW-3 and Sentinel-2 which carries an innovative wide-swath high-resolution multispectral imager at 13 spectral bands). In addition to currently operational programs, there is continuous development of novel observation techniques, methodologies, and technology related to land use and land cover applications at a more scientific level.

Active Systems

Approximately 15 years ago, airborne laser scanning (ALS) generated considerable interest in the forestry sector. Providing 3D information to assess forest conditions has improved from the average forest stand scale to the individual tree scale. In the last decade, ALS has revolutionized the process of automated mapping of forest/non-forest and has been frequently implemented (e.g., Næsset 2007; Straub et al. 2008; Eysn et al. 2015). Recently, full-waveform light detecting and ranging

(LiDAR) systems have opened new perspectives by providing representations of both canopy profiles and surface topography. ALS campaigns have become more feasible at the countrywide scale, and new technologies have facilitated a more precise mapping of forest (Næsset and Gobakken 2005; Koch et al. 2009; Lindberg and Hollaus 2012; Straub et al. 2013). A good overview of individual tree cover and extraction using ALS data can be found in Kaartinen et al. (2012). Promising results for the extraction of woodland areas are obtained by combining LiDAR with multispectral images (Wang et al. 2007; Waser et al. 2015) or hyperspectral images (Dalponte et al. 2014).

Radar systems with synthetic aperture radar (SAR) sensors, such as ERS-1, ERS-2, ENVISAT, and RADARSAT, have increased rapidly over the last few years, and have been used particularly in tropical regions for the detection of forest gaps or for delineation of forest boundaries (Fransson et al. 2007; Hajnsek et al. 2009). Since this method is unaffected by cloud cover and sun illumination, frequent updates of forest conditions of entire regions are possible (Rosenquist et al. 2007). SAR sensors have frequently been used to detect fragmentation (Dong et al. 2014), deforestation, and entire forest areas (e.g., Wagner et al. 2003; Thiel et al. 2006; Rahman and Sumantyo 2010).

Processing of Input Datasets

Prior to mapping woodland and forest, input data has to be prepared. The degree of automation possible and the number of steps required for this preparation will vary depending on the input data used and the chosen mapping approach. The most common steps are briefly described below. From a methodological point of view, improved methods in image processing, automatic generation of DSM using new image matching methods, and image classification based on objects have been developed in recent years. Meanwhile, computer systems and software packages are fast and typically easy to set up so that application can be made to large areas, for instance, when providing wall-to-wall products.

Preprocessing of Image Data

Data preprocessing incorporates atmospheric and radiometric corrections of images, stereo-image matching, and derivation of image features required for forest mapping. Atmospheric and radiometric corrections were originally developed for satellite images, and are frequently and necessarily applied to optimize image quality. The objective of atmospheric correction is to retrieve the surface reflectance (that characterizes surface properties) from image data by removing atmospheric effects that cause absorption and scattering of solar radiation. Radiometric correction is applied to remove radiometric errors or distortions (correction for sun angle and topography). Recently, new software packages also enable radiometric corrections for airborne sensors. For multi-scene applications and for image strips, it is strongly suggested to analyze radiometric corrections at least within—but preferably between as well—image strips from the same acquisition date.

Image-Based Point Clouds

Image-based point clouds from stereo-images of airborne sensors are frequently used to calculate a digital surface model (DSM) in the context of forest mapping. In cases where both a digital surface model (DSM) and a digital terrain model (DTM) (e.g., obtained from LiDAR) are available, the mapping can be based on calculating a potential canopy cover by subtracting the DTM from the DSM (see Figs. 6 and 7). While DSM and DTM on open ground can be derived from both ALS and from stereo-images, DTM generation of forests is restricted to ALS. Point cloud data from images or ALS data have been used operationally to map forest via automated delineation of forest/non-forest vegetation (Straub et al. 2008; Koch et al. 2009; Waser et al. 2011, 2015). Enhanced image matching algorithms, such as semi-global image matching (for example in Hirschmüller 2008) permit the extraction of woodland with high accuracy and degree of detail (e.g., forest borders and small openings between trees).

To separate non-vegetation objects (e.g., buildings) from high vegetation (tall shrubs or trees), spectral information from aerial images (preferably from the near-infrared band) or other types of information extracted from them are used. For example, Ginzler and Hobi (2015) used the normalized difference vegetation index (NDVI) to separate vegetation from non-vegetation areas and produced a wall-to-wall vegetation height model (VHM) for Switzerland in its entirety. Since the near-infrared is crucial for the detection of vegetation, separation problems arise if only true-color images are available. In this case, a canopy height model (CHM) can be used—if stereo-images are available. If only a 3D point cloud from an airborne laser

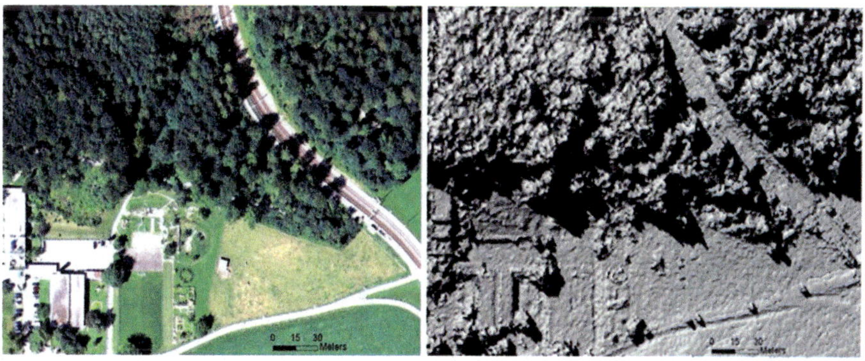

Fig. 6 Example of airborne true-color orthoimage (*left*) with the corresponding hillshade of the potential canopy cover for a study area in Switzerland (*right*)

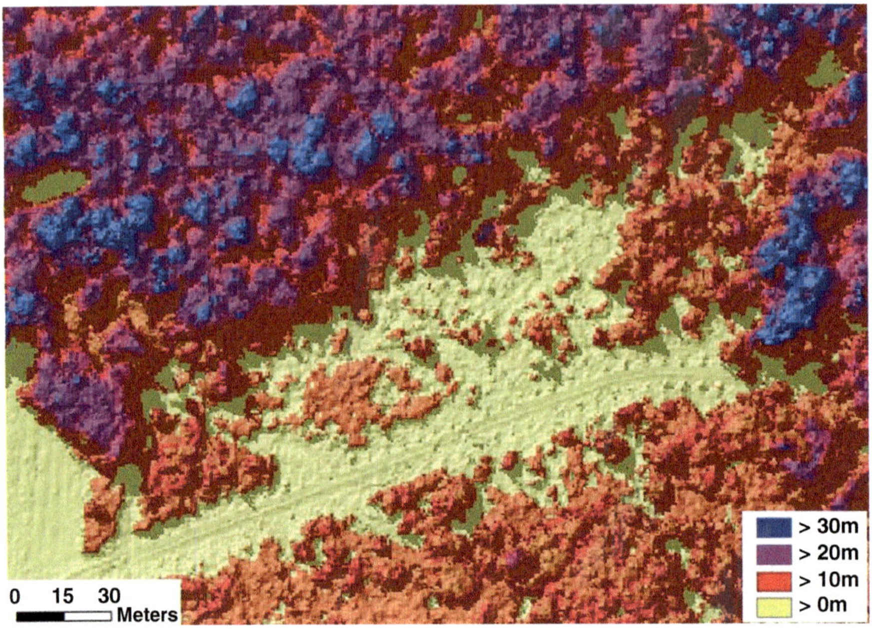

Fig. 7 Hillshade of four different height classes obtained from a normalized DSM

scanning campaign is available, methods of echo-ratio are used to separate vegetation from artificial objects (Eysn et al. 2012).

To summarize, image matching provides essential information—such as height information as shown in Fig. 7—about the surface of vegetation. Since it can be done using commercial software packages and carried out semiautomatically, it requires relatively little expert knowledge. More expertise may be necessary to fully understand the algorithms employed.

Image Segmentation

The extraction of individual trees or groups of tree clusters is needed in the workflow of many highly automated mapping approaches and is often based on image segmentation. In this step, homogeneous image parts are subdivided into smaller patches or partitions. Nowadays, commercially available software packages are widely used making it possible to obtain optimal (according to the requirements of the mapping approach) image segments by iteratively adapting parameters such as the degree of homogeneity and the shape of image objects (Baatz and Schäpe 2000). Successful segmentation is followed by adjusting the size of objects to the scale of the assessment. For example, tree crowns and tree clusters must be adjusted differently when the mapping is set to the individual tree level or to the forest stand level. While image segmentation is carried out with a high degree of automation and is

Fig. 8 True-color aerial image of different forest stands in a mixed temperate forest in Switzerland

relatively simple using most software packages, problems occur when the area of interest increases. Depending on the level of detail and the spatial resolution of the images, these problems may already arise for areas of a few square kilometers. In such cases, subdividing the area into smaller parts can resolve the issue, but this will increase the required effort in terms of time and handling and consequently result in a loss of automation.

Thus, from a practical point of view, operational use on larger areas must be a compromise between the size of the area to be investigated and the level of detail. For example, it may be necessary to proceed at the stand-level scale rather than at the individual tree level. Figures 8 and 9 illustrate aerial image segmentation at the individual tree level for mixed temperate stands.

Image Classification

There are several other methods—with varying degrees of automation—of classifying images that are appropriate for mapping woodland areas and forest. Since image classification is crucial to the mapping process, important principles and background information are briefly outlined below.

Over the past decade, a number of significant developments in object-based image analysis—such as multi-resolution image segmentation and identification of object relationships—have become available for classification purposes. According to Jensen (2005), the general objective of image classification is the automatic allocation of all pixels to land cover classes or specific themes. The most appropriate

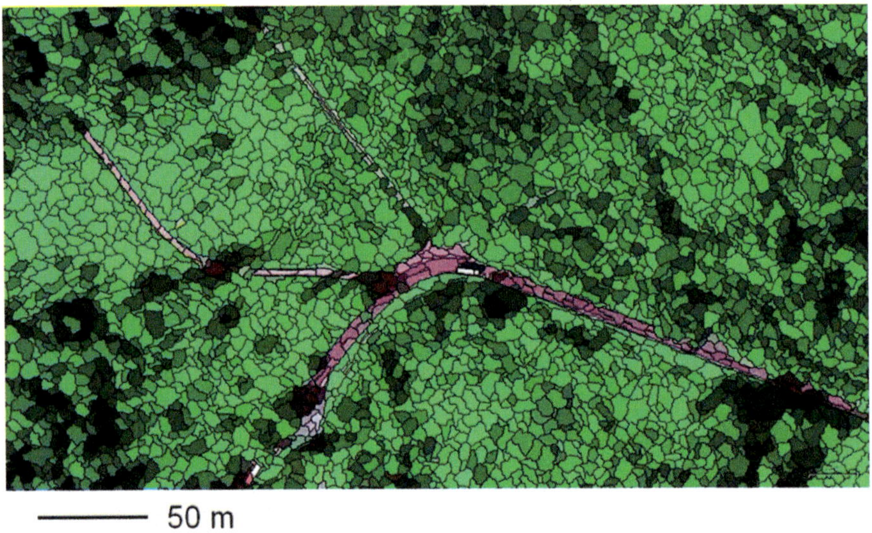

Fig. 9 The same area with the corresponding image objects (different shades of *green*) after segmentation at the individual tree crown level

classification strategy depends on the biophysical characteristics of the research area. These include the topography or heterogeneity of the land cover, the homogeneity of the remote sensing data, the illumination level, the date of acquisition, the training data, the representative samples, and *a priori* knowledge.

Broadly, classification techniques can be categorized as follows:

- Supervised: Imposing our perceptions on the spectral data. Pixels or image objects are assigned to classes by matching them with spectral properties found in training datasets. Examples include minimum distance algorithms, maximum likelihood, more advanced techniques such as neural networks, fuzzy logic (see Chapter "Fuzzy Classification of Vegetation for Ecosystem Mapping"), support vector machine, and regression techniques.
- Unsupervised: Spectral data imposes constraints on our interpretation and involves grouping data into categories based on some measure of inherent similarity. Separability of image clusters must be maximized. Class names are assigned to image clusters after classification (for example, *k*-means).
- Parametric: Based on probability density functions and the assumption that the data has come from a type of probability distribution (for example, discriminant analysis).
- Non-parametric: Based on probability density functions and no assumptions regarding the distribution of the data (for example, nearest neighbor).

Mapping woodland and forest is either carried out on entire forest patches or at the individual tree level (Leckie et al. 2003; Chubey et al. 2009). Several studies highlight the advantages of combining multi-resolution segmentation with object-based classification (Laliberte et al. 2004; Lamonaca et al. 2008; Waser 2012).

In some studies, simple regression techniques were used to extract tree areas (Næsset and Gobakken 2005) or generalized linear models to assess tree and shrub probability (Waser et al. 2008a, b, 2011), since it is well known that a deterministic representation expressed in terms of a limited number of land cover categories (e.g., trees, non-trees) leads to a loss of information. Still other studies are based on local maxima algorithms (Koch et al. 2009; Dalponte et al. 2014; Eysn et al. 2015) or in combination with a maximum likelihood classifier (Leckie et al. 2003). Examples of other techniques include the use of fuzzy algorithms at the individual tree level (Brandtberg 2002), nearest-neighbor techniques (Laliberte et al. 2004), *k*-means clustering (Reitberger et al. 2006), and region merging techniques (Wang et al. 2007).

An overview of studies that have successfully used remotely sensed data with the methods applied for mapping woodland or forest is provided in Table 1.

Mapping Woodland

This section illustrates three highly automated approaches to mapping woodland in different bio-geographical regions in Switzerland. For each approach, the basics of each method, data, results (tables, but especially figures and illustrations), and a discussion of pros and cons are given.

A Hierarchical Segmentation Approach for Mapping Woodland

The principles of image segmentation are given in the section *Image segmentation*. Mean shift—a non-parametric statistical method—is a special form of segmentation which was originally presented by Fukunaga and Hostetler (1975) and generalized by Cheng (1995) 20 years later. It is a robust and highly automated algorithm, and is based on a simple iterative procedure that shifts each data point in the feature space (*n*-dimensions where the imagery variables or canopy height are) to the average of data points in its neighborhood. The algorithm is frequently applied to color images because most of the generated regions can be delineated semantically. The algorithm starts by converting the image data into a feature space. For each data point, the mean shift algorithm defines a window around it and computes the mean of the data point. The algorithm associates each data point with the nearby peak as identified by the dataset's probability function. The center of the window then shifts to the mean. The algorithm is repeated until it converges. Finally, each pixel is grouped together based on all corresponding convergence points. Figure 10 shows an example of the resulting segmentation of one false-color-infrared (CIR) aerial image with the segmented region borders indicated with yellow lines.

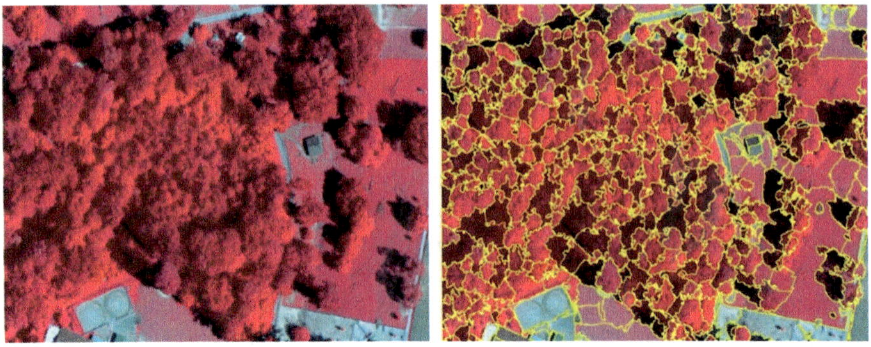

Fig. 10 False-color-infrared aerial image of woodland (*left*) and the segmented crowns obtained by mean shift segmentation (*right*)

This relatively simple technique for most pixel-based tree canopy mapping creates thresholds for either single bands of the image or derived data such as NDVI. Similarly, there are two important thresholds, one of which is the NDVI value and the other is the CHM value. In general, the rule is set whereby a segment is identified as woodland if the mean values of NDVI and CHM of all pixels within one segment are higher than the predefined thresholds. Segments containing shadows, for example, along the boundaries of a woodland, may also include groups of trees. Since these segments can be extracted as woodland, a slight overestimation of woodland is thus obtained. Consequently, a hierarchical segmentation approach is implemented because of the advantages of color-infrared and NDVI images. Figure 11 shows a hierarchical segmentation method flowchart. The two thresholds ψ are applied to NDVI and λ to CHM data. The segment is classified as wooded area if its mean values of NDVI and CHM are higher than the predefined thresholds. Woodland is then classified using the segments as obtained in *woody1* and *woody2*.

The results produced provide the basis for various forest applications, including the calculation of canopy cover, forest area delineation using explicit definitions, and extraction of gaps in forest masks. Due to the highly automated processing chain, further extended study areas would need to be investigated in order to verify the different threshold combinations. The resulting mapped woodland is illustrated for close forest, open forest, and a mixture of the two in Fig. 12.

Individual Tree and Tree Crown Detection

Tree height and crown geometry are two basic woodland features from a remote sensing perspective and complement spectral information with rigid geometry. Numerous highly automated approaches based on ALS or image matching have been proposed to detect individual trees (ITD) and individual tree crowns (ITC).

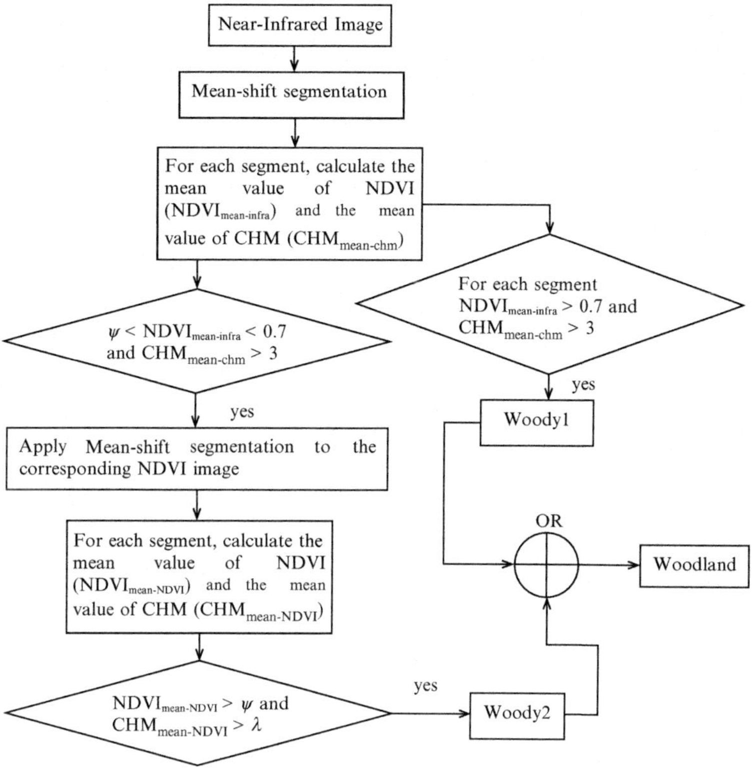

Fig. 11 Flowchart of the hierarchical segmentation method implemented to map woodland based on NDVI and CHM thresholds

Most of these focus on the reconstruction of the DSM or CHM, which provides a representation of the outer geometry of tree canopies. DSMs from ALS data or from image matching are the starting point for the detection of individual tree crowns and the calculation of tree parameters, such as height and crown diameter. Depending on the available point density of the DSM, the location of individual trees, the shape of tree crowns, or the canopy cover can be estimated with decreasing point density. The average point density mainly controls the selection of an appropriate smooth or coarse shape model. Robust ITD using ALS data requires a point density of 2 points/m^2 or higher. Most approaches start by finding a local height maxima in the DSM. For subsequent processing of local neighborhood region-growing (Solberg et al. 2006), height histogram (Kaartinen and Hyyppä 2008), watershed (Straub and Heipke 2001), or cluster analysis (Kaartinen et al. 2012) methods are used.

In the case of coarser CHMs, tree crown or canopy shape models can be applied, of which the most common shape approximations use an ellipsoid of rotation

Fig. 12 Examples of CIR images with different woodland and different types of mapped forests

(Nilson and Peterson 1991; Li and Strahler 1992; Li et al. 1995; Kuusk and Nilson 2000; Garcia-Haro and Sommer 2002; Gerard and North 1997; Gerard 2003). Slightly refined models use a combination of a cone on top of a cylinder (Nilson and Peterson 1991; Chen and Leblanc 1997; Kuusk and Nilson 2000), implement a fixed shape model with adaptive diameter (Vosselman 2003), or use a predefined set of fixed shape models (Rutzinger et al. 2010). For low point densities—which are often the case with large DSMs—less sophisticated shape models are required. A very common terrain shape indicator is curvature (second derivative, and has been used in many applications for terrain analysis—Zeverbergen and Thorne 1987). The most prominent disadvantages of the curvature feature are the limited neighborhood (3×3 box), the requirement of an interpolated grid, and the fact that it has no fitting with a shape model. Curvature processing often results in a highly overestimated number of tree crowns and is consequently only of limited use.

For example, in Switzerland, 40% of the coverage area of the Swiss National DSM has a point density of less than one point per square meter. As a result, robust detection of ITCs is not possible with the shape models discussed above. Consequently, another model is required, which combines local maxima value with a shape fitting approach. The so-called Spider model (Fig. 13) implements a non-parametric shape description with few global spatial constraints, mainly to avoid finding a suitable parameter set for the processing of a national dataset. In the first step, a local maxima must exist within a 3×3 neighborhood. In the second step, for

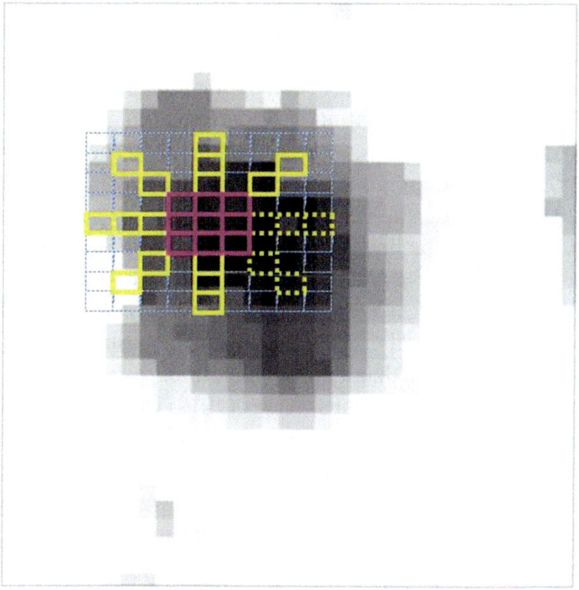

Fig. 13 The Spider model which combines local maxima and a minimal shape fitting of an individual tree crown. The height classes of an individual tree are indicated in *gray* to *black*. The six canopy legs are *yellow*, of which the two dashed legs are invalid (the cells of these legs are not within the 5°–85° limits)

each of the eight directions from the center, the slope values are accumulated for each direction independently. Each slope accumulator contains $(n - 1)/2$ values, i.e., four slope values for each leg of a 9×9 box. Each slope value in the accumulator must be within the limits of 5°–85° to be considered valid. In this way, horizontal and vertical slope extremes are rejected. If all slope values are valid, the corresponding direction is classified as a canopy leg candidate (invalid legs are dashed in Fig. 13). If six out of eight canopy legs are valid and radially contiguous, then the area of the eight slope accumulators will be defined as a canopy patch. The robustness of shape matching is achieved with the nonparametric evaluation of the slope accumulator and the concept of allowed limited local distortions (invalid legs). Using a moving window, the processing of the DSM results in a point cloud of canopy patch centers. The result of this detection process is defined as a canopy patch, mainly due to the given Swiss DSM's coarse resolution of 1 m.

The presented shape model requires only a few parameters compared to other shape models, which is a clear advantage with regard to the degree of automation, robustness, and parameter evaluation. With the increasing spatial resolution power of future aerial cameras, the spider-shaped model is still appropriate to extract single tree crowns from low-quality digital surface models. On the other hand, the lack of a rigid shape model can be seen as a conceptual disadvantage, which occasionally results in the matching of distorted tree canopies. This compromise, however, achieves the desired degree of robustness.

Fractional Tree Cover Approach

The fractional tree cover approach has been applied to several forest ecosystems in central Europe and involves the continuous representation of probabilities (*P* woodland) that each image segment is a tree/shrub or not. The fractional tree cover approach is based on logistic regression models and is described in detail by Waser et al. (2008a, 2010). The logistic regression model is a special case of the generalized linear model (GLM). The linear result is run through a *logistic function*, which runs from 0 (negative infinity) and rises monotonically to 1 (positive infinity). Tree probability (*P* woodland) ranges between 0 and 1 for each pixel that belongs to the class *woodland* (for more details, see Hosmer and Lemeshow 2000). Thus, woodland is continuously mapped in detail and is, therefore, particularly useful for detecting single trees, small trees, and shrubs. Waser (2012) reported that a high degree of automation was achieved with the developed methods, and that trees and shrubs were extracted with a correct classification rate (CCR) of 96%–99%. Visual image inspection revealed that the fractional cover approach was also accurate in areas with complex forest structures, such as open forest and gaps, afforestation, and fuzzy borders—especially in steep terrain. The fractional tree cover approach consists of four main steps. Figure 14 gives an overview of the methodological workflow.

First, potential tree crowns are segmented (as described in Image Segmentation) and a high-resolution CHM is generated by subtracting the DTM (derived from LiDAR) from the DSM (derived from image matching). Second, geometric features (height, slope) are generated from the CHM, and spectral features from digital aerial images (original image bands, color transformation, and NDVI). Digitized polygons

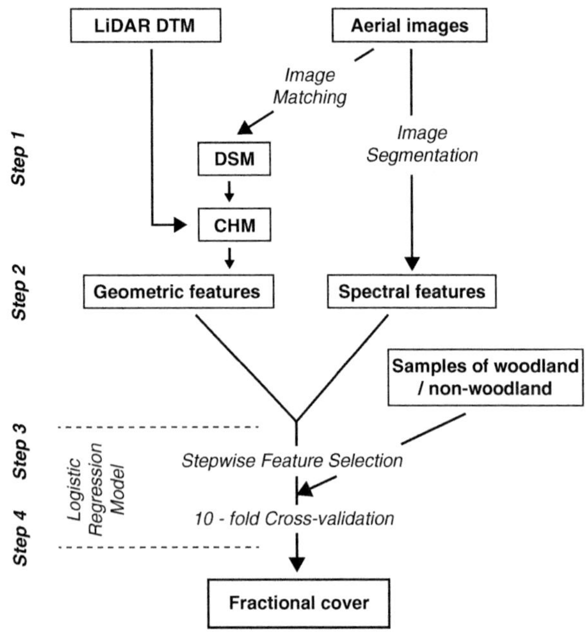

Fig. 14 Methodological workflow of extracting woodland using logistic regression models

of woodland/non-woodland samples (as described above in the section Reference Data) are used as training and validation data. Third, stepwise feature selection is used as an analytical tool to find redundant features which are then excluded from the model. Fourth, the predictive power of the models is then verified by a tenfold cross-validation process using different combinations of selected features.

Usually, good mapping results are obtained if all tree segments with *P* woodland probabilities of greater than 0.5 are included. In the case of overestimated woodland—for example because of grassland on boulders which may be wrongly mapped as trees because of their similarity to real trees and shrubs—the *P* woodland threshold can be adapted accordingly by lowering it, for example, to 0.3. This must be done after the first run and is strongly dependent on the individual terrain and vegetation characteristics of the area to be investigated. Thus, the fractional tree cover approach enables the averaging out of possible over- and underestimations, and has high potential to successfully extract shrubs and small single trees. The limitations of this approach include constraints regarding full automation, and the potential lack of sufficient reference data, in particular when mapping large areas with time restrictions or when reference data simply does not exist. Applying the latest image matching algorithms may improve accuracy at forest borders and small openings between them by improving the generation of the geometric features. An example is provided from a mountainous test site in central Switzerland (Fig. 15: photo of the test site; Fig. 16: true-color orthoimage; Fig. 17: estimated woodland probabilities).

Fig. 15 Example of mountain woodland where the fractional cover is tested from the center of Fig. 16 to the south west. Larch, spruce, and shrubs dominate the area

Forest Mapping

As stressed several times in this chapter, in contrast to mapping woodland, forest mapping implies an unambiguous definition for *forest*. Two highly automated approaches are presented here for forest mapping which are based on merging the wooded areas with *forest* areas using distance and height criteria. Thus, they can be easily adapted to any forest definition. Each approach includes a short description of method, data, results (tables, but especially figures and illustrations), use of different definitions for *forest*, and a discussion of pros (potential) and cons. The impact of minimum areas, minimum width, and crown coverage; the importance of land use definitions; and their impact on forest areas are also discussed. The results derived from these applications are highly dependent on the fundamental input parameters' size, and position of the delineated forest areas.

Moving Window Approach

A highly automated workflow based on ArcGIS functions implemented in Python scripts was developed to generate a wall-to-wall forest cover map of Switzerland. The approach is based on the four key criteria: (1) minimum tree height; (2) minimum tree crown coverage; (3) minimum width; and (4) land use. The description of this approach below was originally published in Waser et al. (2015) and is only partly modified.

Whereas fairly straightforward and automated methods to obtain criteria 1–3 from remote sensing data exist, the land use criterion is not easily assessable when using remotely sensed data. Due to the limitations explained in the ***Introduction,*** the land use criterion cannot be directly obtained from remotely sensed data. Instead, existing map products must be taken into account, which may have a different scale or level of detail compared to the remote sensing data used for forest mapping. The forest mapping approach is suitable for very-high-resolution data from airborne laser scanning (ALS) or digital stereo aerial images. The moving window approach is highly automated and uses threshold techniques in combination with the three criteria above and a land use definition. Thus, its suitability for NFI purposes is entirely supported by the application of the land use criterion and its high level of detail, especially regarding forest borders and gaps.

An overview of the main steps and input datasets is shown in Fig. 18. The four key criteria of the NFI forest definition are explained separately and in more detail below. Figure 18a shows a vegetation height model (VHM) based on image-based point clouds with a spatial resolution of 1 m—which was derived from aerial image blocks of 0.5 × 0.5 km (Ginzler and Hobi 2015). Figure 18b shows a preliminary forest cover map based on the criteria of the NFI forest definition (minimum tree height, crown coverage, and minimum width). Figure 18c shows the application of the land use criterion in order to remove forest on other land (e.g., orchards, urban

Fig. 16 A true-color aerial image of woodland in the central Alps of Switzerland

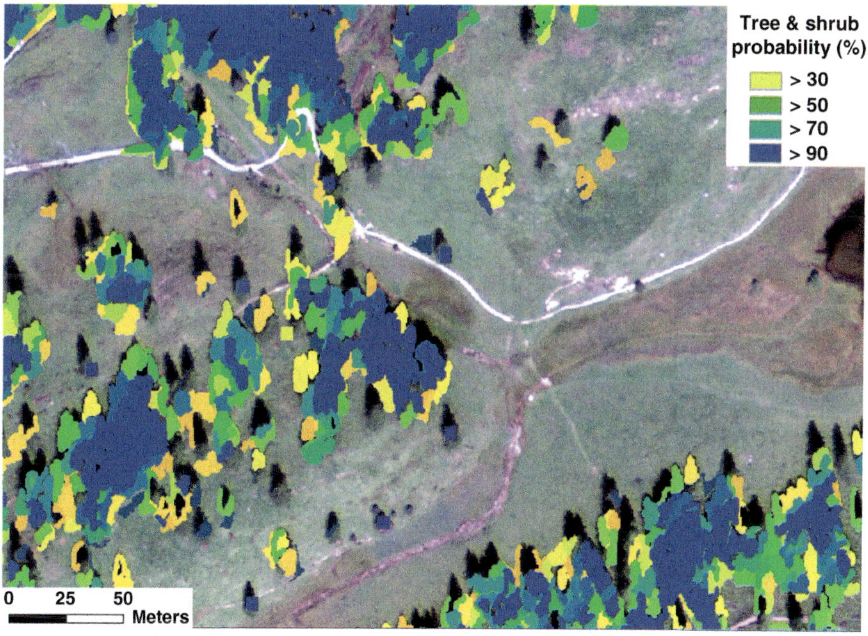

Fig. 17 Mapped woodland (trees and large shrubs) for the same area. The higher the probability for trees and shrubs, the darker the color. Small trees are extracted as well

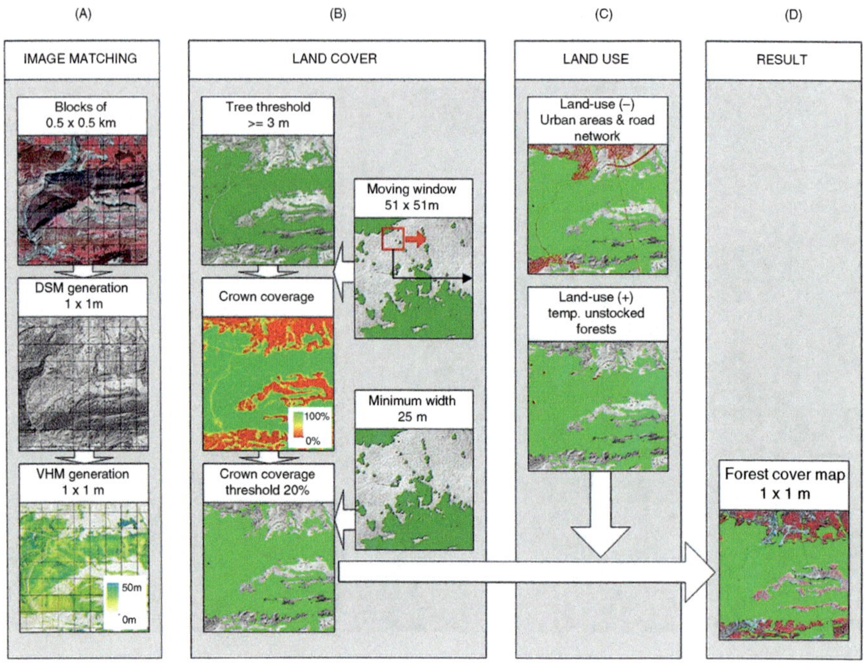

Fig. 18 Workflow of the forest cover map implementing the existing Swiss NFI forest definition. (**a**) Image matching, (**b**) preliminary forest cover map based on land cover and the thresholds and criteria of the NFI forest definition (minimum tree height, crown coverage, and width), (**c**) removal of forest on other land and addition of temporarily unstocked forests, (**d**) final forest cover map (source: Waser et al. 2015)

parks) while adding temporarily unstocked forests (e.g., wind throw, harvesting). Finally, in Fig. 18d, the calculated forest cover map that implements all criteria of the forest definition as used in the stereo-image interpretation of the Swiss NFI is shown.

Height

Terrestrial height measurements are often time consuming for several reasons. In dense forests, it is more challenging to see tree tops from the ground than in open forests. Furthermore, broadleaf trees are even more difficult and ambiguous to measure than coniferous trees, because of the round shape of the crown. Height information—measured remotely using active systems such as ALS or passive systems such as stereo-images—can be applied to large areas with high densities. Given a precise DTM from an ALS campaign, the height of objects can be calculated by subtracting the DTM from the elevation above sea level of the surface model. The result is a CHM with a level of detail which depends on the resolution (point density) of the input DTM and DSM.

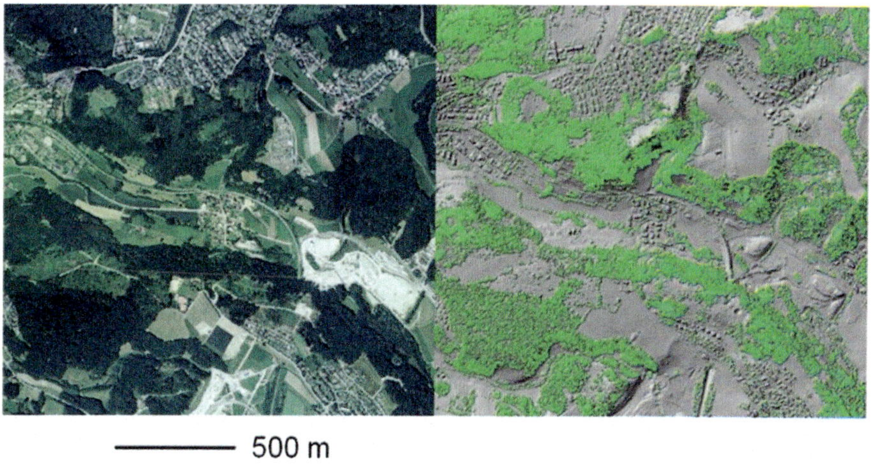

Fig. 19 Stocked areas in the true-color aerial image (*left*). CHM with extracted trees based on the 3-m height criterion as applied in the Swiss NFI (*right*)

To separate trees from other woodland, a height threshold, as specified by a particular forest definition, is used. In the Swiss National Forest Inventory, the minimum tree height is set to 3 m (see Fig. 19). In contrast, in most NFIs and the FAO, the definition of forest specifies a higher minimum tree height of 5 m. Regardless of the NFI, height is always clearly defined as "height in situ," which in fact refers to the height a tree may reach under site-specific conditions in the field and not the actual height. In this way, temporarily unstocked areas (e.g., harvested areas) retain the classification of *forest* in the framework of a forest inventory, but are not detected as forests when using a CHM.

Crown Coverage

For the calculation of crown coverage, a well-defined reference area is essential, which is usually an interpretation area of a fixed size, as used by forest inventories (an area of 50 × 50 m is used by the Swiss NFI). Alternatively, forest stand maps can be used to estimate canopy cover. Crown coverage is calculated using a moving window approach, with a rectangle of 51 × 51 m (odd number in order to obtain the center pixel) which almost corresponds to the interpretation area used in the Swiss NFI terrestrial survey and in the stereo-image interpretation. For each center pixel, the proportion of vegetation greater than or equal to 3 m in height inside the window is calculated and the defined minimum crown coverage of 20% (Swiss NFI definition) is applied. Other definitions of *forest* are in the range of 10%–50% minimum crown coverage. The results vary substantially if different window sizes are applied (see Fig. 20). With this technique, forested area is particularly overestimated at the forest border (see Fig. 21).

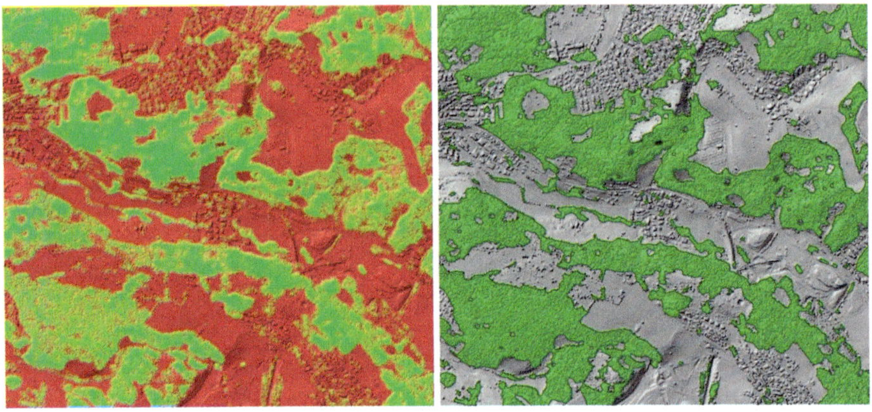

Fig. 20 Extracted tree coverage (crown coverage of 0%—*red*, 100%—*green*) based on the moving window approach as applied in Switzerland (*left*). Example of the same subset after applying a minimal crown coverage threshold of 20% (*right*)

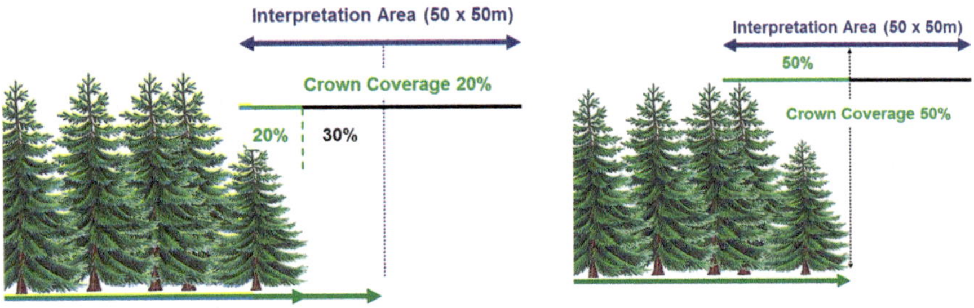

Fig. 21 Problems related to the forest area at forest borders within the interpretation area (*blue line*) when its center (*dashed line*) is clearly outside the forest (*left*) and exactly at the forest border (*right*). Source: Waser et al. (2015)

Figure 21 (left) illustrates that crown coverage is still 20% at the position of the dashed line, even though the center of the interpretation area is clearly outside the desired forested area as defined by the NFI. After the preliminary forest cover map was generated using the 20% crown coverage threshold, the map was shrunk using morphological functions. The number of pixels for the shrinkage process at a specific position *i* (*XY*-coordinates of this pixel) is a function of the size of the interpretation area and the minimum crown coverage threshold as described in Eq. (1).

$$\text{Number of pixels to shrink}_i = \text{Window size} \times (0.5 - \text{Crown coverage threshold}) \times \text{Crown coverage}_i \quad (1)$$

Applying the Swiss NFI forest definition this value is 51 m (window size) × (0.5–0.2) (20% crown coverage threshold) × 1 (100% crown coverage at this location inside the forest) = 15.3 pixels (rounded to 15). For the sake of simplification, the crown coverage inside the forest border at any location can be set to 100%, resulting in Eq. (2).

$$\text{Number of pixels to shrink}_i = \text{Window size} \times (0.5 - \text{Threshold of crown coverage}) \quad (2)$$

Figure 21 (right) illustrates the example for a 50% crown coverage threshold within the interpretation area. In this case, no shrinkage would be applied.

Minimum Area and Width

Minimum area and width are two other essential parameters needed for forest mapping. However, both are handled differently by many NFIs. Minimum width is more frequently used to separate areas of narrow tree elements—which do not fulfill the minimum requirements of the forest definition—from clearly defined forest areas. While minimum area is relatively simple to calculate using standard GIS functions, more effort is needed for minimum width. A frequently used method is to place a circle, with the radius of the minimum width inside the stocked area. In this ideal case, the minimum width criteria is fulfilled. If not, another possibility is to calculate these distances with triangulation by setting the length of the minimum triangle to the required width threshold. In the given example (Fig. 22 on the left), narrow tree groups below the minimum width of 25 m were removed using morphological functions while preserving the shape and size of larger objects. The forest cover map was first shrunk by half of the minimum width of the forest definition and then expanded by half of the minimum width. Parts smaller than the minimum width criterion were assigned the value zero (non-forest) and thus remain non-forest after the expansion of the shrunk forest cover map.

Land Use

Since the land use criterion could not be obtained from the VHM or from aerial images, it was implemented using the respective cover from the Topographic Landscape Model (TLM) (for more details see Swisstopo 2016). It is superior to the currently available CORINE 2006 land cover (CLC2006) with respect to the level of detail and updating, and is the best available proxy for the *forestry* land use. Two classes from the TLM were integrated into the present forest mapping approach: the land cover class *closed forest*—which is actually a land use class—to compensate for temporarily unstocked forests, and the land use class *orchards* and *settlement* to eliminate *forest* area on other land uses. In contrast to the three other criteria, additional manual work was necessary since the exact forest definition applied in the TLM is not entirely clear. Both TLM land layers were visually checked and polygons that were obviously wrong or too generalized were manually deleted.

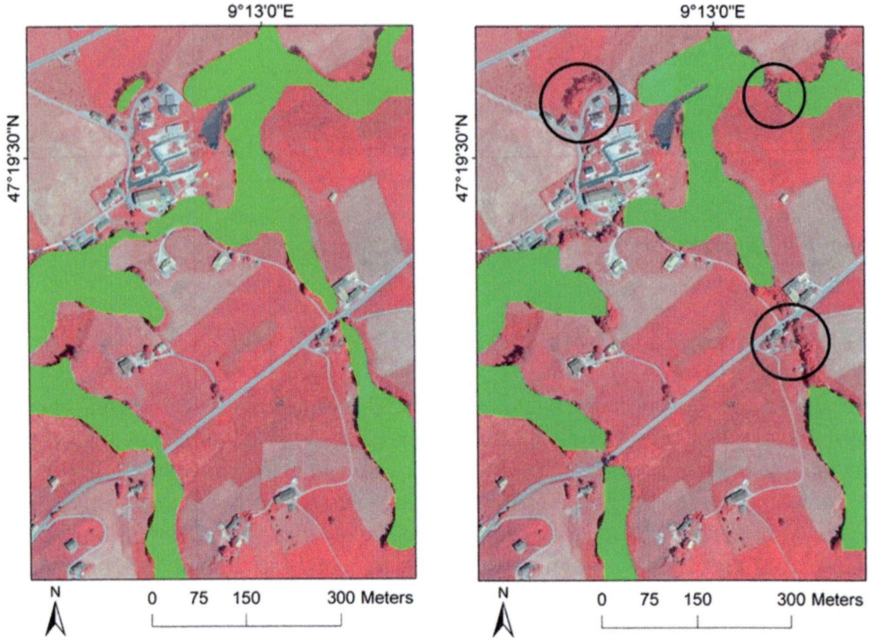

Fig. 22 Removal of narrow tree elements from the forest cover map using morphological filters. Forest cover map after applying the crown coverage threshold of 20% (*left*). Removed narrow elements (*circles*) in the expanded forest cover map with half of the minimum width threshold (*right*) (source: Waser et al. 2015)

Distance Criterion Approach

Another often used approach for forest mapping is based on the distance criterion. The approach is highly automated and aggregates point clouds from both image matching and ALS to forest layers within a spatial grouping process as described below. Similar to the moving window approach, the three criteria of height, minimum width, and minimum crown coverage are used from the forest definition. Thus, in the first step, the height criterion of greater than 3 m based on the VHM according to the Swiss NFI forest definition is applied to the woodland. Figure 23 gives an overview of the distance criterion approach.

In the first step, the two criteria of minimum crown coverage and minimum width/area from the NFI forest definition are obtained indirectly using the distance criterion as described below. The previously extracted centers of individual tree crowns (>3 m) are connected in a non-overlapping mesh using Delaunay triangulation based on a distance criterion (e.g., minimum width of 25 m as defined by the Swiss NFI). In the second step, the shape of the point sample on its surface is reconstructed using alpha shapes. The alpha shapes define a piecewise linear curve around the meshed points. The resulting alpha hull (Fig. 24 on the right) surrounds the tree

Fig. 23 Overview of the methodological workflow of the distance criterion approach based on Delaunay triangulation and alpha shapes

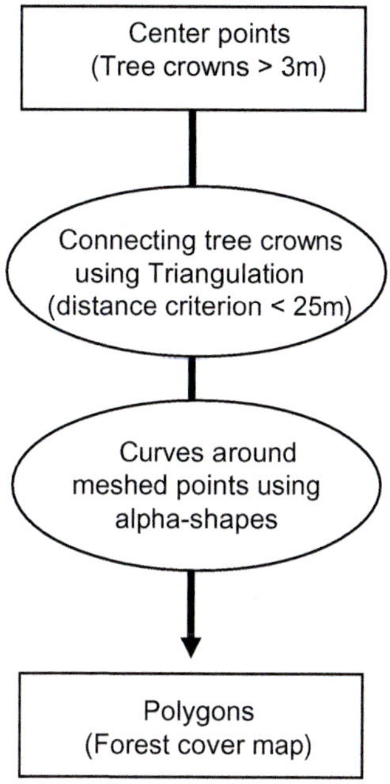

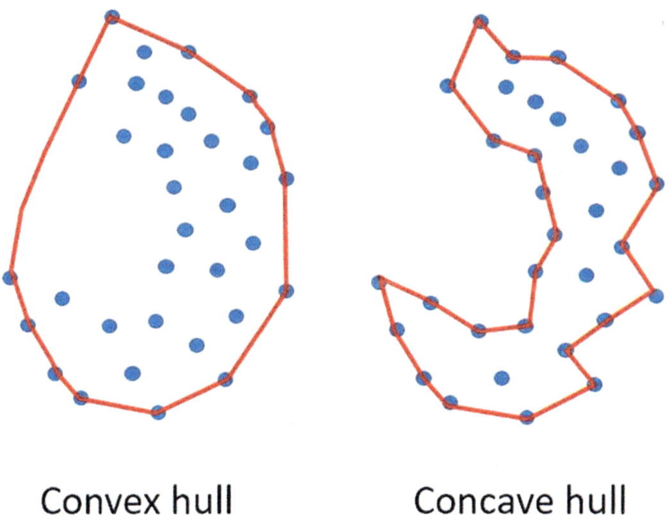

Fig. 24 The principle of convex and alpha hulls (*red line*) of points belonging to tree crown centers. Based on the distance criterion, a tree crown lies within a certain distance from its neighbors

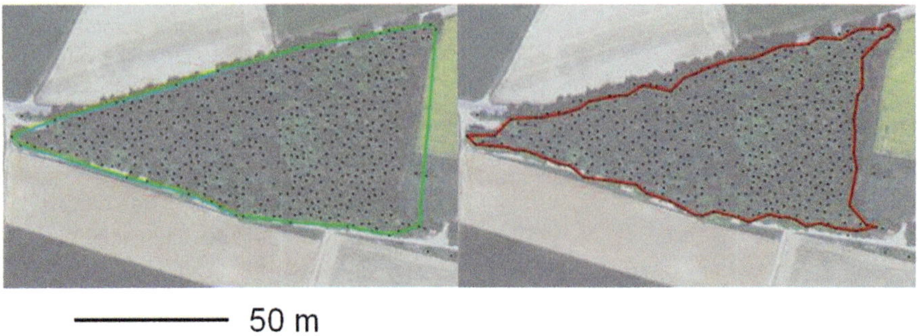

Fig. 25 Convex (*green boundary*) and alpha (*red boundary*) hulls of tree crown centers

crown centers and avoids the overestimation of forest area resulting from the overinclusion of empty areas as in Fig. 24 (on the left) (Mottus et al. 2006; Vauhkonen et al. 2012; Edelsbrunner 1995).

The triangulation of point clouds and the calculation of the alpha hull are closely related to the problem of correct connectivity within a given point cloud (Fig. 25).

The definition of an appropriate connectivity distance is a crucial point. For the Swiss NFI, this distance corresponds to a minimum width of 25 m. The required control of connectivity can be achieved by a preliminary triangulation with a strict distance constraint. Thus, points within a certain distance metric (<25 m) belong to the same object, and only for these points will the alpha hull be calculated (regions A, B, and C in Fig. 26). The maximum distance is required for all three vertices within each triangle. Therefore, linearly aligned points (narrow tree areas such as tree rows, single trees) do not meet the requirements for a spatial object (region E in Fig. 26). If the distance is too great from surrounding points, topological holes will be created (region D in Fig. 25). Thus, forest gaps and clearings are well extracted.

Grouping methods for forest data are strongly related to the application-defined requirements for connectivity and generalization. This approach makes it possible to control the distance- and shape-related grouping process with only two parameters, and can therefore be easily adapted for a wide range of forest applications. Triangulation enables a more appropriate handling of the spatial distribution of point data compared to cluster approaches. For some applications, it could be necessary to use weighted point data, mainly when specific features of forest data are much more important than spatial distributions (e.g., diameter, health status).

Lessons Learned

Nowadays, most existing forest cover maps are still the product of visual interpretation of aerial or satellite images with automation mostly restricted to the processing of the input data. While only a few of these automated methods are currently

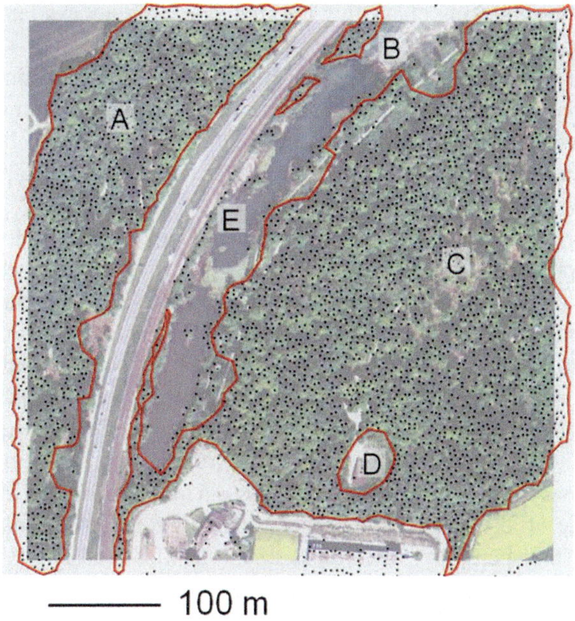

Fig. 26 Triangulation with the distance-constraint and alpha hulls (*red lines*) of tree crown centers (*dots*) with a true-color orthoimage as background. Calculated alpha hulls for points within dense forests (*A*), forest aisles (*B*), and open forest (*C*). The building is extracted as a topological hole (*D*). Tree rows are not extracted due to their linear point alignment

operationalized and applied to large areas, most have been tested within the framework of case studies.

In this chapter, the potential of remote sensing data and techniques towards automated woodland and, in particular, forest mapping has been illustrated. The presented approaches are promising and, because of their automation potential, superior to existing approaches. Because they make use of airborne or spaceborne sensors, and optical datasets can be applied entailing lower costs and less complex data processing and handling, they can be regarded as state of the art, and may be used to gather accurate up-to-date information on forest areas at different scales, including small wooded patterns and single trees. The most likely interest groups that could benefit from the presented mapping approaches are public or private authorities with environmental or forest-related concerns, private environmental agencies, forest districts, private forest owners, and NFIs.

The approaches illustrated here are straightforward from a methodological point of view (well-defined input and output data, with several processing steps), facilitating handling at the operational level and not only at the case study level. They are also highly automated and applicable to larger areas. However, each of the illustrated approaches still consist of steps that need some manual adaptation of parameters. This applies both regarding the processing of datasets, such as the segmentation of images, and the checking of existing training data, or, if not available, the generation of such training data (e.g., digitizing polygon samples).

Automation is reduced when mapping forest is entirely based on a NFI definition, less because of the calculation of geometric parameters (minimum height, minimum crown coverage, and minimum width/area), but more because of the implementation of a land use criterion. The latter may require the appropriate collection of relevant land use information.

From a technical point of view, current handling of large datasets is no longer a problem. In theory, processing and data storage can be done using standard personal computers. However, for wall-to-wall forest cover maps, multi-core computers with large physical memories are recommended. This is not only in order to minimize processing time but also to optimize and speed up each step of the applied mapping approach because most software packages enable the use of clusters for faster calculations.

There are, however, several significant constraints that remain. First, the implicit term *forest* is not always correctly used by the remote sensing community and is confused with *woodland*, since it only comprises a minimum height criterion. Thus, prior to the use of any mapping approach, unambiguous definitions of forest and non-forest are indispensable. Regardless of the applications, the FAO definitions seem to be reliable. Problems regarding automation increase for cross-border mapping approaches because harmonization of existing forest definitions is additionally necessary. Manual interactions and processing steps are also required, such as merging different remote sensing datasets, and harmonizing different forest definition criteria (minimum height, crown coverage, and width/area).

A second area of concern is the implementation of a *land use* criterion, which, in turn, has an impact on the definition of *forest*. Although remote sensing methods enable woodland mapping, the limitations of forest mapping increase if land use is incorporated. Although land cover can be easily and directly assessed using remote sensing data and techniques, for land use it is less straightforward, requiring a combination of image classification expertise and external knowledge. Interviewing foresters or performing *in situ* field visits of the areas under examination are recommended. Problems may also arise if remote sensing-based mapping products are combined with statistically estimated NFI products. In addition to the decreased processing automation, comparisons of the extracted mapped forest areas must be handled with care.

A third concern is that forest mapping may also be challenging because of the huge variety of existing remote sensing data and methods. However, the value of the final map product strongly depends on the requirements of the end user. Thus, in order to maximize automation, a precise understanding of these requirements—that incorporates the end user's needs, budget, availability of training, and validation, quality, and handling of the remote sensing data—must be worked out. In addition, weather conditions (clouds) may also affect the planning process, and a certain degree of flexibility may be needed if alternative data material must be used. Furthermore, the time span between datasets should be kept to a minimum.

A fourth issue is that automation strongly depends on the collection of training data which—although a labor-intensive part of the mapping process—will substantially reduce future work by the implementation of existing NFI sample plot data, such as tree/non-tree information (based on stereo-image interpretation), and terrestrial information (obtained from field visits) within the framework of a NFI.

Future Perspectives

Technical developments in remote sensing data and collection methods in the near future, and the increasing need and demand for long-term forest cover maps, will further push and improve the automation of such products.

Current trends in spaceborne remote sensing also appear promising in a number of ways for the land use and land cover assessment community. The provision of larger swath widths in combination with more bands and high temporal resolution (on the order of a few days) will significantly enhance the operational forest monitoring capabilities of entire countries. Sustainability of such applications is currently supported by the Landsat data continuity mission and systems that are being promoted by the European Space Agency (ESA) and the European Commission (EC) Global Monitoring for Environment and Security (GMES) program. The ESA's Sentinel program—which provides dedicated spaceborne satellite missions that address the operational user and institutional needs within the wider European Copernicus (former GMES) initiative—already represents a major step forward. Dedicated to Copernicus, Sentinel-1 already ensures the continuity of C-band SAR data collection (planned launch of Sentinel-2b in March 2017, building on ESA's and Canada's heritage SAR systems on ERS-1, ERS-2, ENVISAT, and RADARSAT). Full-waveform and multispectral LiDAR data (as presented in Chapter "Airborne LiDAR Applications in Forest Landscapes") also has great potential. These programs provide high-quality 3D information as required by many processing steps in the derivation of CHM or geometric features for forest definitions. Moreover, they are the current standard in many countries and acquired in cycles of 3–6 years.

A higher degree of automation can be achieved by simplifying the methodological workflow using modern machine learning algorithms, scripting, and entire process chains. Focusing on a remote sensing-based derivation of land use by using additional existing information on land use (e.g., from stand maps or forest inventories) can substantially increase the degree of automation for forest mapping. Additionally, newer information and information provided within shorter time spans on the extent of woodland/forest areas by NFIs—which combine terrestrial sampling with remotely sensed estimations—may be used with positive results.

With a higher degree of automation, high-quality forest maps based on remote sensing can be derived more frequently and help provide the latest information on forest resources for entire countries and continents in the future. In turn, retrospective analysis of changes in these areas will become feasible and increasing interest will lead to the development of wall-to-wall forest maps that will become standard in many countries. While future NFIs will be based on both terrestrial surveys and remotely sensed parameter estimations, mapped forest area will play a key role. It is expected that the future of NFIs will be characterized by a complementary combination of diverse types of information and sources of information, such as remotely sensed parameters—e.g., forest cover and trees outside forest (TOF)—enquiries by the local forest service, stereo-image interpretation, and, of course, terrestrial information.

References

Alford J (1993) Towards a new public management model: beyond "managerialism" and its critics. Aust J Publ Admin 52(2):135–148

APEX (2011) Airborne Prism Experiment. http://www.apex-esa.org. Accessed 16 Mar 2017

Baatz M, Schäpe A (2000) Multiresolution segmentation—an optimization approach for high quality multi-scale image segmentation. In: Strobl J, Blaschke T, Griesebner G (eds) Angewandte Geographische Informationsverarbeitung XII. Beiträge zum AGIT Symposium, Wichmann, Heidelberg, pp 12–23

Barrett F, McRoberts RE, Tomppo E, Cienciala E, Waser LT (2016) A questionnaire-based review of the operational use of remotely sensed data by national forest inventories. Remote Sens Environ 174:279–289

Bartholomé E, Belward AS (2005) GLC2000: a new approach to global land cover mapping from Earth observation data. Int J Remote Sens 26:1959–1977

Brandtberg T (2002) Individual tree-based species classification in high spatial resolution aerial images of forests using fuzzy sets. Fuzzy Sets Syst 132(3):371–387

Chen JM, Leblanc S (1997) A four-scale bidirectional reflectance model based on canopy architecture. IEEE Trans Geosci Remote 35:1316–1337

Cheng YZ (1995) Mean shift, mode seeking, and clustering. IEEE Trans Pattern Anal 17:790–799

Chubey M, Stehle K, Albricht R, Gougeon F, Leckie D, Gray S, Woods M, Courville P (2009) Semi-automated species classification in Ontario Great Lakes—St. Lawrence forest conditions. Final report: Great Lakes—St. Lawrence ITC Project (2005/2008). Ontario Ministry of Natural Resources, January 2009, p 71

Dalponte M, Orka HO, Ene LT, Gobakken T, Naesset E (2014) Tree crown delineation and tree species classification in boreal forests using hyperspectral and ALS data. Remote Sens Environ 140:306–317

Darvishsefat A, Kellenberger T, Itten K (2002) Application of hyperspectral data for forest stand mapping. Int Arch Photogram Rem Sens Spatial Inform Sci. 34(4)

Dees M, Koch B, Pelz D (1998) Integrating satellite based forest mapping with Landsat TM in a concept of a large scale forest information system. PFG 4:209–220

Dong J, Xiao X, Sheldon S, Biradar C, Zhang G, Dinh Duong N et al (2014) A 50-m forest cover map in Southeast Asia from ALOS/PALSAR and its application on forest fragmentation assessment. PLoS One 9(1):e85801. doi:10.1371/journal.pone.0085801

Edelsbrunner H (1995) Smooth surfaces for multi-scale shape representation. Lect Notes Comput Sci 1026:391–412

Eysn L, Hollaus M, Schadauer K, Pfeifer N (2012) Forest delineation based on airborne LiDAR data. Remote Sens 4:762–783

Eysn L, Hollaus M, Lindberg E, Berger F, Monnet J, Dalponte M, Kobal M, Pellegrini M, Lingua E, Mongus D, Pfeifer N (2015) A benchmark of lidar-based single tree detection methods using heterogeneous forest data from the alpine space. Forests 6:1721–1747

FAO (1997) Food and Agriculture Organization. *State of the World's Forests*. Food and Agriculture Organization, Rome, Italy, 200 pp

FAO (1999) Food and Agriculture Organization. UNEP terminology for integrated resources planning and management. Food and Agriculture Organization/United nations Environmental Program, Rome, Italy and Nairobi, Kenia
FAO (2000) Food and Agriculture Organization. FRA 2000: on definitions of forest and forest change. Working paper 33, Forest Resource Assessment, Rome, 14 p
FAO (2001) Food and Agriculture Organization. Global Forest Resources Assessment 2000 Main Report. FAO Forestry Paper 140. FAO, Rome, Italy
FAO (2010) Food and Agriculture Organization. Global Forest Resources Assessment 2010 Main Report. FAO Forestry Paper163, Rome
FAO (2012) Food and Agriculture Organization. Global forest land-use change 1990–2005. FAO Forestry Paper 2012, Rome
Fisher P, Comber A, Wadsworth R (2005) Land use and land cover: contradiction or complement. In: Fisher P, Unwin DJ (eds) Re-presenting GIS. Wiley, Hoboken, NJ, pp 199–209
Förster M, Kleinschmit B (2008) Object-based classification of QuickBird data using ancillary information for the detection of forest types and NATURA 2000 habitats. In: Blaschke T, Lang S, Hay G (eds) Object-based image analysis. Springer, Berlin, Heidelberg, pp 275–290
Förster M, Kleinschmit B, Walentwoski H (2005) Monitoring NATURA 2000 forest habitats in Bavaria by the use of ASTER, SPOT5 and GIS data—an integrated approach. In: Proceedings of the ForestSAT 2005 conference, 31th May–2nd June 2005, Borås, Sweden
Fransson JES, Magnusson M, Folkesson K, Hallberg B, Sandberg G, Smith-Jonforsen G, Gustavsson A, Ulander LMH (2007) Mapping of wind-thrown forests using VHF/UHF SAR images. In: Proceedings of the Geoscience and Remote Sensing Symposium, IGARSS 2007, 23rd–28th July 2007, Barcelona, Spain, pp 2350–2353
Fukunaga K, Hostetler LD (1975) Estimation of gradient of a density-function, with applications in pattern-recognition. IEEE Trans Inform Theory 21:32–40
Garcia-Haro FJ, Sommer S (2002) A fast canopy reflectance model to simulate realistic remote sensing scenarios. Remote Sens Environ 81(2–3):205–227
Gebhardt S, Wehrmann T, Ruiz M, Maeda P, Bishop J, Schramm M, Kopeinig R, Cartus O, Kellndorfer J, Ressl R et al (2014) MAD-MEX: automatic wall-to-wall land cover monitoring for the Mexican REDD-MRV program using all Landsat data. Remote Sens 6:3923–3943
Gerard F (2003) Single angle, dual angle and multi-temporal viewing: assessing through modelling the implications for forest structure variable extraction. Int J Remote Sens 24(6):1317–1334
Gerard FF, North PRJ (1997) Analyzing the effect of structural variability and canopy gaps on forest BRDF using a geometric-optical model. Remote Sens Environ 62(1):42–62
GFW (2016) The global forest watch. http://www.globalforestwatch.org/map/3/15.00/27.00/ALL/grayscale/loss,forestgain?begin=2001-01-01&end=2015-01-01&threshold=30. Accessed 23 July 2017
Gillis M, Leckie D (1996) Forest inventory update in Canada. For Chron 72(2):138–156
Ginzler C, Hobi M (2015) Countrywide stereo-image matching for updating digital surface models in the framework of the Swiss national forest inventory. Remote Sens 7:4343–4370
Hajnsek I, Kugler F, Lee SK, Papathanassiou KP (2009) Tropical-forest-parameter estimation by means of Pol-InSAR: the INDREX II campaign. IEEE Trans Geosci Remote 47(2):81–493
Hansen MC, Potapov PV, Moore R, Hancher M, Turubanova SA, Tyukavina A, Thau D, Stehman SV, Goetz SJ, Loveland TR, Kommareddy A, Egorov A, Chini L, Justice CO, Townshend JRG (2013) High-resolution global maps of 21st-century forest cover change. Science 342(6160):850–853
Hirschmugl M, Weninger B, Raggam H, Schardt M (2007) Single tree detection in very high resolution remote sensing data. Remote Sens Environ 110(4):533–544
Hirschmüller H (2008) Stereo processing by semi-global matching and mutual information. IEEE Trans Pattern Anal 30(2):328–341
Hosmer DW, Lemeshow S (2000) Applied logistic regression, 2nd edn. Wiley, New York
Immitzer M, Atzberger C, Koukal T (2012) Tree species classification with random forest using very high spatial resolution 8-band WorldView-2 satellite data. Remote Sens 4(9):2661–2693

JAXA (2016) World's first high-resolution global forest/non-forest map. http://global.jaxa.jp/article/special/geo/shimada_e.html. Accessed 23 July 2017

Jensen JR (2005) Introductory digital image processing: a remote sensing perspective. Prentice Hall, Upper Saddle River, NY

JRC (2016) European Joint Research Center JRC. http://forest.jrc.ec.europa.eu/activities/forest-mapping/forest-cover-map-2006/. Accessed 23 July 2017

Kaartinen H, Hyyppä J (2008) EuroSDR/ISPRS Project, Commission II "Tree Extraction" Final Report. EuroSDR (European Spatial Data Research): Dublin, Ireland, 2008

Kaartinen H, Hyyppä J, Yu X, Vastaranta M, Hyyppä H, Kukko A, Holopainen M, Heipke C, Hirschmugl M, Morsdorf F, Næsset E, Pitkänen J, Popescu S, Solberg S, Wolf BM, J-C W (2012) An international comparison of individual tree detection and extraction using airborne laser scanning. Remote Sens 4(4):950–974

Keil M, Schardt M, Schurek A, Winter R (1990) Untersuchung und Kartierung von Waldschäden mit Methoden der Fernerkundung. DLR-Abschlussdokumentation, Kapitel Auswertung von Satellitendaten, pp 71–131

Kennedy P, Bertolo F (2002) Mapping sub-pixel forest cover in Europe using AVHRR data and national and regional statistics. Can J Remote Sens 28(2):302–321

Koch B, Dees M, van Brusselen J, Eriksson L, Fransson J, Gallaun H, Leblon B, McRoberts RE, Nilsson M, Schardt M, Seitz R, Waser LT (2008) Forestry applications. In: Li Z, Chen J, Baltsavias E (eds) Advances in photogrammetry, remote sensing and spatial information—ISPRS 2008 Congress Book. Taylor & Francis Group, London, pp 439–468

Koch B, Straub C, Dees M, Wang Y, Weinacker H (2009) Airborne laser data for stand delineation and information extraction. Int J Remote Sens 30(4):935–963

Kuusk A, Nilson T (2000) A directional multispectral forest reflectance model. Remote Sens Environ 72(2):244–252

Laliberte AS, Rango A, Havstad KM, Paris JF, Beck RF, McNeely R (2004) Object-oriented image analysis for mapping shrub encroachment from 1937 to 2003 in southern New Mexico. Remote Sens Environ 93(1–2):198–210

Lamonaca A, Corona P, Barbati A (2008) Exploring forest structural complexity by multi-scale segmentation of VHR imagery. Remote Sens Environ 112(6):2839–2849

Leckie DG, Gougeon FA, Walsworth N, Paradine D (2003) Stand delineation and composition estimation using semi-automated individual tree crown analysis. Remote Sens Environ 85(3):355–369

Leckie DG, Tinis S, Nelson T, Burnett C, Gougeon FA, Cloney E, Paradine D (2005) Issues in species classification of trees in old growth conifer stands. Can J Remote Sens 31(2):175–190

Li XW, Strahler AH (1992) Geometric-optical bidirectional reflectance modeling of the discrete crown vegetation canopy—effect of crown shape and mutual shadowing. IEEE Trans Geosci Remote 30(2):276–292

Li XW, Strahler AH, Woodcock CE (1995) A hybrid geometric optical–radiative transfer approach for modeling albedo and directional reflectance of discontinuous canopies. IEEE Trans Geosci Remote 33:466–480

Lindberg E, Hollaus M (2012) Comparison of methods for estimation of stem volume, stem number and basal area from airborne laser scanning data in a hemi-boreal forest. Remote Sens 4(4):1004–1023

Lund HG (2016) rev*. Definitions of forest, deforestation, afforestation, and reforestation. Gainesville, V.A.: Forest Information Services. [Online] Note this paper has been continuously updated since 1998. DOI: 10.13140/RG.2.1.2364.9760

Magdon P, Fischer C, Fuchs H, Kleinn C (2014) Translating criteria of international forest definitions into remote sensing image analysis. Remote Sens Environ 149:252–262

McRoberts RE (2012) Satellite image-based maps: scientific inference or pretty pictures? Remote Sens Environ 115(2):715–724

McRoberts RE, Tomppo EO (2007) Remote sensing support for national forest inventories. Remote Sens Environ 110(4):412–419

McRoberts RE, Liknes GC, Domke GM (2014) Using a remote sensing-based, percent tree cover map to enhance forest inventory estimation. For Ecol Manag 331:12–18
Mottus M, Sulev M, Lang M (2006) Estimation of crown volume for a geometric radiation model from detailed measurements of tree structure. Ecol Model 198:506–514
Næsset E (2007) Airborne laser scanning as a method in operational forest inventory: status of accuracy assessments accomplished in Scandinavia. Scand J For Res 22(5):433–442
Næsset E, Gobakken T (2005) Estimating forest growth using canopy metrics derived from airborne laser scanner data. Remote Sens Environ 96(3–4):453–465
Nilson T, Peterson U (1991) A forest canopy reflectance model and a test case. Remote Sens Environ 37:131–142
Pekkarinen A, Reithmaier L, Strobl P (2009) Pan-European forest/non-forest mapping with Landsat ETM+ and CORINE Land Cover 2000 data. ISPRS J Photogramm 64(2):171–183
Petrie G, Walker AS (2007) Airborne digital imaging technology: a new overview. Photogramm Rec 22(119):203–225
Rahman M, Sumantyo J (2010) Mapping tropical forest cover and deforestation using synthetic aperture radar (SAR) images. Appl Geomat 2:113–121
Reitberger J, Krzystek P, Heurich M (2006) Full-waveform analysis of small footprint airborne laser scanning data in the Bavarian forest national park for tree species classification. In: Koukal T, Schneider W (eds) Proceedings of the International Workshop 3D Remote Sensing in Forestry, 14-15th February 2006. University of Natural Resources and Applied Life Sciences, Vienna, Austria, pp 218–228
Rosenquist Å, Shimada M, Ito N, Watanabe M (2007) ALOS PALSAR: a pathfinder mission for global-scale monitoring of the environment. IEEE Trans Geosci Remote 45(11):3307–3316
Rutzinger M, Pratihast AK, Elberink SO, Vosselman G (2010) Detection And Modelling Of 3D Trees From Mobile Laser Scanning Data. Int Arch Photogram Rem Sens Spatial. Inf Sci 38(5):520–525
Schepaschenko D, See L, Lesiv M, McCallum I, Fritz S, Salk C, Moltchanova E, Perger C, Schepashchenko M, Shvidenko A et al (2015) Development of a global hybrid forest mask through the synergy of remote sensing, crowdsourcing and FAO statistics. Remote Sens Environ 162:208–220
Shimada M, Itoh T, Motooka T, Watanabe M, Shiraishi T, Thapa R, Lucas R (2014) New global forest/non-forest maps from ALOS PALSAR data (2007–2010). Remote Sens Environ 155:13–31
Solberg S, Naesset E, Bollandsas OM (2006) Single tree segmentation using airborne laser scanner data in a structurally heterogeneous spruce forest. Photogramm Eng Remote Sens 72:1369–1378
Spurr SH (1960) Photogrammetry and photo-interpretation, 2nd edn. Ronald Press, New York
Stibig HJ, Achard F, Fritz S (2004) A new forest cover map of continental southeast Asia derived from SPOT-VEGETATION satellite imagery. Appl Veg Sci 7(2):153–162
Stoffels J, Mader S, Hill J, Werner W, Ontrup G (2012) Satellite-based standwise forest cover type mapping using a spatially adaptive classification approach. Eur J For Res 131(4):1071–1089
Straub BM, Heipke, C (2001) Automatic extraction of trees for 3D city models from images and height data. Automatic extraction of man-made objects from aerial and space images, vol 3, pp 267–277
Straub C, Weinacker H, Koch B (2008) A fully automated procedure for delineation and classification of forest and non-forest vegetation based on full-waveform laser scanner data. Int Arch Photogram Rem Sens Spatial Inform Sci 37(8):1013–1019
Straub C, Stepper C, Seitz R, Waser LT (2013) Potential of UltraCamX stereo images for estimating timber volume and basal area at the plot level in mixed European forests. Can J For Res 43(8):731–741
Swisstopo (2016) The topographic landscape model TML. http://www.swisstopo.admin.ch/internet/swisstopo/en/home/topics/geodata/tlm.html. Accessed 23 July 2017
Thiel C, Drezet P, Weise C, Quegan S, Schmullius C (2006) Radar remote sensing for the delineation of forest cover maps and the detection of deforestation. Forestry 79(5):589–597

Tomppo E, Halme M (2004) Using coarse scale forest variables as ancillary information and weighting of variables in k-NN estimation: a genetic algorithm approach. Remote Sens Environ 92(1):1–20

Tomppo E, Haakana M, Katila M, Peräsaari J (2008) Multi-source national forest inventory-methods and applications, Managing Forest Ecosystems, vol 18. Springer, Berlin

Tomppo E, Gschwantner T, Lawrence M, McRoberts RE (eds) (2010) National forest inventories—pathways for common reporting, 1st edn. Berlin, Springer

Vauhkonen J, Seppänen A, Packalén P, Tokola T (2012) Improving species-specific plot volume estimates based on airborne laser scanning and image data using alpha shape metrics and balanced field data. Remote Sens Environ 124:534–541

Vosselman G (2003) 3D reconstruction of roads and trees for city modeling. Int Arch Photogram Rem Sens Spatial Inform Sci. 34(3), W13, Dresden, Germany, pp 231–236

Wagner W, Luckman A, Vietmeier J, Tansey K, Balzter H, Schmullius C, Davidson M, Gaveau D, Gluck M, Le Toan T, Quegan S, Shvidenko A, Wiesmann A, JJ Y (2003) Large-scale mapping of boreal forest in SIBERIA using ERS tandem coherence and JERS backscatter data. Remote Sens Environ 85(2):125–144

Wang Z, Boesch R, Ginzler C (2007) Color and LIDAR data fusion: application to automated forest boundary delineation in aerial images. Int Arch Photogram Rem Sens Spatial Inform Sci 36(1)

Wang Z, Ginzler C, Waser LT (2015) A novel method to assess short-term forest cover changes based on digital surface models from image-based point clouds. Forestry 88:429–440

Waser LT (2012) Airborne remote sensing data for semi-automated extraction of tree area and classification of tree species. Dissertation, ETH Zürich, 153 p. http://e-collection.library.ethz.ch/view/eth:6087

Waser LT, Baltsavias E, Ecker K, Eisenbeiss H, Feldmeyer-Christe E, Ginzler C, Küchler M, Thee P, Zhang L (2008a) Assessing changes of forest area and shrub encroachment in a mire ecosystem using digital surface models and CIR-aerial images. Remote Sens Environ 112(5):1956–1968

Waser LT, Baltsavias E, Ecker K, Eisenbeiss H, Ginzler C, Küchler M, Thee P, Zhang L (2008b) High-resolution digital surface models (DSM) for modeling fractional shrub/tree cover in a mire environment. Int J Remote Sens 29(5):1261–1276

Waser LT, Klonus S, Ehlers M, Küchler M, Jung A (2010) Potential of digital sensors for land cover and tree species classifications—a case study in the framework of the DGPF-Project. PFG 2:141–156

Waser LT, Ginzler C, Kuechler M, Baltsavias E, Hurni L (2011) Semi-automatic classification of tree species in different forest ecosystems by spectral and geometric variables derived from Airborne Digital Sensor (ADS40) and RC30 data. Remote Sens Environ 115(1):76–85

Waser LT, Küchler M, Jütte K, Stampfer T (2014) Evaluating the potential of WorldView-2 data to classify tree species and different levels of ash mortality. Remote Sens 6:4515–4545

Waser LT, Fischer C, Wang Z, Ginzler C (2015) Wall-to-wall forest mapping based on digital surface models from image-based point clouds and a NFI forest definition. Forests 6:4510–4528

Zeverbergen LW, Thorne CR (1987) Quantitative analysis of land surface topography. Earth Surf Processes 12:47–56

Epilogue: Toward More Efficient and Effective Applications of Forest Landscape Maps

Ajith H. Perera and Tarmo K. Remmel

Abstract Significant advances in geospatial sciences are being made, aided by multibillion dollar technological industries that focus on remote sensing, geographic information systems (GIS), global positioning system (GPS), and computing. As a result of these improvements, forest landscape ecologists and forest managers now have access to increasingly comprehensive and accurate georeferenced information that is frequently updated. However, the knowledge of forest landscape patterns and processes and the ability to convey this knowledge for applications have not grown in proportion to the advances in mapping technology. Thus, there are vast opportunities to improve the knowledge of forest landscape patterns, especially in remote regions that are difficult to study in the field. There are numerous avenues to improve the efficacy and efficiency of map use. Here we highlight the importance of resisting the allure of excessive detail and instead focus on identifying the most appropriate scale, which may not be the most detailed resolution available, to support a user's research or management goals. It is also crucial to improve our awareness of the many embedded assumptions that shape how information is mapped and the sources and magnitude of the errors that are inherent to any map, however advanced the science and technology that produced it. We emphasize that ongoing communication and interaction between the communities of map developers and map users will be essential to achieve wise use of forest landscape maps.

Abbreviations

2D	Two-dimensional
3D	Three-dimensional
GIS	Geographic information system

A.H. Perera (✉)
Ontario Forest Research Institute, Ontario Ministry of Natural Resources and Forestry,
1235 Queen Street East, Sault Ste. Marie, ON, Canada, P6A 2E5
e-mail: ajith.perera@ontario.ca

T.K. Remmel
Department of Geography, York University,
4700 Keele Street, Toronto, ON, Canada, M3J 1P3
e-mail: remmelt@yorku.ca

T.K. Remmel, A.H. Perera (eds.), *Mapping Forest Landscape Patterns*,
DOI 10.1007/978-1-4939-7331-6_8

GPS	Global positioning system
LiDAR	Light detection and ranging
UAV	Unmanned aerial vehicle (drone)

Background

Forest landscapes occupy more than 30% of Earth's land surface and occur on every continent except Antarctica. They therefore represent one of the most important global ecosystem resources. Forests are highly valued for ecological reasons (e.g., conservation of biodiversity, protective functions), economic reasons (e.g., supply of wood and other products), and social reasons (e.g., aesthetics, cultural values). There are many management activities that correspond to these roles of forest landscapes. These include developing plans for extracting and conserving resources; detecting and monitoring changes in the composition, functions, and patterns of forests; and assessing the ecological impacts of human activities. Forest landscape maps are essential for planning and executing all such management activities, whether these applications are simple and subjective approaches, or involve complex and more objective methods such as decision support systems. Mapped information about forest landscapes also provides the basis for periodic official reports on the status of a region's forests that are generated by governmental and nongovernmental agencies at global, national, and regional scales.

All of these applications are informed by numerous ongoing research efforts that study all facets of forest landscapes. It is now common for researchers to use forest landscape maps as an essential input for such academic investigations. For example, the mathematical models developed to simulate most forest landscape characteristics—composition, patterns, structure, function, processes, and utility—are based on spatially explicit information about forest landscapes. The output from these models is also presented in the form of maps that aid exploration, discovery, prediction, and discovery. The demand for such spatially explicit information about forest landscapes is increasing globally, at a rapidly increasing rate, and both the developers and users of this information must be aware of the potential and the limitations of the knowledge and of the technologies and methods used to generate it.

This demand is being met by concurrent developments in several theoretical and technological fields during the last several decades. First, data collection techniques have advanced rapidly since the widespread adoption of coarse-resolution aerial photography in the 1950s, and the launch of the Landsat satellite in 1972. Now, there are many global platforms that support an array of sensors capable of detecting a wide range of signals from forest landscapes. With these tools, it has become possible not only to map the most remote forest landscapes, but also to make these methods and the resulting data readily accessible and affordable to a wide audience of potential users. Second, researchers in the fields of geography and spatial statistics have developed a variety of methods for data analysis, most notably geographic information systems—the ubiquitous GIS. Because these techniques for analysis of spatial data are now provided by GIS software, sophisticated and comprehensive spatial pattern analyses are increasingly available to an increasingly broad spectrum of individuals who are interested in the mapping of forest landscapes. Third, continuous advances in computer science and technology have made it feasible to store and consistently analyze massive quantities of spatial data. The data storage and analysis capabilities that were only available to supercomputer users as little as a decade ago are now available in inexpensive personal computers.

The net result of these developments is the availability of highly advanced techniques for the capture, storage, and analysis of spatial data, combined with readily available and relatively cheap access to high-resolution thematic, spatial, and temporal data about forest landscapes. Furthermore, the detection, analysis, and visualization of forest landscape characteristics have become a global enterprise for both the academic and the commercial sectors.

Despite the expanding demand for this data, and the ready and advanced supply of mapped information, our knowledge about forest landscape patterns and processes and our wisdom about how to use this knowledge appear to be lagging behind. In fact, the gap between the supply of information and its judicious application in forest landscape management may even have widened over time.

Goals of This Chapter

Our goal in this concluding chapter is to offer a narrative on how to improve the application of forest landscape maps that bridges the intent of the first chapter and the contents of the other chapters. It is mostly based on our observations, interactions, and experiences with applications and uses of forest landscape maps for more than five decades. This is not a review based on a compilation of details from an exhaustive search of the scientific literature; the topic that we address is rarely written about by scientists for other scientists, and typically does not appear in the scientific literature in relation to forest landscape mapping applications. Scientists who develop forest landscape maps appear to implicitly assume that map users, whether they are researchers or forestry practitioners, understand the limitations and the scope of the utility of their maps. In our experience, this assumption does not appear to be valid.

Here, we also contemplate the future of forest landscape mapping by focusing on improved efficiency and effectiveness of map use. By *use*, we mean the application of mapped information to support forest landscape management by professionals (the end users), although researchers and the public also use maps. The maps we address here are not only primary cartographic products, but also include spatially portrayed secondary information that is generated by intermediate applications such as simulation models and predictive models. By *efficiency* we mean the ratio of the useful information content to the total cost (including time) of acquiring and utilizing that information, and by *effectiveness*, we mean that the map provides the optimal information content to support the goal of using the map. Both efficiency and effectiveness are judged from the perspective of users of the maps.

In brief, efficient and effective applications of forest landscape maps involve three activities: generating a demand among practitioners for forest landscape maps that are relevant and pertinent to real-world questions, meeting that demand by supplying mapped information with the highest possible reliability, and applying the mapped information to achieve real-world goals with minimum effort. Though these may appear to be simple tasks, in practice they involve complex and fuzzy decisions, perhaps made more complex by ambiguities in the user's goals, miscomprehension of the mapped products, and (most of all) poor communication between the developers and users of maps.

Our discussion in the rest of this chapter involves considerations that could help improve the efficiency and effectiveness of forest landscape maps. These may be implicitly obvious to the scientists and technicians who develop maps, and may even be considered mundane and trivial. However, our experience suggests that these considerations are rarely explicit for most users, or may not be readily evident in many applications. We anticipate that users of forest landscape maps will benefit directly from the contents of this chapter by familiarizing themselves with the points we discuss, and will benefit indirectly by learning to improve communication with those who generate the information.

Considerations in Forest Landscape Mapping

The Community of Map Developers and Users is Broad

It is perhaps not readily apparent that the overall forest landscape mapping process includes multiple participants. In practice, map developers and users form a broad community that represents the confluence of expertise from several academic fields (e.g., remote sensing, GIS, spatial statistics, and ecology), several professional groups (e.g., natural resource policy developers and managers, professionals employed by the natural resource extraction sector, environmental monitoring and conservation researchers, educators), governments, and even the general public. Mapped information about forest landscapes is used directly by the research community and by professionals who manage the land, and (to a smaller extent) indirectly by the public, who may rely on the secondary information generated by researchers and land managers (Fig. 1). Even though the public is not generally considered an important user group, they play a significant role in shaping forest landscape policies and management, both of which are informed by the use of forest landscape maps. Representatives of all these domains of expertise interact to create a dynamic demand and supply for forest landscape maps.

A key to successful development and application of these maps is communication—a clear, two-way flow of relevant information between those who obtain and

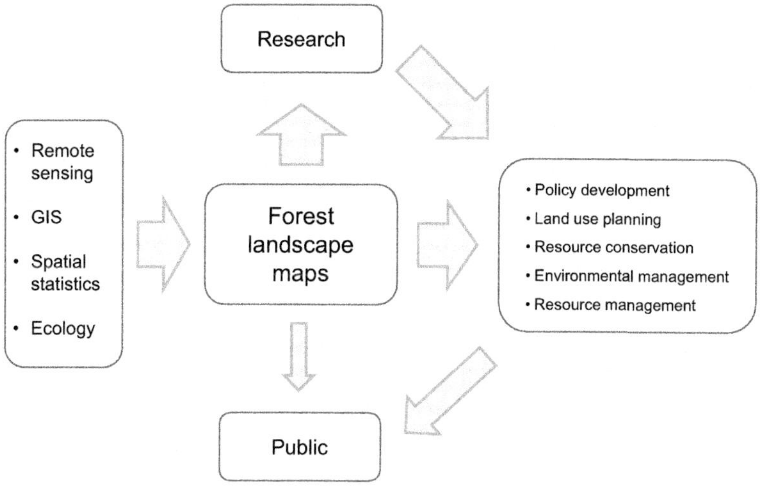

Fig. 1 Illustration of the overall process of forest landscape mapping, which includes both the demand for and the supply of forest landscape maps. Various user groups (researchers, land managers, the public) generate the demand and use the information from forest landscape maps, and that supply is met by a group of interacting specialists (e.g., remote-sensing specialists, geographers, spatial statisticians, and ecologists). The width of the *arrows* is roughly proportional to the relative importance of the information flow. The *arrows* are depicted as one-way for simplicity, but in reality, they may consist of many feedback loops

present the information and those who use the information. This involves continuous and active engagement of all parties, although only pairs of these parties may be involved in any particular situation. Communication and information exchanges between some groups are well established, and function effectively, with feedback loops to improve the maps. For example, such exchanges occur between the communities of researchers and map developers, who converse both via the research literature and in the many forums for communication of scientific knowledge (e.g., conferences, online discussion groups). Though land managers represent the crucial end users of forest landscape maps, the flow of information they receive is less formal and explicit, and is often ad hoc (i.e., inspired by a specific need rather than being part of an ongoing process). Information exchanges with the public are mostly ad hoc, and feedback loops may not exist; that is, researchers and map developers rarely ask the public about their needs.

Maps are Model Outputs

All forest landscape maps are abstractions of reality, and not just the first maps that were produced, since any map is a representation of one or more implicit mental models. This is true even for the most "realistic" modern maps that have been

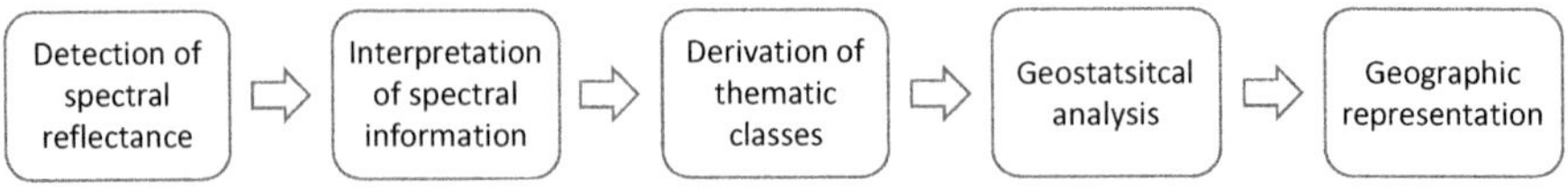

Fig. 2 Some major steps involved in producing a forest landscape map. This example uses remotely sensed data from a satellite as the starting point. All subsequent steps involve mathematical and statistical models, each with specific assumptions and methods. The final map product embodies the assumptions and limitations of all models employed in each step of the process

generated by applying explicit methods of gathering, processing, and portraying spatial data based on sophisticated technology and advanced scientific methods (Fig. 2). For example, a map of a forest landscape generated two centuries ago would probably portray navigability and extractable values such as timber volumes based on the mental model of the explorer who created the map. The corresponding map of the same landscape generated today may represent the age and species composition of the forest stands, tree heights and wood volumes, estimates of the best route for accessibility without damaging the site, and even economic and ecological optima for timber extraction—all based on data gathered by optical sensors, georeferenced by GPS, and then interpreted, analyzed, and predicted by an array of mathematical models. Forest landscape maps still remain (mathematical and statistical) simplifications and are only estimates of the vast complexities of nature, spatially portrayed as simulations and predictions. Maps are therefore method specific. That is, maps may differ in their information content based on the procedures used for data collection, analysis, and spatial portrayal.

As in the case of all quantitative models, forest landscape maps are based on many assumptions; some are inherent to the scientific and statistical methods used by the developer, whereas others stem from limitations in technological, scientific, and expert knowledge. To follow the example of good science, these assumptions should be made explicit for each method that is applied. Applications of the model's outputs (in this case, maps) are valid only if the assumptions are appropriate for the intended use, and this consideration must be made explicit. Some map production methods are more robust than others, but the assumptions and methods of any method will create method- and assumption-specific limits to the scope of their applications.

Maps are Probabilistic

The contents of forest landscape maps, however rigorously obtained, processed, and represented, are probabilistic; that is, they represent some level of confidence, not absolute certainty. This probability arises, in part, from the many assumptions, errors, and uncertainties in the methods used to obtain, process, and represent the

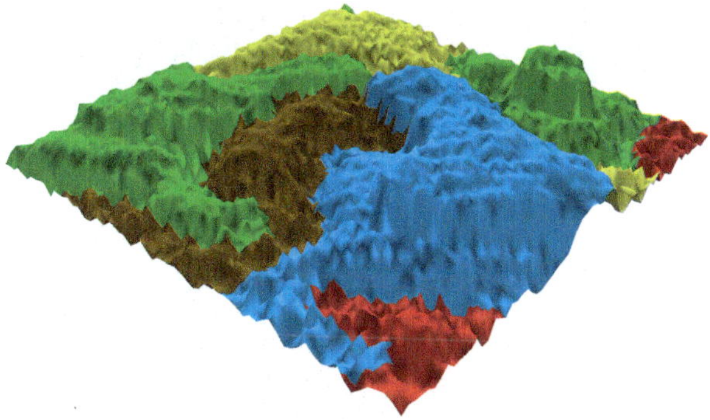

Fig. 3 An example of a 3D rasterized map of forest cover types (*colors*), with changes in elevation of each point above the plane representing spatial variability of the probabilities of each pixel belonging to the indicated forest type: the greater the elevation, the greater the probability of being assigned to the correct thematic class. These probabilities vary among the pixels within a cluster, among clusters with the same theme, and among clusters with different themes. Such explicit expressions of uncertainty illustrate the degree of confidence in all mapped thematic classes and pixels within those classes

data. The problem of probabilistic mapped information also arises because natural objects and their patterns (in this case, the characteristics of forest landscapes) are inherently variable, and our perceptions of them are fuzzy. Consequently, the assignment of thematic classes to the parts of a landscape and georeferencing the objects to mapped features of the forest landscape are not about establishing absolute certainty, but rather about maximizing the probability of being correct. For example, a feature mapped as "old coniferous forest" may not have old conifers throughout each entity that has been assigned that classification, and entities within such a thematic class may not be identical because of such variations. This classification means only that the classified entities have a high probability of containing old conifers and of being more similar to each other than to entities in other thematic classes. Explicit articulation of the spatial variability of the probabilities associated with thematic classifications is uncommon, but is necessary, especially if the mapped information will be used in subsequent models and decision support systems (Fig. 3).

The probabilistic nature of maps is even more evident around boundaries and transitions between mapped entities. Because discrete boundaries are relatively rare in natural forest landscapes, thematic classifications could be less certain at the boundaries within vectorized maps (Fig. 4).

Furthermore, the probability of an entity belonging to a specific class could also vary over time in the case of dynamic maps, in which the mapped information (e.g., vegetation cover) explicitly changes over time. As we discuss later, in section "Map Contents are Scale Related", the probabilities of mapped information are also scale dependent.

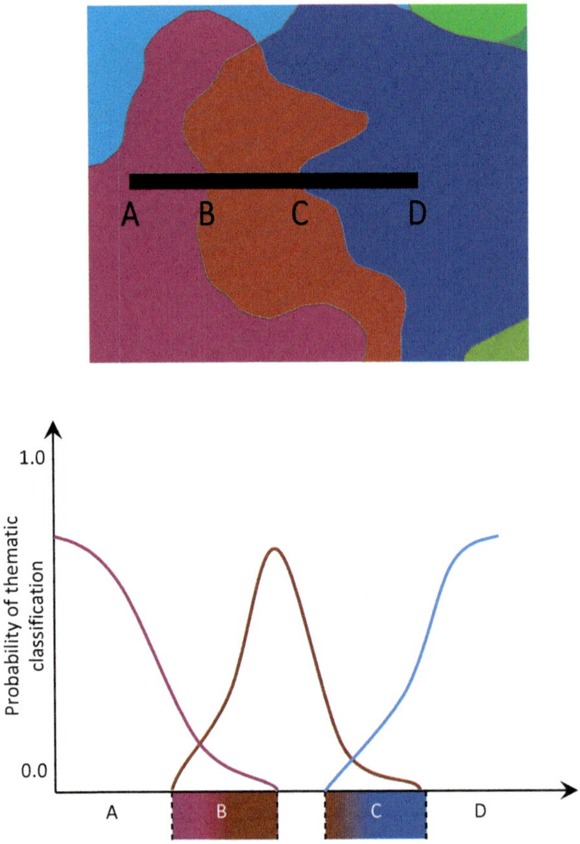

Fig. 4 (*Top*) A typical vectorized forest cover classification map with a hypothetical transect drawn across the boundaries between three polygons. (*Bottom*) The probability of correct thematic classification along the transect. The boundaries (at points *B* and *C*) are not discrete; rather, they represent a transition zone in which both thematic classes could occur. In reality, the probability distribution may not be smoothly unimodal, as depicted in the example, but could be a multimodal distribution like the one illustrated in Fig. 3

Maps Contain Errors

Even forest landscape objects that have been mapped with the most rigorous and advanced methods have some misclassification or location errors, and such inaccuracies are commonplace in maps. When the errors are detected, typically through a validation process (e.g., ground-truthing), specific errors can be corrected and details about the possibility of these errors elsewhere in the map can be included in the methodological descriptions. A common means of expressing thematic errors is by cross-tabulating mapped versus validated entities for a sample of landscape positions to create a table known as a "confusion matrix" (Fig. 5). This table describes the matches and mismatches, which can be further analyzed quantitatively. The

		Validation Categories			
		Water	**Forest**	**Bedrock**	**Total**
Mapped Categories	**Water**	65	5	7	**77**
	Forest	2	94	3	**99**
	Bedrock	12	0	37	**49**
	Total	**79**	**99**	**47**	**225**

Fig. 5 An example of a confusion matrix that summarizes the match and mismatch frequencies between mapped categories and their corresponding validated categories. The diagonal entries are the matches and off-diagonal entries are mismatches. Column and row totals are the "marginal distributions" and are used for computing overall, user, and producer accuracy values and nonspatial accuracy summaries. The total number of observations (i.e., 255 in this example) is a function of the desired level of sampling locations to validate

Table 1 A summary of mapping errors that illustrates partitioning of the error sources into either thematic (categorization) or positional (georeferencing) errors, with specific examples of each case

Error type	Example errors
Thematic	Wrong label (e.g., forest rather than shrubs)
	Wrong label level (e.g., dense rather than open conifers)
	Validation data incorrect (e.g., false-positive or false-negative accuracy assessment)
Positional	Inappropriate vector representation for a given resolution (point, line, polygon)
	Inappropriate spatial resolution for the minimum mappable unit
	Features mapped to incorrect coordinates (e.g., either systematic or nonsystematic coordinate errors)
	Boundaries mapped at incorrect locations (e.g., either systematic or nonsystematic coordinate errors)
	Coordinate precision not appropriate for a given resolution (e.g., too much detail for coarse-resolution mapping)
	Boundary complexity not matched to actual complexity (e.g., boundary too simple or too complex)
	Boundary width not characterized properly (e.g., ecotones not recognized)

expression of georeferencing errors is not common in mapping, perhaps because it is sometimes difficult to separate thematic errors from location errors during assessment of the final map product (i.e., empirical map validation), and necessitates sophisticated evaluations of detection methods and instruments. Consequently, classification errors and georeferencing errors are typically not separated in accuracy assessments (Table 1).

Another aspect of these errors that affects map use is related to precision. The concept of map precision is often confused with map accuracy. Precision refers to variability of the prediction of a given mapped property (i.e., the ability to repeatedly assign the same label to pixels in a specific thematic class), whereas accuracy

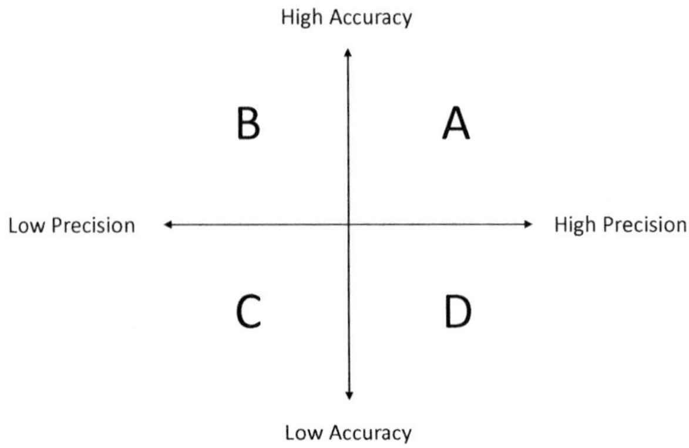

Fig. 6 Map errors can be described in terms of the accuracy (correctness) of the classification and the precision (repeatability) of the classification. Domains *A*, *B*, *C*, and *D* represent different combinations of accuracy and precision, which have different implications for the efficiency and effectiveness of the map

refers to the correctness of the labeling. They are independent (perhaps even orthogonal) concepts, and express two different types of mapping error. For example, a mapped entity could be very precisely expressed (i.e., in minute detail and low variability in prediction), but inaccurately classified (i.e., incorrect labeling), as in the case of domain D in Fig. 6, or can be imprecisely expressed (i.e., described in broad terms), but accurately classified (i.e., correct), as in the case of domain B in Fig. 6. It is also possible that increasingly fine resolution can lead to a false sense of precision, as is the case when the measurements become finer than the phenomenon being measured. Domain A in Fig. 5 represents an ideal situation (i.e., the optimal balance between the effectiveness and efficiency of information): it represents the most accurate and most precise mapped information. Domain B will be efficient, but perhaps less effective than domain A. Domain D is less accurate, and represents a common trap for map users, who are attracted by high precision. As with all aspects of mapping, errors are scale dependent. Even though spatial portrayals of error estimates offer important information, equivalent in importance to estimates of the mapped themes, they are rarely reported in real-world applications. As in the case of probability surfaces (e.g., Fig. 3), communicating the spatial variability of errors in the estimates would provide additional evidence about the reliability of mapped information.

Map Contents are Scale Related

By "scale," we refer here to the level of detail (i.e., resolution of information) in the maps—not the ecological scale that refers to the hierarchy of ecological organization or the cartographic scale that refers to the ratio between distances portrayed on

Table 2 Relationships between the scale (resolution) of the map contents and the relative level of detail with respect to the thematic, spatial, and temporal information. Here, we have illustrated these relationships using the example of a boreal forest landscape

	Map resolution	
Map content	Coarse	Fine
Thematic classification	Broader forest cover classes (e.g., dense coniferous forest)	Detailed forest community types (e.g., an overstorey of 40% black spruce, 20% white spruce, 20% jack pine, 15% poplar, and 5% larch, with an understorey of Labrador tea, sheep-laurel and blueberry, and herbaceous ground cover)
Spatial differentiation	Large pixel size (e.g., 1 ha) or a large minimum mappable unit (e.g., 20 ha)	Small pixel size (e.g., 1 m^2) or a small minimum mappable unit (e.g., 1 ha)
Temporal interval	Long frequency of map updates (e.g., decadal)	Short frequency of map updates (e.g., annual)

a map relative to their real-world distances on the Earth's surface. As described with examples in Table 2, a coarse-scale map will have less detail (lower resolution in thematic, spatial, and temporal information) and a fine-scale map will have more detail (higher resolution in thematic, spatial, and temporal information).

The map resolution (high or low) is related to the thematic information (e.g., classification details), spatial information (e.g., pixel size), or temporal information (e.g., interval between changes). Mapped information could be gathered, analyzed, and portrayed with higher resolution (e.g., more thematic classes, smaller pixel size, more frequent updates) or at a lower resolution (e.g., fewer thematic classes, larger pixel size, and less frequent updates). This concept of resolution is imperative when considering all aspects of map contents and use, but is often ignored, misunderstood, or incorrectly applied. High-resolution maps with more thematic detail, finer spatial resolution, and shorter temporal intervals are often considered to be more informative and better than low-resolution maps with less thematic detail, coarser spatial resolution, and longer temporal intervals. This is not universally true. Fine-resolution maps may be more precise, but may have lower accuracy, higher uncertainty, and lower predictive capability than the corresponding coarse-resolution map. Coarse-resolution maps are often more accurate, more certain, and more valid, and may have higher predictive ability than fine-resolution maps that contain the same information, despite the lower precision of the coarser maps (Fig. 7). Because of technological advances and the decreasing cost of production, maps of forest landscapes are increasingly produced at finer resolutions—with smaller features, more thematic detail, and more frequent updates. However, this should not mislead their users into believing that these maps are inherently superior to coarser maps. Maps with finer resolution may include unnecessary information that requires the user to ignore or exclude the information (i.e., a higher cognitive burden), may be more expensive to produce (without providing benefits that justify the additional

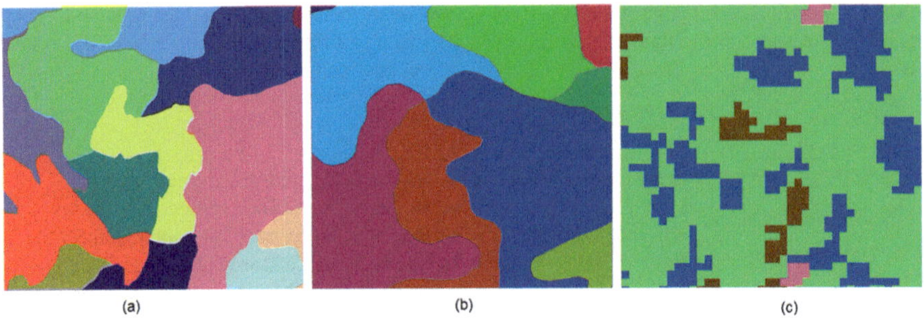

Fig. 7 Three 1 km × 1 km maps of the same forest landscape, obtained from different sources at different resolutions: (**a**) a recent forest inventory from the Leica ADS 40 airborne sensor at 0.4 m spatial resolution and vectorized; (**b**) an old forest inventory map created from 1:20,000 aerial photographs; and (**c**) a land cover map from the Landsat TM sensor at 30 m spatial resolution. The thematic details differ among the maps: (**a**) 14 forest community types based on tree, shrub, and ground vegetation species composition, basal area, age, and height of the canopy species; (**b**) 9 forest stand types based on tree species composition, basal area, tree age, and height of the tree canopy; (**c**) 4 generalized forest cover types based on the overstorey composition. The minimum mappable unit sizes also vary: 10 ha in (**a**) and (**b**), and 1 ha in (**c**). The temporal resolution also varies: 10 years for (**a**) and (**b**), and 16 days for (**c**)

cost), and may mislead users into focusing on smaller details than would be appropriate (e.g., varying the width of a riparian buffer strip in 1-m intervals instead of using a single width for the whole length of a river).

Map Applications are Scale Related

Each map application dictates a specific optimal resolution for its different purposes (i.e., a range of thematic, spatial, and temporal resolutions). This is true whether the maps are used to support management activities (e.g., policy development, land use planning, timber extraction, biodiversity conservation) or academic pursuits (e.g., simulating environmental impacts, predicting ecosystem processes). Using maps with resolutions coarser than that optimal resolution will not be effective because users will lack sufficient detail on certain characteristics of the mapped area. Conversely, use of maps with resolutions finer than the optimal resolution will not be efficient (or potentially effective) because there will be too much information, thereby requiring the user to ignore or eliminate the information they cannot use.

Both of these instances of resolution mismatches have adverse consequences for users of the maps. The consequences of using maps with too coarse resolution and inadequate information to support the intended use are generally recognized by users of forest landscape maps: they risk missing small but important areas that require different treatment. However, the consequences of using maps with too fine resolution are less commonly recognized. Indeed, there appears to be a general belief that increasing map resolution will always better inform the user. It is less obvious that, in addition to the higher cost of acquisition, storage, analysis, and updating, finer map resolutions have a hidden cost: they overinformed users and complicate use of the map. For example, the information content and relative accuracy and certainty of that information (as well as the cost) differ among the three maps in Fig. 7. Of these, the best suited map is the one that matches the resolution (thematic, spatial, and temporal) of the application. The key to selecting the optimal map resolution lies in determining the level of detail required for a given application right from the start; only then should the user determine the map resolution that can provide that level of detail. We do not mean to imply that choosing the optimal resolution is a simple task, nor that this resolution will be readily evident. Researchers have been examining the problem of defining an optimal resolution for studies of ecosystem structure and processes for many decades, without achieving consensus on the best solution, and land managers continue to struggle with choosing the optimal resolution to support planning, decision making, monitoring, and assessment. Nonetheless, determination of the optimal map resolution to support real-world applications is essential, especially when multiple options are readily available.

Mapping Methods are Advancing Rapidly

Sensors have improved, allowing narrower regions of the electromagnetic spectrum to be measured at higher spatial, temporal, and radiometric resolutions than ever before. Software for handling and manipulating these new data sources (e.g., GIS and remote sensing software) permits efficient analysis of these data and their conversion into increasingly informative maps. Delivery and sharing of these data sources and maps online are increasingly bringing maps into the hands of nontechnical users, and anyone with a smartphone, tablet, or computer and a connection to the Internet can now view, interpret, and enjoy rich map data more easily than ever before.

Laser 3D scanning and light detection and ranging (LiDAR) are making new dimensions of data collection available to map developers. Low-altitude images obtained from unmanned aerial vehicles (UAVs, which are also called "drones") using custom cameras and sensors are creating a resurgence in photogrammetric technologies, assisted by powerful computers and software capable of processing vast quantities of 2D and 3D data. The rapid advances that are currently under way in 3D printing technology will make it possible to convert maps into tactile and multidimensional representations that will provide new opportunities for visualization of landscape data. We imagine that in time, with increasing miniaturization of technology, forest landscape mapping will become increasingly rapid, informative, readily available, and affordable from these new lightweight UAV platforms.

A Brief List of Best Practices for Using Forest Landscape Maps

In this section, we provide a short list of suggestions for best practices that would improve the efficiency and effectiveness of forest landscape map applications. Many other practical and conceptual considerations could have been included, but we chose to focus on those that we deemed most important but that have been most overlooked in practice. Even though we primarily address the users of mapped information in this section, developers of maps must also pay attention to these considerations. They also have a responsibility to ensure that the users are aware of these and other considerations that affect how they will use maps.

- *Determine the resolution required to provide enough detail for the application before acquiring the mapped information.*

  Ideally, whether by means of formal sensitivity analyses and risk assessments or based on expert knowledge, users should determine the levels of thematic and spatial detail necessary to support their use of the eventual map. They also should assess the expected life span of the map application, and determine the interval (temporal resolution) required for information updates. If the applications span a range of resolutions (thematic, spatial, and temporal), then users should assess the best method for hierarchical nesting of the required levels of thematic, spatial, and temporal details (i.e., to allow users to efficiently increase and decrease the resolution of the information).

- *Select the maps with optimum level of detail for the application.*

  Based on the resolution required by the application, users should select a map resolution that will provide the optimum level of detail and information. Maps that are too coarse will not be effective because they will not provide enough detail to support the application, whereas maps whose resolution is too fine will not be efficient. In our experience, most map users understand the negative consequences of maps that are too coarse, but incorrectly assume that maps with finer resolution are a better option. This results in higher cost of acquisition, processing, interpreting, and updating of the maps while creating a false sense of the true precision and accuracy of the resulting products.

- *Scrutinize and assess the methods involved in generating maps.*

  Users can discern the quality of maps by evaluating the methods used to create them and the assumptions those methods rely upon; these will affect the level of confidence that users can place in the maps. Even though the mapping process may be complicated and may involve sophisticated techniques, all map development procedures, including assumptions and terminology, must be made explicit so their consequences can be accounted for. The repeatability of all methodological steps (including data acquisition, analysis, interpretation, and verification) is as important to assure the reliability of a map as the empirical validation of the results that is typically carried out as a means of quality assurance.

- *Acquire estimates of all sources of uncertainty associated with maps.*

  The many sources of uncertainty included in mapped information must be made explicit to the users. The spatial distribution of errors (e.g., noise, classification inaccuracy, incorrect georeferencing), probabilities, and confidence limits of the predictions provides important insights into the reliability of mapped information and helps users to use mapped products more prudently. A 3D map (with error estimates and confidence limits represented by elevations) will be more illustrative and valuable than a typical 2D map that assumes a homogeneous distribution of errors and uncertainties.

- *Interact and communicate with map developers.*

  All of these best practices require continuous engagement and dialogue between map developers and map users. This communication must not end after maps have been produced and delivered to their users. Interactions between developers and users of maps must continue through all stages of development and application of the map products. With rapid advances in mapping processes, maps are progressively becoming better, less expensive, and more accessible. However, because the range of options available is becoming broader, clarity of information exchange is mandatory to support wise use of maps. Efficient and effective use of forest landscape maps will occur only through active, continued, and open dialogue between those who generate maps and those who use them.

Conclusions

There have been several significant advances in forest landscape mapping over the past 100 years; these include the advent of aerial photography, launch of satellites such as Landsat, emergence of GIS technology, expansion of GPS use, and arrival of new sensing technologies such as LiDAR. Our recent ability to map the most remote regions of the Earth with high spatial and thematic resolutions, and to monitor changes through frequent reimaging of an area, is increasingly commonplace. These scientific and technological advances will soon make it possible to rapidly portray forest landscapes in three dimensions, accompanied by real-time information on their dynamics in composition, structure, and even functions. Though the technologies of forest landscape mapping are advancing rapidly, it is not yet clear that applications of the mapped information are advancing in parallel. Our focus in this chapter has been to reveal ways to improve these applications by capitalizing on the rapid progress in mapping capability.

We contend that an important new role of the professionals responsible for developing forest landscape maps (i.e., geographers, cartographers, landscape ecologists, statisticians, software developers) will be to interact and communicate with all users of their products to ensure that the mapped information is understood and applied correctly. These professionals must communicate that maps, however advanced, are only abstractions and are, at best, approximations of the complex, fuzzy, nonlinear, and dynamic patterns that exist in nature; the simplicity, certainty, linearity, and self-similarity that may exist in most maps of the human world cannot be expected from forest landscape maps. Furthermore, maps depend on many assumptions, and include multiple sources of error; awareness of these problems and cautious use of maps are therefore essential. Map users must also understand that there are optimal thematic, spatial, and temporal resolutions for any application, and that the finest and most detailed map is not always the most suitable. This point must be emphasized repeatedly to avoid the trap of providing more information than is appropriate; this leads to high costs of acquiring, storing, and updating maps, and the excess information can overwhelm both users of the map and the processing systems used to create it. Goal- and resolution-specific maps are the most efficient solution, and are most informative in revealing and understanding relevant spatial patterns.

Finally, we reiterate that all professionals involved in the process of developing and using maps must seek ways to ensure that the enormous richness of spatial information that is available to us will translate into an equivalent wealth of advanced and reliable knowledge. Only then will we achieve the true potential of forest landscape maps as indispensable tools in advancing ecological sciences and discoveries to support the development of forest management strategies and decisions and to educate the general public.

Index

T.K. Remmel, A.H. Perera (eds.), *Mapping Forest Landscape Patterns*,
DOI 10.1007/978-1-4939-7331-6

Printed by Printforce, the Netherlands